# Bioinformatics and Functional Genomics

# Bioinformatics and Functional Genomics

Editor: Christina Marshall

R CALLISTO
REFERENCE

www.callistoreference.com

**Callisto Reference,**
118-35 Queens Blvd., Suite 400,
Forest Hills, NY 11375, USA

Visit us on the World Wide Web at:
www.callistoreference.com

ISBN: 978-1-64116-073-5 (Hardback)

**Cataloging-in-Publication Data**

Bioinformatics and functional genomics / edited by Christina Marshall.
    p. cm.
Includes bibliographical references and index.
ISBN 978-1-64116-073-5
1. Bioinformatics. 2. Genomics. 3. Genetics--Data processing. I. Marshall, Christina.
QH324.2 .B56 2019
570--dc21

# Table of Contents

# Preface

Bioinformatics is a rapidly growing branch of science, which integrates the concepts of biology, engineering, mathematics and computer science in order to develop software tools. These tools are used in analyzing and interpreting biological data. Functional genomics is a sub-field of molecular biology, which uses the tools of bioinformatics to understand the diverse aspects of genes such as regulation of gene expression, DNA sequencing, gene transcription, protein-protein interactions, etc. There has been rapid progress in these fields and their applications are finding their way across multiple industries. This book is compiled in such a manner, that it will provide in-depth information about the theory and practice of bioinformatics and functional genomics. Students, researchers, experts, geneticists, biologists and biological engineers will benefit alike from this book.

The information contained in this book is the result of intensive hard work done by researchers in this field. All due efforts have been made to make this book serve as a complete guiding source for students and researchers. The topics in this book have been comprehensively explained to help readers understand the growing trends in the field.

I would like to thank the entire group of writers who made sincere efforts in this book and my family who supported me in my efforts of working on this book. I take this opportunity to thank all those who have been a guiding force throughout my life.

**Editor**

# Genomic prediction using subsampling

Alencar Xavier[1], Shizhong Xu[2], William Muir[3] and Katy Martin Rainey[1*]

## Abstract

**Background:** Genome-wide assisted selection is a critical tool for the genetic improvement of plants and animals. Whole-genome regression models in Bayesian framework represent the main family of prediction methods. Fitting such models with a large number of observations involves a prohibitive computational burden. We propose the use of subsampling bootstrap Markov chain in genomic prediction. Such method consists of fitting whole-genome regression models by subsampling observations in each round of a Markov Chain Monte Carlo. We evaluated the effect of subsampling bootstrap on prediction and computational parameters.

**Results:** Across datasets, we observed an optimal subsampling proportion of observations around 50% with replacement, and around 33% without replacement. Subsampling provided a substantial decrease in computation time, reducing the time to fit the model by half. On average, losses on predictive properties imposed by subsampling were negligible, usually below 1%. For each dataset, an optimal subsampling point that improves prediction properties was observed, but the improvements were also negligible.

**Conclusion:** Combining subsampling with Gibbs sampling is an interesting ensemble algorithm. The investigation indicates that the subsampling bootstrap Markov chain algorithm substantially reduces computational burden associated with model fitting, and it may slightly enhance prediction properties.

**Keywords:** Genome-wide selection, Bayesian analysis, Bootstrapping

## Background

The use of genomic tools has become important for the genetic improvement of complex traits in plants and animals through genome-wide prediction (GWP). GWP provides an interesting solution for the selection of traits with low heritability, such as grain yield in crops and milk production in dairy cattle, as well as for traits that present challenging or expensive phenotyping.

Over the past decade, researchers have tried to overcome the pitfalls of increased computational burden associated with gains in prediction accuracy from GWP of complex traits. Increases in predictive ability (and computational burden) are often associated with better statistical learning properties, such as regularization and variable selection [1]. Hence models with an improved ability to identify patterns provide more robust predictions, but computational costs are involved.

In statistical learning, resampling techniques are common approaches used to turn weak learners into strong learners [2]. Gianola et al. [3] showed that bootstrapping aggregation could improve prediction accuracy of kernel-based genomic best linear unbiased prediction (GBLUP) model in genomic prediction of plant and animals. We hypothesized that a similar approach could apply to whole-genome regression methods, often referred to as the Bayesian alphabet [4].

Besides computational advantages offered by some resampling methods, these techniques may also help to overcome theoretical shortcomings of some of these Bayesian methods, such as the bias of BayesA [5]. The objective of this study was to evaluate the predictive and computational outcomes from the application of a resampling technique ensemble with the Gibbs sampler to a Bayesian ridge regression model.

### Sampling procedures

In addition to the increasing number of markers available over time due to higher density single nucleotide polymorphism (SNP) arrays and even resequencing, computation challenges include the large number of samples from which those genotypes are taken [6]. The computational burden associated with large population

* Correspondence: krainey@purdue.edu
[1]Department of Agronomy, Purdue University, 915 W. State St., Lilly Hall, West Lafayette, IN 47907, USA
Full list of author information is available at the end of the article

sizes is more evident in plant breeding, where hundreds of crosses with large offspring are genotyped and selected every season using GWP. Sampling methods are often necessary to enable such complex statistical procedures in large datasets. Among those, two main classes of sampling techniques are Markov chain Monte Carlo (MCMC) and Bootstrapping.

The MCMC method is possibly the most popular Monte Carlo algorithm with application to linear models, providing a feasible framework to resolve high-dimensional problems (i.e., more parameters than observations) with moderate computer power [7]. Likewise, bootstrapping also provides an interesting framework for solving large-scale problems [8, 9], particularly a method known as subsampling [10] used to reduce data dimensionality.

## Gibbs sampling

Gibbs sampling is a widely used MCMC technique, applied in conjunction with Bayesian methods to generate the posterior distribution of the parameters. The posterior distribution is denoted as $p(\Theta|X)$, where $\Theta$ represents the set of unknown parameters $\Theta = \{\theta_1, , \theta_2, ...., , \theta_r\}$, and $X$ represents the data. The Gaussian model described in the following section, unknown parameters include the intercept ($\mu$), the vector of regression coefficients ($\mathbf{b}$) and variance components, as $\Theta = \{\mu, \mathbf{b}, \sigma_b^2, \sigma_e^2\}$, whereas the observed data comprises the genotypic information ($\mathbf{X}$) of individuals and phenotype ($\mathbf{y}$), as $X = \{\mathbf{X}, \mathbf{y}\}$.

Gibbs sampling algorithms are based on updating each parameter with samples drawn from the full-conditional posterior distribution, one parameter at a time while holding every other parameter constant. Each parameter $\theta$ is sampled from

$$p(\theta|X) \propto f(X|\theta)\pi(\theta), \forall \theta \in \Theta, \tag{1}$$

where $p(\theta|X)$ denotes the posterior distribution of $\theta$, the likelihood is expressed as $f(X|\theta)$ and the prior distribution of $\theta$ is $\pi(\theta)$.

In most implementations, regression coefficients are sampled individually from normal distributions whereas variance components are sampled from scaled inverse chi-squared distributions [4, 5]. Every time a parameter (i.e., regression coefficients and variance components) or a conjugated prior is updated, its value is stored as samples of the posterior distribution. The final Bayesian estimator is the expectation of the posterior distribution, obtained as the mean of the posterior distribution.

## Bootstrapping aggregation

A natural strategy to increase prediction accuracy is to build and combine multiple prediction models generated from samples of a large dataset, averaging the outcome predictor [11]. Bootstrapping aggregation, or simply

'bagging', is implemented in linear models by fitting the function $f_1(x), f_2(x), ..., f_B(x)$ with $B$ bootstrapped samples of the dataset and the final model, with reduced variance, will be given by

$$\hat{f}_{avg}(x) = \frac{1}{B}\sum_{b=1}^{B} \hat{f}_b(x), x \subset X. \tag{2}$$

Regression coefficients are stored each time the model is fitted, hence generating an empirical distribution of each parameter. Bagging parameters are obtained as the mean of this distribution.

With bootstrapping, when samples are obtained with replacement, the number of observations sampled is commonly the same size as the initial dataset, recognizing that some observations may be sampled more than once. When bootstrapping is performed with fewer samples than the original number of observations, sampling can proceed either with or without replacement. The latter case is known as subsampling.

## Subsampling bootstrap Markov chains

MCMC and Bootstrap are usually implemented separately, such that some studies have attempted to compare the performance of these samplers [12]. Nevertheless, both methods can be co-implemented. A co-implementation that is becoming popular in the context of big data is a technique known as subsampling bootstrap Markov chain (SBMC). In this algorithm, the Markov chain update mechanism is performed upon a subset ($x$) of the whole data ($X$) and a different subset is used to update the parameters in each round of MCMC. Therefore, each parameter is sampled from the posterior distribution

$$p(\theta|x) \propto f(x|\theta)\pi(\theta), \forall \theta \in \Theta, x \subset X. \tag{3}$$

The concept of subsampling Gibbs sampler was first presented by Geyer [13] and some predictive properties were further investigated by MacEachern and Peruggia [14]. Regarding the applications to genome-wide prediction of complex traits, SBMC can be used to update the regression coefficients [15], hence increasing the computational performance of model fitting.

## Methods
### Statistical model
The family of whole-genome regression methods is a standard set of models widely applied for genomic prediction [4]. Among these, Bayesian ridge regression is a regularized model that assigns the same variance to every marker. The linear model is described as follows:

$$\mathbf{y} = 1\mu + \mathbf{Xb} + \mathbf{e} \tag{4}$$

where $\mathbf{y}$ is the response variable (i.e., the phenotypic information), $\mu$ is a scalar representing the intercept, $\mathbf{X}$ is the genotypic matrix coded as {0,1,2} for {AA, Aa, aa} where rows correspond to the genotypes and columns correspond to the molecular markers, $\mathbf{b}$ is a vector of regression coefficients that represents the additive value of allele substitutions, and $\mathbf{e}$ is the vector of residuals. In this model, both regression coefficients and residuals are assumed to be normally distributed as $\mathbf{b} \sim N(0, I\sigma_b^2)$ and $\mathbf{e} \sim N(0, I\sigma_e^2)$. The variances are assumed to follow a scaled inverse chi-squared distribution with a given prior shape (S) and prior degrees of freedom (ν), thus $\sigma_b^2 \sim \chi^{-2}$ $(S_b, , \nu_b)$ and $\sigma_e^2 \sim \chi^{-2}(S_e, , \nu_e)$.

High-dimensional methods are regularized to enable fitting the model without losing predictive properties [2]. The regularization of linear models occurs by shrinking regression coefficients, which also biases predictions downwards [1]. The Bayesian ridge regression attempts to estimate regression coefficients with the minimum bias necessary for a satisfying prediction (i.e., minimum variance), a solution referred to as best linear unbiased predictor [4, 5]. As an optimization problem, the loss function to be minimized by the model (equation 4) that balances variance and bias is described as

$$L_2 = (\mathbf{y}-\mu-\mathbf{Xb})^{'}(\mathbf{y}-\mu-\mathbf{Xb}) + \lambda(\mathbf{b}^{'}\mathbf{b}) \quad (5)$$

where $\lambda$ is the regularization parameter, the ratio between the residual variance and the genetic variance of marker effects, as $\lambda = \sigma_e^2/\sigma_b^2$. For the model in consideration, the regularization parameter assumes a single value for all regression coefficients.

## Coefficient update

Sorensen and Gianola [16] show that the full conditional distribution of regression coefficients for Gibbs sampling from a normal distribution has a closed form. The expectation is derived from the Cholesky decomposition of the left-hand side (LHS) of the mixed model equation. The computational cost of operations such as solving the mixed model equation is described in terms of $n$ observations and $p$ parameters. The cost associated with the Cholesky decomposition is $p^3$, making it computationally unfeasible for high-dimensional problems ($p \gg n$), such as whole-genome regression methods. On the other hand, the Gauss-Seidel residual updating (GSRU) algorithm [15] has a computational cost of $3pn$, which is much lower than for the Cholesky decomposition in this scenario. A Gibbs sampler based on GSRU updates the $j^{th}$ regression coefficient as

$$b_{j^{t+1}}\big|* \sim N\left(\frac{x_j^{'}e^t + x_j^{'}x_j b_j^t}{x_j^{'}x_j + \lambda_j}, \frac{\sigma_e^2}{x_j^{'}x_j + \lambda}\right) \quad (6)$$

where $x_j$ is the vector corresponding to the $j^{th}$ marker and $*$ represents the data and all parameters other

than the one being updated. The coefficient update is followed by update of the vector of residual

$$\mathbf{e}^{t+1} = \mathbf{e}^t + \mathbf{x}_j(b_j^{t+1}-b_j^t). \quad (7)$$

The greatest advantage of GSRU comes from the low computational cost of updating the right-hand side (RHS) of the mixed model equation [15], solving the linear system one parameter at a time without computing $\mathbf{X'X}$. Subsequently, variance components are updated as

$$\sigma_b^2\big|* \sim \frac{\mathbf{b}^{'}\mathbf{b} + S_b\nu_b}{\chi_{p+\nu_b}^2} \text{ and } \sigma_e^2\big|* \sim \frac{\mathbf{e}^{'}\mathbf{e} + S_e\nu_e}{\chi_{n+\nu_e}^2}. \quad (8)$$

where $S_e$, $\nu_e$, $S_b$, and $\nu_b$ correspond to the prior parameters "shape" and "degrees of freedom" of the residual and genetic variance, respectively.

## SBMC extension

We here propose incorporating subsampling into the Gibbs sampler. This variation implies sampling a $\psi$ fraction of the data ( $\psi \in [0, 1]$ ) to update regression coefficients and residual variance in each round of MCMC.

For a matter of notation, let $\mathbf{X}^{\sim}$ and $\mathbf{e}^{\sim}$ represent the bagged subsamples, in other words, a fraction of $\mathbf{X}$ and $\mathbf{e}$ that contains $\psi$ percent of observations sampled at random in a given round of MCMC. This modified GSRU would have an expected computational cost of $3pn\psi$. To accommodate bagged samples, sampling algorithms of regression coefficients and residual variance undergo a slight modification. Regression coefficients are updated or sampled as

$$b_j^{t+1}\big|* \sim N\left(\frac{\tilde{x}_j^{'}\tilde{e}^t + \psi x_j^{'}x_j b_j^t}{\psi x_j^{'}x_j + \lambda_j}, \frac{\sigma_e^2}{\psi x_j^{'}x_j + \lambda_j}\right) \quad (9)$$

with subsequent residual update

$$\mathbf{e}^{\sim t+1} = \mathbf{e}^{\sim t} + \mathbf{x}_j(b_{j^{t+1}}-b_j^t). \quad (10)$$

The entire $k^{th}$ round of MCMC is updated using the subsampled dataset $x^k = \{\mathbf{X}^{\sim}, \mathbf{e}^{\sim}\}$. Since the residual variance is a function of the number of observations, its update is slightly modified from equation 8 as

$$\sigma_e^2\big|* \sim \frac{\tilde{e}^{'}\tilde{e} + S_e\nu_e}{\chi_{\psi n+\nu_e}^2}. \quad (11)$$

The sampling procedure above assumes that the variance associated to markers in the subsamples are approximately the same as in the whole data ($\sigma_x^2 \sim \approx \sigma_x^2$). That is, the marker sum of squares ($x'x$) is expected to reduce linearly according to the proportion of bag samples ($\psi x'x$) to avoid recalculating the sum of squares of bagged markers $(\mathbf{x}^{\sim'}\mathbf{x}^{\sim})$ for each round of MCMC. In genetic terms, the subset is assumed to have the same allele frequencies as the whole set.

The SBMC algorithm is implemented in the R package bWGR [17] using the $R^2$ rule proposed by Pérez and de Los Campos [18] to estimate prior shapes using the whole data, based on $R^2 = 0.5$, with the values of prior degrees of freedom set as $\nu_e = 5$ and $\nu_b = 5$. In the $R^2$ rule [18], prior shapes are estimated as

$$S_b = R^2 \times \sigma_y^2 \times \frac{(\nu_b + 2)}{\sum_j \sigma_{x_j}^2} \tag{12}$$

and

$$S_e = \left(1 - R^2\right) \times \sigma_y^2 \times (\nu_e + 2). \tag{13}$$

### Dataset

Three datasets available on R packages [18, 19] were chosen to demonstrate the effect of bagging on genomic prediction, including a wheat panel from the International Maize and Wheat Improvement Center (CIMMYT), as the median of grain yield observed in four environments [20]; a mouse panel designed to study body mass index [21] but using only half the SNP panel obtained by skipping every other marker; a soybean panel with eight bi-parental families with elite parents from the SoyNAM project [19] with phenotypes observed in eighteen environments; and a simulated $F_2$ population with 10 chromosomes of 50 cM each, genotyped at density of 1 SNP/cM, and with one QTL every 10 cM placed between markers. Heritability of traits was computed by restricted maximum likelihood (REML) upon the animal model with additive kernel [22]. Markers with minor allele frequency below 0.05 were removed. Datasets are summarized in Table 1.

### Prediction metrics

Prediction statistics were obtained with a 10-fold cross validation scheme. We fitted the Bayesian ridge regression model using subsampling from 25 to 100%, by increment of 1%, with and without replacement. We set the algorithm for 4000 MCMC iterations to ensure convergence [16], with 500 of burn-in [18].

To determine the efficacy of subsampling, we evaluated the mean square prediction error (MSPE), prediction bias as the slope of linear regression between predictions and observations ($\beta_{y,y}$), computation time in minutes, and predictive ability as the Pearson's correlation between predictions and observations ($Cor_{y,y}$).

### Results

The mean outcome of prediction metrics across datasets is presented in Fig. 1. The results by individual dataset are presented in the Additional file 1. Numeric results for some proportions of subsampling are presented in Table 2.

### Computational improvement

The computational time had a linear response to subsampling (Fig. 1d). As expected, subsampling is clearly beneficial to speed up the computation of model fitting. The same trend was observed for individual datasets (Additional file 1). Although our evaluation of the improvement of computational performance used relatively small datasets, we believe the results must hold for larger datasets.

In comparison to fitting the model with whole data (Table 2), the computation time to fit the model at 50% subsampling was 33.6% faster with replacement and 58.3% faster without replacement. Yet, the computational cost was less than expected, once $3pn\psi$ with $\psi = 0.5$ should provide a model fitting 100% faster. This difference can be attributed to the computational cost of the sampling process along with the fixed cost of the initial problem settings. Computationtime 100% faster was achieved for subsampling 33% (or less) without replacement.

Interestingly, subsampling with replacement presented a slightly higher computational cost, also presenting worse predictive properties for subsampling lower than 40% or higher than 60%.

### Implications of subsampling on prediction parameters
*Bias*
The use of the complete dataset was nearly unbiased (Table 2). Subsampling with replacement was biased downwards, presenting the least bias at 40% replacement ($\beta_{y,y} = 0.824$). Subsampling without replacement presented slight upward bias, being 1.8 and 5.8%

**Table 1** Summary of datasets used in this study

| Species | Population type | Trait | n | p | $h^2$ | Source |
|---|---|---|---|---|---|---|
| Mouse | Heterogeneous stock | Body mass index | 1814 | 5173 | 0.146 | Legarra et al. [21] |
| Soybean | Nested Ass. Panel | Grain yield | 1079 | 4307 | 0.345 | Xavier et al. [19] |
| Wheat | Diverse panel | Grain yield | 599 | 1209 | 0.434 | Crossa et al. [20] |
| Simulation | Experimental F2 | Simulated | 400 | 500 | 0.516 | Technow [29] |

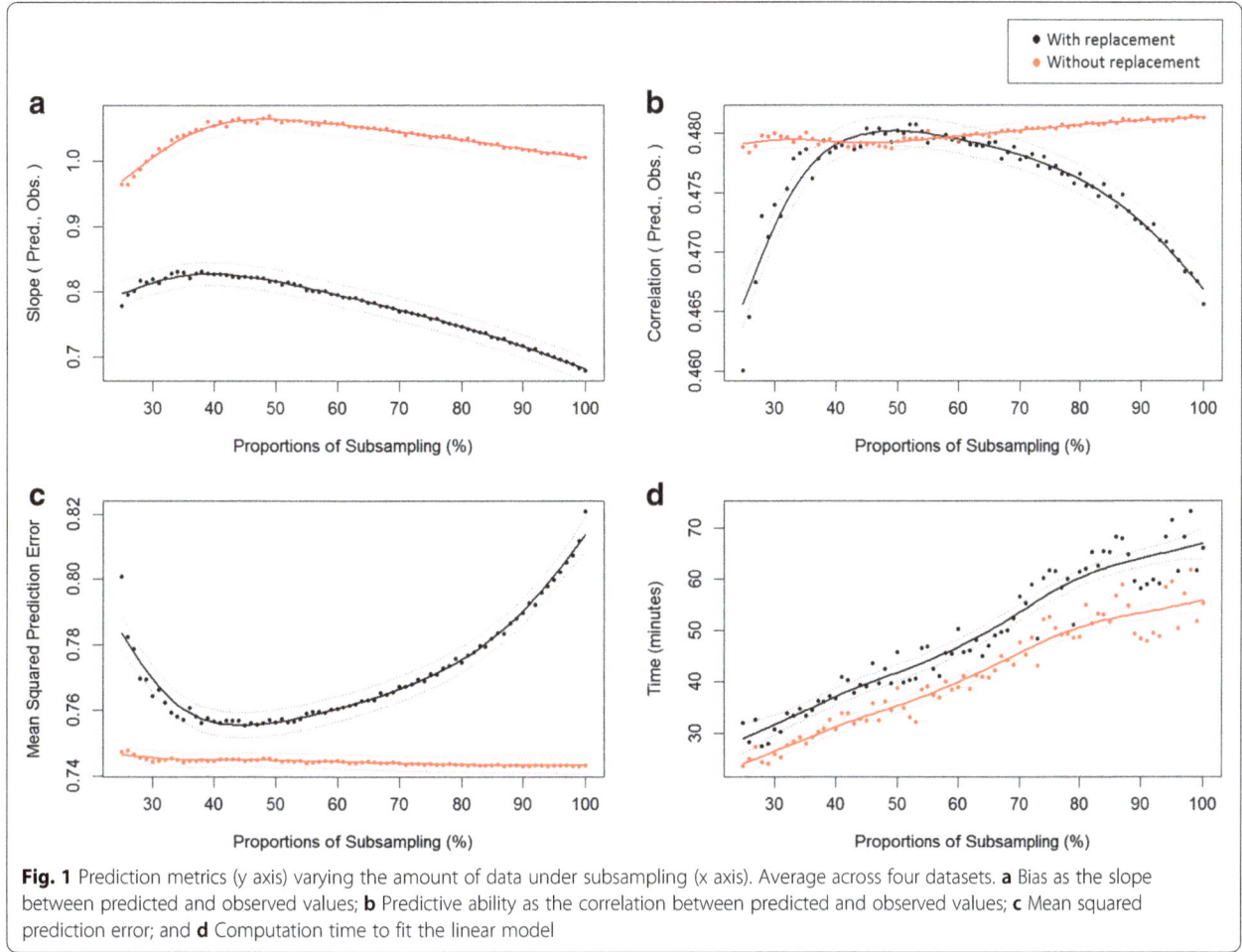

**Fig. 1** Prediction metrics (y axis) varying the amount of data under subsampling (x axis). Average across four datasets. **a** Bias as the slope between predicted and observed values; **b** Predictive ability as the correlation between predicted and observed values; **c** Mean squared prediction error; and **d** Computation time to fit the linear model

more biased than the complete dataset at 33 and 50% subsampling, respectively.

### Predictive ability

Across datasets (Table 2), the loss in predictive ability was negligible. Correlation between predictions and observations decreased 0.2% by subsampling with replacement at 50% subsampling, and 0.4% without replacement at both 33 and 50% subsampling.

**Table 2** Summary of prediction metrics with for the complete dataset (Complete), and subsampling 50% with replacement (wR), and 33 and 50% without replacement (woR)

|          | Time (min.) | $Cor_{y,y}$ | MSPE   | $\beta_{y,y}$ |
|----------|-------------|-------------|--------|---------------|
| Complete | 55.90       | 0.4814      | 0.7431 | 1.0058        |
| woR 33%  | 27.90       | 0.4794      | 0.7454 | 1.0239        |
| woR 50%  | 35.32       | 0.4794      | 0.7447 | 1.0642        |
| wR 50%   | 41.84       | 0.4802      | 0.7562 | 0.8161        |

$Cor_{y,y}$, correlation between observed and predicted value; MSPE, mean squared prediction error; $\beta_{y,y}$, Prediction bias

### MSPE

The negative impact on MSPE due to subsampling was also negligible. An increase of 0.3 and 0.2% were observed at 33 and 50% subsampling without replacement (Table 2). The impact of subsampling on MSPE was slightly higher with replacement, increasing 1.76% at 50% subsampling.

### Dataset specific analysis

Although negligible, we observed a slight improvement in predictive ability and MSPE for all datasets at some optimal subsampling point. The optimal subsampling and respective improvement in predictive ability and MSPE are presented in Table 3.

## Discussion

### Prediction machinery

Any algorithm that enhances prediction or computation performance is valuable for machine learning. At its optimal utilization, SBMC has the potential of improving prediction while reducing the computational cost [14]. However, reported results vary regarding any prediction

**Table 3** Optimal sampling observed for individual datasets to enhance predictive ability (PA) and mean squared prediction error (MSPE). Subsampling performed with (wR) and without replacement (woR)

|  | Optimal PA | Increase in PA | Optimal MSPE | Decrease in MSPE |
|---|---|---|---|---|
| Mouse | wR 66% | 2.5% | woR 32% | <0.1% |
| Soybean | woR 25% | 0.1% | woR 25% | 0.1% |
| Wheat | woR 34% | 0.7% | woR 33% | 0.5% |
| Simulated $F_2$ | wR 87% | 0.1% | wR 66% | 0.2% |

improvement provided by subsampling [8, 23]. Subsampling has not been investigated in big data, for neither large $n$ nor large $p$, and that is a specific niche where subsampling may work best.

Previous studies indicate that there are no guarantees that SBMC will improve prediction, but it at least provides results equivalent to the whole dataset; however, we showed that subsampling can also provide a positive outcome for genomic prediction besides the computational aspects (Table 3), where the improvement reached 2.5% for the mouse data. We recommend including a bagging WGR with 50% subsampling without replacement in cross-validation studies looking for the most accurate prediction model.

### Random data

An interesting statistical property provided by SBMC is that data is sampled from a larger set, which is associated with that definition of a random term. This occurs because the observations used to update parameters are sampled from the empirical distribution of the data. This property violates the Bayesian assumption that data are *given*.

In classical Bayesian analysis, inferences are made based upon the posterior distribution of *parameters given data*, whereas random data implies that the parameters are sampled from the distribution of parameters given the current state of data. MCMC drives the posterior towards a relative entropy, possibly with larger sample variance associated with the continuous resampling used to update parameters with different subsets of data, but without obvious implications for the interpretation of the results [24].

### Incompleteness of data

Geyer [25] discussed the issue of subsampling Markov chains concluding "one does not get a better answer by throwing away data." Nevertheless, he emphasizes that the value of the technique is 1) the reduction of dimensionality of $n$, and 2) the reduction of auto-correlation among chains.

Our counterargument is that the all data are used in the course of model fitting, although not simultaneously. In addition, accurate estimates are obtained when the subsampling strategy is used correctly [14]. We show

that subsampling is a valid approach for genomic prediction purposes to fit high-dimensional models ($p \gg n$).

### Future directions

Subsampling uses only part of the data to fit the model in each MCMC round, that enables the computation of prediction statistics with the subset left out, which is referred to as out-of-bag statistics (OOB) [26]. The information provided by OOB is similar to the outcome of a cross-validation, with the advantage of being computed during the model fitting. Therefore, OOB could be used to re-weight observations (i.e., boosting). Another possibility is to adapt SBMC to other learning methods, such as elastic net [27], where OOB statistics could be utilized in the search for the tuning parameters without having to perform explicit cross-validation [28].

### Conclusion

SBMC decreases computation time without compromising prediction properties. We observed that subsampling approximately 33–50% without replacement and 40–60% with replacement in each round of MCMC is advantageous for fitting the model. Subsampling can dramatically reduce computational burden with little reduction in accuracy and, in some cases, enhanced predictive properties. This study provides insight into a general method for incorporating a particular type of bagging ensemble into the Gibbs sampling of whole-genome regressions.

### Additional file

**Additional file 1:** Results presented by individual dataset **Figure S1.** Time to fit the model (y axis) varying the subsampling method (x axis). **Figure S2.** Prediction ability (y axis) varying the subsampling method (x axis). Methods include Bayesian ridge regression (BRR) with regular sampler, and SBMC subsampling from 25 to 100%, with and without replacement. **Figure S3.** Mean squared prediction error (y axis) varying the subsampling method (x axis). Methods include Bayesian ridge regression (BRR) with regular sampler, and SBMC subsampling from 25 to 100%, with and without replacement. **Figure S4.** Bias (y axis) varying the subsampling method (x axis). Methods include Bayesian ridge regression (BRR) with regular sampler, and SBMC subsampling from 25 to 100%, with and without replacement.

### Abbreviations

CIMMYT: International Maize and Wheat Improvement Center; GBLUP: Genomic best linear unbiased prediction; GSRU: Gauss-Seidel residual updating; GWP: Genome-wide prediction; LHS: Left-hand side; MCMC: Markov chain Monte

Carlo; MSPE: Mean square prediction error; OOB: Out-of-bag statistics;
REML: Restricted maximum likelihood; RHS: Right-hand side; SBMC: Subsampling
bootstrap Markov chain; SNP: Single nucleotide polymorphism

## Acknowledgments
Not applicable.

## Funding
This study did not receive any specific funding.

## Authors' contribution
AX wrote the manuscript. SX revised the mathematical notation and
theoretical basis of subsampling. WM and KMR provided insight for the
method evaluation, predictive metrics and the suggested the datasets. All
authors read and approved the final manuscript.

## Competing interests
The authors declare that they have no competing interests.

## Author details
[1]Department of Agronomy, Purdue University, 915 W. State St., Lilly Hall,
West Lafayette, IN 47907, USA. [2]Department of Plant Science, University of
California, 3134 Batchelor Hall, Riverside, CA 92521, USA. [3]Department of
Animal Science, Purdue University, 915 W. State St., Lilly Hall, West Lafayette,
IN 47907, USA.

## References
1. Okser S, Pahikkala T, Airola A, Salakoski T, Ripatti S, Aittokallio T. Regularized machine learning in the genetic prediction of complex traits. PLoS Genet. 2014;10(11):e1004754.
2. Hastie T, Tibshirani R, Friedman J. Elements of statistical learning. New York: Springer; 2009.
3. Gianola D, Weigel KA, Krämer N, Stella A, Schön CC. Enhancing genome-enabled prediction by bagging genomic BLUP. PLoS One. 2014;9(4):e91693.
4. de los Campos G, Hickey JM, Pong-Wong R, Daetwyler HD, Calus MP. Whole-genome regression and prediction methods applied to plant and animal breeding. Genetics. 2013;193(2):327–45.
5. Gianola D. Priors in whole-genome regression: the Bayesian alphabet returns. Genetics. 2013;194(3):573–96.
6. Misztal I. Inexpensive computation of the inverse of the genomic relationship matrix in populations with small effective population size. Genetics. 2016;202(2):401–9.
7. Scott SL, Blocker AW, Bonassi FV, Chipman HA, George EI, McCulloch RE. Bayes and big data: the consensus monte carlo algorithm. Int J Manag Sci Eng Manag. 2016;11(2):78–88.
8. Flegal JM. Applicability of subsampling bootstrap methods in Markov chain monte Carlo. In: Monte Carlo and quasi-monte Carlo methods. Heidelberg: Springer; 2012. p. 363–72.
9. Kleiner A, Talwalkar A, Sarkar P, Jordan M. The big data bootstrap. arXiv preprint arXiv;1206.6415;2012.
10. Politis DN, Romano JP, Wolf M. On the asymptotic theory of subsampling. Statistica Sinica. 2001;11(4):1105–24.
11. James G, Witten D, Hastie T, Tibshirani R. An introduction to statistical learning. New York: Springer; 2013. p. 331.
12. Alfaro ME, Zoller S, Lutzoni F. Bayes or bootstrap? a simulation study comparing the performance of Bayesian Markov chain monte Carlo sampling and bootstrapping in assessing phylogenetic confidence. Mol Biol Evol. 2003;20(2):255–66.
13. Geyer CJ. Practical Markov chain monte Carlo. Stat Sci. 1992;7(4):473–83.
14. MacEachern SN, Peruggia M. Subsampling the gibbs sampler: variance reduction. Stat Probab Lett. 2000;47(1):91–8.
15. Legarra A, Misztal I. Technical note: Computing strategies in genome-wide selection. J Dairy Sci. 2008;91(1):360–6.
16. Sorensen D, Gianola D. Likelihood, Bayesian, and MCMC methods in quantitative genetics. Springer Science & Business Media, New York. 2002.
17. Xavier A, Muir W, Rainey KM. bWGR: Bagging Whole-Genome Regression. CRAN, version 1.3.1. 2016.
18. Pérez P, de Los Campos G. Genome-wide regression & prediction with the BGLR statistical package. Genetics. 2014;198(2):483–95.
19. Xavier A, Beavis WD, Specht JE, Diers BW, Howard R, Muir WM, Rainey KM. SoyNAM: Soybean Nested Association Mapping Dataset. CRAN, version 1.2. 2015.
20. Crossa J, de Los Campos G, Pérez P, Gianola D, Burgueño J, Araus JL, Makumbi D, Singh RP, Dreisigacker S, Yan J, Arief V. Prediction of genetic values of quantitative traits in plant breeding using pedigree and molecular markers. Genetics. 2010;186(2):713–24.
21. Legarra A, Robert-Granié C, Manfredi E, Elsen JM. Performance of genomic selection in mice. Genetics. 2008;180(1):611–8.
22. Xu S. Mapping quantitative trait loci by controlling polygenic background effects. Genetics. 2013;195(4):1209–22.
23. Brooks S, Gelman A, Jones G, Meng XL. editors. Handbook of Markov Chain Monte Carlo. CRC Press. 2011.
24. Shalizi CR. Dynamics of Bayesian updating with dependent data and misspecified models. Electron J Stat. 2009;3:1039–74.
25. Geyer CJ. Introduction to Markov chain Monte Carlo. Handbook of Markov Chain Monte Carlo. Chapman and Hall/CRC. 2011;10:3–48.
26. Breiman L. Out-of-bag estimation. Technical Report, Statistics Department, University of California Berkeley, Berkeley CA 94708. 1996b;33,34; 1996.
27. Zou H, Hastie T. Regularization and variable selection via the elastic net. J R Stat Soc Series B Stat Methodology. 2005;67(2):301–20.
28. Xavier A, Muir WM, Craig B, Rainey KM. Walking through the statistical black boxes of plant breeding. Theor Appl Genet. 2016;129(10):1933–49.
29. Technow F. hypred: Simulation of Genomic Data in Applied Genetics. CRAN, version 0.5. 2014.

# 2

# NoGOA: predicting noisy GO annotations using evidences and sparse representation

Guoxian Yu[*] 🆔, Chang Lu and Jun Wang

## Abstract

**Background:** Gene Ontology (GO) is a community effort to represent functional features of gene products. GO annotations (GOA) provide functional associations between GO terms and gene products. Due to resources limitation, only a small portion of annotations are manually checked by curators, and the others are electronically inferred. Although quality control techniques have been applied to ensure the quality of annotations, the community consistently report that there are still considerable noisy (or incorrect) annotations. Given the wide application of annotations, however, how to identify noisy annotations is an important but yet seldom studied open problem.

**Results:** We introduce a novel approach called *NoGOA* to predict noisy annotations. NoGOA applies sparse representation on the gene-term association matrix to reduce the impact of noisy annotations, and takes advantage of sparse representation coefficients to measure the semantic similarity between genes. Secondly, it preliminarily predicts noisy annotations of a gene based on aggregated votes from semantic neighborhood genes of that gene. Next, NoGOA estimates the ratio of noisy annotations for each evidence code based on direct annotations in GOA files archived on different periods, and then weights entries of the association matrix via estimated ratios and propagates weights to ancestors of direct annotations using GO hierarchy. Finally, it integrates evidence-weighted association matrix and aggregated votes to predict noisy annotations. Experiments on archived GOA files of six model species (H. sapiens, A. thaliana, S. cerevisiae, G. gallus, B. Taurus and M. musculus) demonstrate that NoGOA achieves significantly better results than other related methods and removing noisy annotations improves the performance of gene function prediction.

**Conclusions:** The comparative study justifies the effectiveness of integrating evidence codes with sparse representation for predicting noisy GO annotations. Codes and datasets are available at http://mlda.swu.edu.cn/codes.php?name=NoGOA.

**Keywords:** Gene ontology, GO annotations, Evidence codes, Sparse representation

## Background

With the influx of biological data, it is difficult for researchers to collect and search functional knowledge of gene products (including proteins and RNAs), as different databases use different schemas to describe gene functions. To overcome this problem, Gene Ontology Consortium (GOC) collaboratively developed Gene Ontology (GO) [1]. GO has two components: GO and GO annotations (GOA) files. GO uses structured vocabularies to annotate molecular function, biological roles and cellular location of gene products in a taxonomic and species-neutral way. Particularly, GO arranges GO terms into three branches: molecular function (MF), biological process (BP) and cellular component (CC). Each branch organizes terms in a direct acyclic graph to reflect hierarchical structure relationship among them. GOA files store functional annotations of gene products, which associate gene products with GO terms. Each annotation encodes the knowledge that the relevant gene products carry out the biological function described by the associated GO term. Hereinafter, for brevity, we abuse annotations of gene products as annotations of genes.

GO annotations are originally extracted from published experimental data by GO curators. These annotations provide solid, dependable sources for function inference

*Correspondence: gxyu@swu.edu.cn
College of Computer and Information Sciences, Southwest University, Chongqing, China

[2], and are also biased by the research interests of biologists [3]. With the development and application of high-throughput technologies, accumulated large volume of biological data enable to computationally predict gene functions. Various computational approaches have been proposed to predict gene function without curator intervention [4, 5]. Manually checking these electronically predicted annotations is low throughput and labor-intensive.

Electronically inferred annotations provide a broad coverage and have a significantly larger taxonomic range than manual ones [6, 7]. On the one hand, since these annotations are not checked by curators, they may have lower reliability than manual ones [8]. On the other hand, curated annotations are restricted by experiment protocols and contexts [3]. Therefore, both inferred and curated annotations include some incorrect annotations [9]. As we known, GO is regularly updated with some terms obsolete or appended as the updated biological knowledge. Similarly, annotations of genes are also updated as the accumulated biological evidences and evolved GO. However, we want to remark that the removed annotations in archived GOA files, from our preliminary investigation, do not solely result from updated GO terms and structure. For example, in an archived (date: May 9th, 2016) GOA file of S. cerevisiae, 'AAC1' (ADP/ATP Carrier) was annotated with a GO term 'GO:0006412' (translation), but 'AAC1' was not annotated with 'GO:0006412' in a recently archived (date: September 24th, 2016) GOA file. Further investigation using QuickGO [10] shows this removed annotation is not caused by the change of GO. In fact, annotations in archived GOA files have already underwent several quality control measures to ensure consistency and quality [7]. Gross et al. [11] studied the evolution and (in)stability of GO annotations and found that there were evolution operations for annotations. These instable annotations are not only caused by the changes of gene products or ontology, but also by the incorrect (or inappropriate) annotations. Gross et al. [12] further found that past changes in the GO and GOA are non-uniformly distributed over different branches of the ontology. Gillis et al. [13] also showed instabilities of annotation data and detected that 20% annotations of the genes could not be mapped to themselves after a two year interval. Clarke et al. [14] investigated annotations and structural ontology changes from 2004 to 2012, and found that annotation changes are largely responsible for the changes of enrichment analysis on angiogenesis and the most significant terms. These observations suggest that there are some incorrect annotations in GOA files. Hereinafter, we call these incorrect annotations as *noisy* annotations. These noisy annotations can mislead the downstream analysis and applications, such as GO enrichment analysis [14, 15], diseases analysis [16], drug repositioning [17] and so on.

Some researchers tried to improve annotation quality using association rules. Faria et al. [18] summarized that erroneous annotations, incomplete annotations, and inconsistent annotations affect the annotation quality, and introduced a association rule learning method to evaluate inconsistent annotations in the MF branch. Agapito et al. [19] considered different GO terms have different information contents, and proposed a weighted association rule solution based on the information contents to improve annotation consistencies. This solution only uses one ontology. Agapito et al. [20] extended this solution to mine cross-ontology association rules, i.e., association rules whose terms belong to different branches of GO. Despite these efforts to avoid errors and inconsistencies, most groups are more concerned with replenishing (or predicting) new GO annotations of genes than removing noisy ones [5, 7], and how to predict noisy annotations is a rarely studied but essential problem.

Each GO annotation is tagged with an evidence code, recording the type of evidence (or source) the annotation extracted from [1, 8]. GO currently uses 21 evidence codes and divides them into four categories, which are shown in Table 1. All these evidence codes are reviewed by curators, except IEA (Inferred from Electronic Annotation). There are several studies on assessing GO annotation quality with evidence codes. Thomas et al. [21] recommended to use evidence codes as indicator for the reliability of annotations. They investigated annotations of different species and categorized homology-based, literature-based and other annotations, and found that literature-based (experimental and author statement) annotations are more reliable than others. Clark et al. [22] investigated the quality of NAS (Non-traceable Author Statement) and IEA annotations, and found IEA annotations were much more reliable in MF branch than NAS ones. Gross et al. [11] estimated stability and quality of different evidence codes by considering evolutionary changes. Buza et al. [23] took advantage of GO annotation quality score based on a ranking of evidence codes to assess the quality of annotations available for specific biological processes. Jones et al. [24] found that electronic annotators that using ISS (Inferred from Sequence or structural Similarity) annotations as the basis of predictions are likely to have higher false prediction rates, and suggested to consider avoiding ISS annotations where possible. All these methods just analyze the quality of annotations for different evidence codes. However, none of them pay attention to automatically predicting noisy GO annotations.

Evidence codes are also adopted to measure the semantic similarity between genes [25, 26]. Benabderrahmane et al. [25] assigned different weights to GO annotations based on the evidence codes tagged with these annotations, and used a graph-based similarity measure to compute the semantic similarity between genes. They

**Table 1** Four categories of evidence codes used in GO and their meanings

| Experimental | Computational | Author | Curatorial |
|---|---|---|---|
| EXP: inferred from experiment | ISS: inferred from sequence or structural similarity | TAS: traceable author statement | IC: inferred by curator |
| IDA: inferred from direct assay | ISO: inferred from sequence orthology | NAS: non-traceable author statement | ND: no biological data available |
| IPI: inferred from physical interaction | ISA: inferred from sequence alignment | | |
| IMP: inferred from mutant phenotype | ISM: inferred from sequence model | | |
| IGI: inferred from genetic interaction | IGC: inferred from genomic context | | |
| IEP: inferred from expression pattern | IBA: inferred from biological aspect of ancestor | | |
| | IBD: inferred from biological aspect of descendant | | |
| | IKR: inferred from key residues | | |
| | IRD: inferred from rapid divergence | | |
| | RCA: inferred from reviewed computational analysis | | |
| | IEA: inferred from electronic annotation | | |

observed this evidence weighted semantic similarity was more consistent with the sequence similarity between genes than the counterpart without considering the evidence codes. Semantic similarity is found to be positively correlated with the sequence similarity between genes, protein-protein interactions and other types of biological data [27, 28]. Given that, it has been applied to predict the missing annotations of incompletely annotated genes and to validate protein-protein interactions [29–31]. Lu et al. [32] pioneered noisy annotations prediction and suggested a method called NoisyGOA. NoisyGOA firstly computes a vector-based semantic similarity between genes, and a taxonomic similarity between terms using GO hierarchy. Then, it aggregates the maximal taxonomic similarity between terms annotated to a gene and terms annotated to neighborhood genes. After that, it takes terms with the smallest aggregated scores as noisy annotations of the gene. However, NoisyGOA is still suffered from noisy annotations in measuring the semantic similarity between genes, and it does not differentiate the reliability of different annotations.

There are more than 43,000 terms in GO and each gene is often annotated with dozens or several of these terms. From this perspective, the gene-term association matrix, encoding GO annotations of genes, is sparse with some noisy entries. To accurately measure the semantic similarity between genes, we use sparse representation [33], which has been extensively applied in image and signal de-noising, sparse feature learning [34]. When the input signals are sparse with some noises, sparse representation shows superiority in capturing the ground-truth signals. Motivated by these observations, we advocate to integrate sparse representation with evidence codes to predict noisy annotations and introduce an approach called *NoGOA*. NoGOA applies sparse representation on the gene-term matrix to compute the sparse representation coefficients and takes the coefficients as the semantic similarity between genes. Then, it votes noisy annotations of a gene based on annotations of its neighborhood genes. Next, it estimates ratios of noisy annotations for each evidence code based on archived GOA files in different releases, and weights each entry of the gene-term matrix by estimated ratios and GO hierarchy. The final prediction of noisy annotations is obtained from the integration of the weighted gene-term matrix and the aggregated votes from neighborhood genes.

There are no off-the-shelf noisy annotations to quantitatively study the performance of NoGOA in predicting noisy annotations. For this purpose, we collected GOA files archived on four different periods, May 2015, May 2016, September 2015 and September 2016. For each year, we call the GOA file archived in May as the *historical* one, and the GOA file archived in September as the *recent* one. We take the annotations available in the historical GOA file but absent in the recent one as noisy annotations. Based on this protocol, we conducted experiments on archived GOA files of six model species (H. Sapiens, A. thaliana, S. cerevisiae, G. gallus, B. Taurus and M. musculus). Comparative study shows that noisy annotations are predictable and NoGOA outperforms other related techniques in predicting noisy annotations. The empirical study also demonstrates removing noisy annotations can significantly improve the performance of gene function prediction.

## Methods

Let $\mathbf{A} \in \mathbb{R}^{N \times |\mathcal{T}|}$ be a gene-term association matrix, $N$ is the number of genes, $\mathcal{T}$ is the set of GO terms and $|\mathcal{T}|$ is the cardinality of $\mathcal{T}$. $\mathbf{A}$ is defined as follows:

$$\mathbf{A}(i,t) = \begin{cases} 1, & \text{if gene } i \text{ is annotated with} \\ & \text{term } t \text{ or } t\text{'s descendants} \\ 0, & \text{otherwise} \end{cases} \quad (1)$$

The objective of NoGOA is to identify noisy annotations in $\mathbf{A}$ and update corresponding entries from 1 to 0. Although identifying noisy annotations can be viewed as a different face of gene function prediction, we still would like to remark that identifying noisy annotations is different from replenishing missing annotations of incompletely annotated genes [29, 31], which updates some entries of $\mathbf{A}$ from 0 to 1. It is also different from negative examples selection [35, 36], which updates some entries of $\mathbf{A}$ from 0 to -1 and indicates that the relevant genes are clearly not annotated with the given GO terms.

### Preliminary noisy annotations prediction using sparse representation

In this section, we firstly compute the semantic similarity between genes, and then use this similarity to select neighborhood genes of a gene and to preliminarily infer noisy annotations. There are some noisy annotations in the GOA files. In other words, there are some noisy entries in $\mathbf{A}$. Although various semantic similarity measures have been proposed and widely applied, most of them are still suffered from shallow and incomplete GO annotations of genes [27, 28, 37, 38]. Sparse representation has been widely and successfully applied to handle images with blurs, speech data with noises and to recover samples with noisy features [33, 34]. Actually, the portion of non-zero entries in $\mathbf{A}$ is no more than 2%. Therefore $\mathbf{A}$ is a sparse matrix with some noisy entries. Given the characteristics of $\mathbf{A}$ and of sparse representation, we resort to sparse representation on $\mathbf{A}$ to measure the semantic similarity between genes. In this paper, we use an $l_1$ norm regularized sparse representation objective function as follows:

$$\hat{\gamma}_i = \arg \min_{\gamma_i} ||\mathbf{A}(i,\cdot) - \gamma_i^T \bar{\mathbf{A}}^i||_2 + \lambda ||\gamma_i||_1, s.t. \, \gamma_i \geq 0 \quad (2)$$

The target of sparse representation is to find a sparse coefficient vector $\gamma_i \in \mathbb{R}^{(N-1)}$, with $\mathbf{A}(i,\cdot) \approx \gamma_i^T \bar{\mathbf{A}}^i$ and $||\gamma_i||_1$ is minimized. $||\gamma_i||_1$ is the $l_1$ norm that sums the absolute values of $\gamma_i$, and minimizing $||\gamma_i||_1$ can enforce $\gamma_i$ to be a sparse vector. $\lambda(> 0)$ is a scalar regularization parameter that balances the tradeoff between reconstruction error and sparsity of coefficients [34]. $\bar{\mathbf{A}}^i \in$

$\mathbb{R}^{(N-1) \times |\mathcal{T}|}$ is a sub-matrix of $\mathbf{A}$ with the $i$-th row removed. In this way, $\mathbf{A}(i,\cdot)$ is linearly reconstructed by other rows of $\mathbf{A}$, instead of itself. $\gamma_i(j)$ can be seen as the reconstruction contribution of $\mathbf{A}(j,\cdot)$ to $\mathbf{A}(i,\cdot)$. In other words, the larger the semantic similarity between $\mathbf{A}(i,\cdot)$ and $\mathbf{A}(j,\cdot)$, the larger the $\gamma_i(j)$ is. Here, we solve the optimal $\gamma_i$ using the sparse learning with efficient projection package [39]. To further explain the usage of sparse representation to measure the semantic similarity between genes, we provide a simple workflow in Additional file 1: Figure S1.

Next, we employ $\gamma_i$ to define the semantic similarity between the $i$-th gene with respect to other genes, and use $\mathbf{S} \in \mathbb{R}^{N \times N}$ to store the semantic similarity between $N$ genes. $\mathbf{S}(i,\cdot)$ stores the similarity of the $i$-th gene with other genes, and it is defined as follows:

$$\mathbf{S}(i,j) = \begin{cases} \gamma_i(j), & \text{if } j < i \\ \gamma_i(j-1), & \text{if } j > i \\ 0, & \text{otherwise} \end{cases} \quad (3)$$

By iteratively applying Eqs. (2–3) for $N$ genes, we can sequentially fulfil each row of $\mathbf{S}$. The similarity between a gene and itself is set as 0, since noisy annotations of a gene are predicted based on the annotations of semantic similar genes of that gene, instead of itself. To make $\mathbf{S}$ being a symmetric matrix, we set $\mathbf{S} = (\mathbf{S}^T + \mathbf{S})/2$. In fact, various approaches [34] utilize Eq. (3) to measure the similarity between samples, and find this similarity often performs better than many other widely-used similarity metrics, and is robust to noisy features.

A simple and intuitive idea to predict noisy annotations of a gene is to select neighborhood genes of a gene based on the semantic similarity between them and regard these genes as voters, and then to vote whether a term should be removed or not, based on the term's association with these voters. The fewer votes the term obtains, the more likely the term as a noisy annotation of the gene is. In fact, this idea is widely used to aggregate annotations and to solve the disagreement between annotators [40, 41], and also adopted by NoisyGOA [32]. However, this idea does not differentiate varieties of neighborhood genes. To take into account these varieties, we use the semantic similarity derived from sparse representation to predict noisy annotations. If $t$ is annotated to gene $i$, namely $\mathbf{A}(i,t) > 0$, the aggregated vote of $t$ for the gene is counted as follows:

$$\mathbf{V}_{SR}(i,t) = \sum_{j=1}^{N} \mathbf{S}(i,j) \times \mathbf{A}(j,t) \quad (4)$$

Equation (4) is similar to a weighted $k$ nearest neighborhood ($k$NN) classifier [42], since $\mathbf{S}(i,\cdot)$ is a sparse vector with most entries as (or close to) zeros and neighborhood genes of gene $i$ are automatically determined by these nonzero entries. Equation (4) can be regarded as a weighted voting method and the weights are specified by the semantic similarity between them. If a term is annotated to a gene, but this term is not (or less frequently) annotated to that gene's neighborhood genes than other terms, then this term has a larger probability as a noisy annotation of that gene than other terms. Here, we want to remark that if gene $i$ has few similar genes, then all entries in $\mathbf{S}(i, \cdot)$ will be equal or close to zeros. Consequently, terms annotated this gene are more likely to receive lower voting scores and to be identified as noisy annotations. Indeed, this extreme case is worthwhile for future pursue.

### Weighting annotations using evidence codes

Using aggregated votes to predict noisy annotations is a feasible solution [32, 41], but it does not take into account the differences among annotations. Evidence codes, attached with GO annotations, illustrate the sources where these annotations collected from. Some researchers used GO annotations archived on different periods to analyse the quality of annotations under different evidences codes [11, 21, 24], and found the quality varying among different branches and evidence codes. Motivated by these analysis, we estimate the ratios of noisy annotations for each evidence code in each branch and then employ the ratios to weight the gene-term association matrix $\mathbf{A}$. Here, we collected two GOA files that archived on different months, then we take the annotations available in the former month but absent in the latter month as noisy annotations of the former GOA file. To account for GO change and its cascade influence on GO annotations, we only use the shared GO hierarchy in the two contemporary GO files. Let $N^m(c)$ be the number of annotations attached with evidence code $c$ in the $m$-th version GOA file, and $\bar{N}^m(c)$ be the number of noisy annotations tagged with evidence code $c$ in that GOA file. The estimated ratio of noisy annotations for $c$ can be approximated as:

$$r_{ec}^m(c) = \frac{\bar{N}^m(c)}{N^m(c)} \tag{5}$$

To more accurately estimate the ratio of noisy annotations for the $m$-th version, we sum up the ratios estimated from its $l$ previous versions as follows:

$$\tilde{r}_{ec}^m(c) = \frac{1}{l} \sum_{l'=m-l+1}^{m} r_{ec}^{l'}(c) \tag{6}$$

Obviously, a large $\tilde{r}_{ec}^m(c)$ indicates annotations tagged with $c$ are unstable and more likely to contain noisy annotations, since they change frequently in the previous versions. Based on $\tilde{r}_{ec}^m(c)$, we set different weights to different evidence codes as follows:

$$w_{ec}(c) = \begin{cases} 1, & \text{if } \tilde{r}_{ec}^m(c) < \tau \\ \theta, & \text{otherwise} \end{cases} \tag{7}$$

$\tau$ is a threshold and set as the average value of $\tilde{r}_{ec}^m$ with respect to different evidence codes. Annotations tagged with evidence codes whose $\tilde{r}_{ec}^m(c) \geqslant \tau$ are unstable and likely to be noisy annotations. Therefore, we set $w_{ec}$ of these annotations as $\theta(< 1)$, and others as 1. Other specifications of $\theta$ and $\tau$ is postponed to be discussed in the next section.

GOC follow a convention to annotate genes with the appropriate and as well as specific terms that correctly describe the biology of the genes. The annotations stored in the GOA files are called *direct* annotations, and each of them is tagged with an evidence code. To make use of these direct annotations and evidence codes, if $\mathbf{A}^d(i, t)$ is tagged with evidence code $c$, we update the gene-term association matrix $\mathbf{A}^d \in \mathbb{R}^{N \times |\mathcal{T}|}$ as follows:

$$\mathbf{A}_{ec}^d(i, t) = \mathbf{A}^d(i, t) \times w_{ec}(c) \tag{8}$$

where $\mathbf{A}^d$ is initialized by direct annotations only. If there are multiple evidence codes for the same gene-term association $\mathbf{A}^d(i, t)$, we set the maximal weight of these involved evidence codes to $\mathbf{A}_{ec}^d$.

Annotated with a term implies the gene also annotated with its ancestor terms via any path of GO hierarchy. In other words, if a gene is annotated with term $t$, this gene is inherently annotated with all the ancestors of $t$. This rule is called *true path rule* [1, 43]. To make use of this rule, we propagate the weights and extend $\mathbf{A}_{ec}^d$ to ancestor annotations of direct ones as follows:

$$\mathbf{A}_{ec}(i, s) = max \left\{ \mathbf{A}_{ec}^d(i, t) | s \in anc(t) \right\} \tag{9}$$

where $anc(t)$ includes all ancestors of $t$. If ancestor annotation $s$ is propagated from two or more direct annotations, we take maximal value of these direct annotations as the weight of $\mathbf{A}_{ec}(i, s)$. This setting ensures the weights of ancestor annotations equal (or larger) than descendant annotations, since a descendant term describes more specific biological function than its ancestor terms and annotations with respect to ancestor terms are generally more easier to be verified than descendant ones. Another reason for this maximal setting is motivated by accumulated evidences from different sources. If the weight for an ancestor annotation is smaller than its descendant ones, the relevant term will be more likely to be identified as a noisy annotation than its descendants. This setting is not desirable. From the true path rule, if the ancestor term is not annotated to a gene, then all its descendants are not annotated to that gene, too.

### Noisy annotations prediction

To this end, we integrate the evidence weighted annotations in Eq. (9) and aggregated votes in Eq. (4) to predict noisy GO annotations of genes as follows:

$$\mathbf{V}(i, t) = \alpha \times \mathbf{V}_{SR}(i, t) + (1 - \alpha) \times \mathbf{A}_{ec}(i, t) \tag{10}$$

where $\alpha$ is a scalar parameter to adjust the contribution of $\mathbf{V}_{SR}$ and $\mathbf{A}_{ec}$. If both $t$ and $s$ are annotated to the $i$-the gene and $\mathbf{V}(i,t) < \mathbf{V}(i,s)$, then $t$ is more likely to be a noisy annotation than $s$. Eq. (10) is motivated by the observation that if a term is annotated to a gene, but this term is not (or rarely) annotated to neighborhood genes of the gene and the evidence code attached with this annotation has a large estimated ratio of noisy annotations, then the annotation is more likely to be a noisy one. One shortcoming of Eq. (10) is that if a noisy annotation appears in successive GOA files and its relevant GO term is frequently annotated to neighborhood genes of the gene, this noisy annotation is difficult to be identified by NoGOA. This kind of noisy annotations are more challenging and remain for future pursue. To select a reasonable value for $\alpha$, we can adjust it in the range [0, 1] by taking GOA files archived prior to the historical GOA files to train NoGOA and use the GOA files archived no late than the historical GOA files to validate the prediction. After that, we can select the optimal $\alpha$ to train NoGOA on the historical GOA files. Fortunately, our following empirical parameter sensitivity analysis shows that it is easy to select a reasonable and consistent $\alpha$ for NoGOA on GOA files of different species.

To predict noisy annotations, NoGOA not only takes advantage of sparse representation to reduce the interference of noisy annotations and of aggregated votes from neighborhood genes, but also weights annotations based on the estimated ratios of noisy annotations with respect to different evidence codes. Therefore, NoGOA has the potential to achieve better performance than using sparse representation or evidence codes alone. Our following experimental study corroborates this advantage and shows evidence codes can be used as a plugin with other semantic similarity based methods to improve the performance in predicting noisy annotations.

## Results and discussion
### Experimental protocols and comparing methods
We downloaded four versions of GOA files (archived in May and September) of six model species [44], *H. sapiens, A. thaliana, S. cerevisiae, G. gallus, B. Taurus* and *M. musculus* to comparatively study the performance of NoGOA and of other comparing methods in two successive years (2015 and 2016), respectively. To mitigate the impact of GO change in long intervals, we use the GO annotations archived in the first four months of the year (2015 or 2016) to estimate the ratio of noisy annotations for each evidence code and the annotations archived in May for prediction. We then validate the prediction based on annotations archived in September of the same year. Accordingly, we also downloaded contemporary GO files [45], which were archived on the same

date as GOA files. To reduce the impact of evolved GO and annotations for evaluation, similar to the 2nd CAFA (Critical Assessment of protein Function Annotation algorithms) [5], we retain the terms that are included both in the historical and recent GO files, and filter out terms that are absent in historical or recent GO files. Next, these retained terms, direct annotations in the GOA files and the inherited ancestor annotations of these direct ones, are used to initialize the historical (archived in May) gene-term association matrix $\mathbf{A}^h$ and recent (archived in September) gene-term matrix $\mathbf{A}^r$, respectively. We consider the annotations available in $\mathbf{A}^h$ but absent in $\mathbf{A}^r$ as noisy annotations. To be honest, this consideration is not very good, because of the complicated evolutionary mechanism of GO and GO annotations [7, 11]. However, since noisy annotations are not readily available, we regard these removed annotations as 'noisy annotations' and use them to validate the predicted noisy annotations made by the comparing methods. The statistics of genes and annotations in 2015 and 2016 are listed in Tables 2 and 3. For instance, in 2016, there are 18,932 genes in H. sapiens and these genes are annotated with 13,172 BP GO terms. These genes in total have 1,141,456 annotations in BP branch, among them there are 22,706 noisy annotations.

To comparatively study the performance of NoGOA, we take eight related methods as comparing methods. The details of these methods are introduced as follows:

(i) *Random* randomly chooses a term annotated to a gene as the noisy annotation of that gene.

(ii) *LF* randomly selects the term annotated to a gene but with the Lowest Frequency among $N$ genes as the noisy annotation of the gene.

(iii) *SR* is solely based on Sparse Representation [34] in Eq. (4) to predict noisy annotations.

(iv) *EC* is solely based on Evidence Code to predict noisy annotations. More specifically, it chooses the term annotated to the $i$-th gene but with lowest weight in $\mathbf{A}_{ec}(i, \cdot)$ as a noisy annotation of the gene.

(v) *NtN* is a semantic similarity based approach that can be adopted to predict noisy annotations [46]. It views each gene as a document and terms annotated to the gene as words of that document. It firstly utilizes the term-frequency, inverse document frequency in vector space model [47], and GO hierarchy to weight annotations located at different locations. Next, it employs singular value decomposition on the weighted gene-term association matrix and then chooses the term annotated to a gene but with lowest entry value in the decomposed matrix as a noisy annotation of that gene.

**Table 2** Statistics of GO annotations of *H. sapiens, A. thaliana, S. cerevisiae, G. gallus, B. Taurus* and *M. musculus* (archived date: May, 2015)

| | Branch($|\mathcal{T}|$) | Annotations | Noisy annotations |
|---|---|---|---|
| | BP (13875) | 1183415 | 23143 |
| H. sapiens(18939) | CC (1672) | 375982 | 2770 |
| | MF (4244) | 234599 | 2322 |
| | BP (5132) | 794092 | 2651 |
| A. thaliana(24377) | CC (848) | 222465 | 498 |
| | MF (2684) | 197422 | 2301 |
| | BP (4768) | 244374 | 898 |
| S. cerevisiae(5887) | CC (931) | 104831 | 87 |
| | MF (2282) | 65745 | 338 |
| | BP (11783) | 572194 | 19603 |
| G. gallus(12782) | CC (1451) | 201471 | 3859 |
| | MF (3350) | 144112 | 2345 |
| | BP (11783) | 768861 | 20788 |
| B. Taurus(17316) | CC (1521) | 272289 | 3745 |
| | MF (3350) | 189509 | 2371 |
| | BP (13744) | 1036467 | 15376 |
| M. musculus(21188) | CC (1621) | 356694 | 1603 |
| | MF (4148) | 231078 | 2195 |

The data in the parentheses of the 1st column is the number of genes, data in the 2nd column is the number of involved GO terms ($|\mathcal{T}|$), the 3rd column is the number of annotations for a particular branch, and the last column is the number of noisy annotations, which were available in the GOA file archived in May, but absent in the GOA file archived in September of the same year

**Table 3** Statistics of GO annotations of *H. sapiens, A. thaliana, S. cerevisiae, G. gallus, B. Taurus* and *M. musculus* (archived date: May, 2016)

| | branch($|\mathcal{T}|$) | Annotations | Noisy annotations |
|---|---|---|---|
| | BP (13172) | 1141456 | 22706 |
| H. sapiens(18932) | CC (1707) | 385525 | 3141 |
| | MF (4345) | 243928 | 4660 |
| | BP (4157) | 243249 | 15918 |
| A. thaliana(6931) | CC (750) | 97616 | 2937 |
| | MF (2271) | 81318 | 3554 |
| | BP (4385) | 222754 | 13647 |
| S. cerevisiae(6719) | CC (990) | 108186 | 2768 |
| | MF (2379) | 65032 | 4394 |
| | BP (10643) | 244374 | 898 |
| G. gallus(10912) | CC (1429) | 177491 | 4448 |
| | MF (3298) | 124997 | 2130 |
| | BP (11724) | 753976 | 6541 |
| B. Taurus(17886) | CC (1550) | 281284 | 2244 |
| | MF (3298) | 194425 | 1396 |
| | BP (13141) | 481417 | 18182 |
| M. musculus(21279) | CC (1686) | 367461 | 3917 |
| | MF (4238) | 239664 | 2705 |

The data in the parentheses of the 1st column is the number of genes, data in the 2nd column is the number of involved terms ($|\mathcal{T}|$), the 3rd column is the number of annotations for a particular branch, and the last column is the number of noisy annotations, which were available in the GOA file archived in May, but absent in the GOA file archived in September of the same year

(vi) *NoisyGOA* is originally proposed for predicting noisy annotations by our team [32]. It was elaborated in the last part of the 6th paragraph of Introduction section.

(vii) *NtN+EC* integrates the predictions from evidence code updated gene-term association matrix $\mathbf{A}_{ec}$ (see Eq. (9)) and those from NtN (similar as Eq. (10)) to predict noisy annotations.

(viii) *NoisyGOA+EC* integrates the predictions from $\mathbf{A}_{ec}$ and those from NoisyGOA (similar as Eq. (10)) to predict noisy annotations.

$\lambda = 0.5$ is used in Eq. (2), and the parameters of NtN and NoisyGOA are fixed as the authors suggested in their original papers. In practice, we conducted experiments to study the sensitivity of $\lambda \in [0.1, 1]$ (as suggested by the package provider) [39] and found that NoGOA has stable performance in this range, so we use the median value $\lambda = 0.5$ for experiment. In the following experiments, we denote the number of noisy annotations for gene $i$ as $q$, and then take $q$ entries with nonzero values in $\mathbf{A}(i, \cdot)$ but with the smallest values in $\mathbf{V}(i, \cdot) \in \mathbb{R}^{|\mathcal{T}|}$ (see Eq. (10)) as the predicted noisy annotations of that gene. In this

way, we can avoid genes having fewer neighborhood genes to receive systematically lower voting scores, since we determine noisy annotations by referring to $\mathbf{A}(i, \cdot)$ and $\mathbf{V}(i, \cdot)$, instead of all entries in $\mathbf{V}$. To reach fair comparison, NoGOA and all other comparing methods use the same protocol to select $q$ noisy annotations. This adopted protocol may affect the prediction of noisy annotations. Other more appropriate protocols are interesting future pursue. From the true path rule, if a term is not annotated to a gene, its descendant terms are also not annotated to this gene. To ensure consistency, if the descendant terms of the predicted $q$ terms are annotated to the $i$-th gene, all the comparing methods will take descendant terms of these $q$ terms as predicted noisy annotations of the gene, too.

To quantitatively analyze the performance of noisy annotations prediction, three metrics are adopted: *Precision, Recall* and *F1-Score*. The formal definitions of these metrics are provided as follows:

$$p_i = \frac{TP_i}{TP_i + FP_i}, \; r_i = \frac{TP_i}{TP_i + FN_i} \tag{11}$$

$$\text{Precision} = \frac{1}{N} \sum_{i=1}^{N} p_i, \ \text{Recall} = \frac{1}{N} \sum_{i=1}^{N} r_i \qquad (12)$$

$$\text{F1-Score} = \frac{1}{N} \sum_{i=1}^{N} \frac{2 \times p_i \times r_i}{p_i + r_i} \qquad (13)$$

where $TP_i$ is the number of correctly predicted noisy annotations of the $i$-th gene, $FP_i$ is the number of wrongly predicted noisy annotations, and $FN_i$ is the number of noisy annotations not predicted by the predictor. $p_i$ and $r_i$ are the precision and recall on the $i$-th gene, they evaluate the fraction of predicted noisy annotations that are true noisy annotations and the fraction of noisy annotations that are correctly predicted, respectively. F1-Score firstly computes individual precision and recall for each gene, and then takes the average of harmonic mean of individual precision and recall of $N$ genes.

### Results of predicting noisy annotations

In this section, we predict noisy annotations of genes based on the annotations in the historical GOA files, and then use the annotations in the recent GOA files to validate the predicted noisy annotations. Similar to CAFA2 [5], to get reliable and repeatable experimental results, we use bootstrapping to randomly take 85% genes and their annotations in the recent GOA files to validate the predicted noisy annotations. We independently repeat the above bootstrapping 500 times to avoid random effect. In these experiments, $\alpha$ in Eq. (10) is set as 0.2, and $\theta$ in Eq. (7) is set as 0.5. Other input values of $\alpha$ and $\theta$ will be discussed later. The recorded experiments results (average and standard deviation) on a particular species for a particular branch are revealed in Table 4 and Tables S1-S11 of the supplementary file. We use pairwise $t$-test at 95% significant level to check the difference among these comparing methods and highlight the best (or comparable best) performance in **boldface**.

From these tables, we can easily observe that NoGOA achieves the best (or comparable best) performance among these comparing algorithms in most cases in terms of Precision and F1-score. NoisyGOA or Noisy-GOA+EC get better performance than NoGOA on some species (such as *A. thaliana* in the BP branch (archived in May, 2015), and *G. gallus* in the BP branch (archived in May, 2016)), but NoGOA still obtains better results than other comparing approaches (Random, LF, NtN, EC and NtN+EC). This global observation validates the effectiveness of NoGOA in identifying noisy annotations. Both NoGOA and SR employ sparse representation to define the semantic similarity between genes and then use a $k$NN style algorithm to predict noisy annotations. SR often loses to NoGOA. This is principally because NoGOA additionally takes advantage of evidence codes to set different weights to different annotations. Similarly, NoGOA always gets better Precision and F1-score than EC, which predicts noisy annotations by only utilizing the evidence code weighted gene-term association matrix. This observation shows that integrating sparse representation with evidence code can generally improve the performance of noisy annotation prediction.

We adopt Wilcoxon signed rank test [48, 49] to assess the difference between NoGOA and these comparing algorithms with respect to F1-score on multiple species across three GO branches, and observe that NoGOA significantly works better than them with all the $p$-value smaller than 0.001. From these results, we can draw a conclusion that it is necessary and effective to integrate evidence codes with sparse representation for identifying noisy annotations. However, the F1-Score is between 34% and 74%, which means only a portion of noisy annotations can be correctly predicted and there is much space for future pursue.

Another observation from these tables is that EC has larger Recall than SR and NoGOA in most cases. The reason is that EC picks up terms with the lowest values in $\mathbf{A}_{ec}(i, \cdot)$ as noisy annotations, without considering the terms' association with other genes. EC also takes

**Table 4** Performance of predicting noisy annotations in GOA files of *H. sapiens* (archived date: May, 2016)

|    |           | Random | LF | NtN | NoisyGOA | SR | EC | NtN+EC | NoisyGOA+EC | NoGOA |
|----|-----------|--------|-----|------|----------|-----|-----|--------|-------------|-------|
| BP | Precision | 23.99 ± 0.49 | 29.50 ± 0.57 | 23.71 ± 0.47 | 33.98 ± 0.67 | 35.24 ± 0.56 | 29.43 ± 0.56 | 26.30 ± 0.51 | 38.55 ± 0.72 | **41.14 ± 0.76** |
|    | Recall    | **57.75 ± 1.00** | 29.58 ± 0.57 | 55.84 ± 0.87 | 41.08 ± 0.76 | 35.67 ± 1.48 | 49.04 ± 0.86 | 52.52 ± 0.89 | 44.82 ± 0.81 | 41.45 ± 0.76 |
|    | F1-Score  | 31.51 ± 0.60 | 29.54 ± 0.57 | 30.94 ± 0.55 | 36.63 ± 0.70 | 35.44 ± 0.69 | 35.04 ± 0.64 | 33.24 ± 0.61 | **40.93 ± 0.75** | **41.28 ± 0.76** |
| CC | Precision | 19.34 ± 0.52 | 28.62 ± 0.77 | 17.75 ± 0.52 | 36.41 ± 0.89 | **41.41 ± 1.01** | 17.40 ± 0.45 | 18.00 ± 0.48 | 36.13 ± 0.88 | **41.34 ± 0.97** |
|    | Recall    | 50.62 ± 1.12 | 28.69 ± 0.77 | 49.68 ± 1.18 | 44.45 ± 1.02 | 41.91 ± 1.02 | **79.22 ± 1.40** | 44.80 ± 1.07 | 44.15 ± 1.02 | 41.85 ± 0.98 |
|    | F1-Score  | 25.98 ± 0.65 | 28.65 ± 0.77 | 24.22 ± 0.65 | 38.79 ± 0.93 | **41.63 ± 1.02** | 25.34 ± 0.58 | 24.34 ± 0.61 | 38.50 ± 0.92 | **41.56 ± 0.97** |
| MF | Precision | 27.74 ± 0.39 | 23.60 ± 0.38 | 36.43 ± 0.45 | 38.16 ± 0.48 | 46.18 ± 0.54 | 41.25 ± 0.50 | 49.90 ± 0.55 | 52.18 ± 0.57 | **58.92 ± 0.60** |
|    | Recall    | 41.94 ± 0.50 | 23.63 ± 0.38 | 48.83 ± 0.57 | 46.41 ± 0.55 | 46.57 ± 0.54 | **60.46 ± 0.64** | 56.80 ± 0.60 | 58.26 ± 0.62 | 59.47 ± 0.60 |
|    | F1-Score  | 30.35 ± 0.41 | 23.61 ± 0.38 | 38.82 ± 0.47 | 39.44 ± 0.48 | 46.34 ± 0.54 | 44.45 ± 0.51 | 51.75 ± 0.56 | 53.23 ± 0.58 | **59.14 ± 0.60** |

The numbers in **boldface** denote the best performance

descendant terms of these picked up terms as noisy annotations of the $i$-th gene and results in a large number of predicted noisy annotations. For this reason, it gets larger Recall but lower Precision than NoGOA, and loses to NoGOA on F1-score.

NtN also weights the gene-term association matrix by employing the GO hierarchy, but it does not consider the evidence codes attached with annotations. It frequently has large Recall but low Precision and F1-score. That is because NtN sets larger weights to specific terms (or annotations) than general ones, and the terms corresponding to general annotations are ranking ahead of specific ones as candidate noisy annotations. Because of true path rule, all the annotations with respect to descendant terms of these general terms are also deemed as noisy annotations by NtN. For this reason, NtN often gets larger Recall but much lower Precision and F1-score than other comparing methods.

Similar as SR, NtN and NoGOA, NoisyGOA also utilizes the semantic similarity between genes and it additionally uses taxonomic similarity between GO terms. NoisyGOA outperforms NtN, Random, and LF in many cases. This fact indicates taxonomic similarity is helpful for predicting noisy annotations. However, NoisyGOA is frequently outperformed by SR. This observation suggests that semantic similarity contributes much more than taxonomic similarity in predicting noisy annotations. NoisyGOA often loses to NoGOA. The reason is threefold: (i) NoGOA differentially treats neighborhood genes to aggregate votes, whereas NoisyGOA equally treats neighborhood genes; (ii) NoGOA takes advantage of evidence codes of annotations, while NoisyGOA does not; (iii) NoGOA adopts sparse representation to measure the semantic similarity between genes, which is less suffered from noisy annotations than the Cosine similarity adopted by NoisyGOA.

LF selects terms annotated to a gene but with the lowest frequency among $N$ genes as noisy annotations of the gene. It frequently gets larger Precision and F1-score than Random and NtN. This observation indicates that the frequency of terms can be used as an important feature for predicting noisy annotations. In fact, NoGOA, SR and NoisyGOA also take advantage of this feature. More specifically, to determine whether a term should be annotated to a gene or not, they count how many times the term annotated to neighborhood genes of the gene.

Random randomly selects terms from all the terms annotated to a gene, and took these selected terms and their descendant terms as noisy annotations of that gene. It sometimes can get the largest Recall. That is principally because these randomly selected terms often have many descendants, which are also annotated to the same gene. Given the superior results of NoGOA to Random,

LF and EC, we can conclude that noisy annotations are predictable.

To further study the rationality of using evidence codes, we also report the results of NoisyGOA+EC and NtN+EC in Table 1 and Additional file 1: Tables S1–S11. With the help of evidence codes, NoisyGOA+EC has improved performance than NoisyGOA, and NtN+EC also shows this pattern. These results show evidence codes can be used as a plugin to improve the performance of noisy annotation prediction. NoGOA performs significantly better than NoisyGOA+EC and NtN+EC. The fact again justifies the rationality of synergy SR with EC for predicting noisy annotations.

### Parameter sensitivity analysis

NoGOA are involved with three parameters $\alpha$ (in Eq. (10)), $\tau$ and $\theta$ (in Eq. (4)). We conduct additional experiments on GOA files of *H. sapiens*, *A. thaliana* and *S. cerevisiae* to study the sensitivity of NoGOA to these parameters and report the results in Fig. 1 (for $\alpha$), Additional file 1: Figure S2 (for $\theta$) and Additional file 1: Tables S12–S17 (for $\tau$). When $\alpha = 0$, NoGOA is equivalent to EC. Likewise, when $\alpha = 1$, NoGOA is equivalent to SR.

In Fig. 1, we set $\theta$ as 0.5 and $\tau$ as the average of $r_{ec}^m$. There are 18 broken lines, and each of them denotes the change of F1-Scores under different input values of $\alpha$. With the increase of $\alpha$, these lines rise at first and then decrease (14 of 18) or keep stable. NoGOA always gets better results than the special case $\alpha = 0$ (or EC), and it also performs better than the special case $\alpha = 1$ (or SR). When $\alpha \in [0.1, 0.3]$, NoGOA generally achieves better (or similar) performance than EC and SR across GOA files of different species archived in different years, so we set $\alpha$ as 0.2 for experiments. The sensitivity analysis of $\alpha$ further corroborates the necessity and advantage of integrating sparse representation with evidence codes. In some branches, F1-Scores remains relatively stable when $\alpha \in [0.1, 1]$. That is because SR plays a major role in noisy annotation prediction in these branches.

### Removing noisy annotations improves gene function prediction

To further study the influence of removing noisy annotations, we downloaded protein-protein interactions (PPI) network of *H. sapiens*, *A. thaliana* and *S. cerevisiae* from BioGrid [50] (archived date: 2016-05-01) for experiments. We take annotations whose aggregated scores $\mathbf{V}(i, t)$ smaller than 0.45 as predicted noisy annotations, and then update the gene-term association matrix $\mathbf{A}$. From Eq. (10), for $\alpha = 0.2$ and $\theta = 0.5$, $\alpha \times \mathbf{V}_{SR}(i, t) \in [0, 0.2]$ and $(1 - \alpha) \times \mathbf{A}_{ec}(i, t) \in [0.4, 0.8]$. So we take the annotations with the lowest $\mathbf{A}_{ec}(i, \cdot)$ and $\mathbf{V}_{SR}(i, \cdot) < 0.25$ as noisy

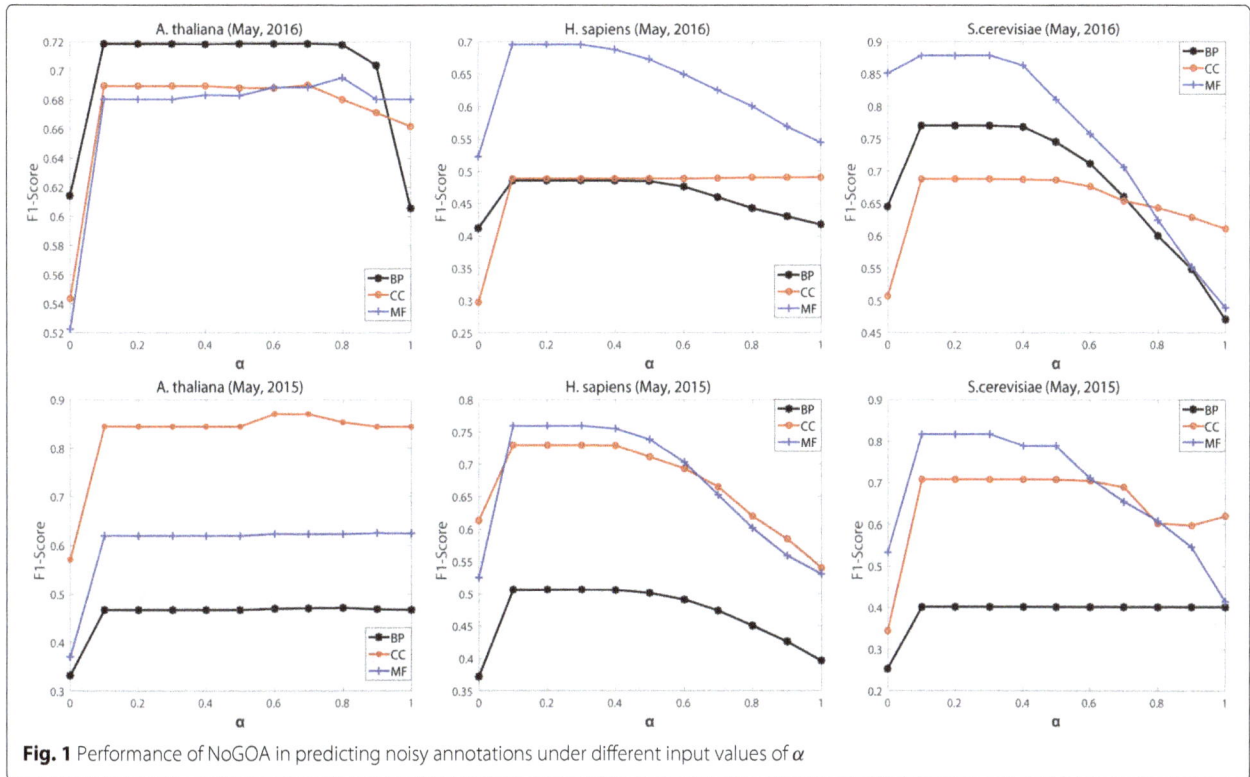

**Fig. 1** Performance of NoGOA in predicting noisy annotations under different input values of $\alpha$

annotations of the $i$-th gene. Next, we apply a majority vote based function prediction model [51], which predicts GO annotations of a gene using the annotations of its interacting partners based on updated **A**. After that, we use the annotations in the recent GOA files to validate the predicted annotations. For comparison, we also apply the majority vote model on the same PPI network and the original **A**, and then follow the same protocol to evaluate the predictions. We label the latter method as 'Original'.

To reach a comprehensive evaluation of gene function prediction, we use six evaluation metrics, namely *MicroAvgF1*, *MacroAvgF1*, *AvgPrec*, *AvgROC*, *Fmax* and *Smin*. These metrics have been applied to evaluate the results of gene function prediction [5, 36]. Except *Smin*, the higher the value of these metrics is, the better the performance is. These metrics measure the performance from different aspects, it is difficult for a method consistently better than others across all the metrics. The formal definitions of these metrics are provided in the supplementary file. The results with respect to *H. sapiens*, *A.thaliana* and *S. cerevisiae* are included in Table 5 and Additional file 1: Tables S18-S19.

From the results in Table 5 and Additional file 1: Tables S18-S19, we can see that NoGOA has improved performance in gene function prediction than Original in most cases. We use Wilcoxon signed rank test to check the difference between the results of NoGOA and Original on

these three model species, and find the $p$-value is smaller than 0.003.

From these results, we can draw a conclusion that removing noisy annotations improves the performance of gene function prediction.

**Real examples**

To further investigate the ability of NoGOA in predicting noisy annotations of genes, we firstly study the number of predicted noisy annotations of *H. sapiens*, *A. thaliana* and *S. cerevisiae* for each evidence code. Since

**Table 5** Results of gene function prediction on *H. sapiens* (archived date: May, 2016)

|  | BP | | CC | | MF | |
|---|---|---|---|---|---|---|
|  | Original | NoGOA | Original | NoGOA | Original | NoGOA |
| MicroAvgF1 | **92.85** | 92.64 | 93.72 | **93.92** | 93.10 | 93.10 |
| MacroAvgF1 | 89.04 | **90.05** | 88.06 | **89.96** | 89.55 | **90.30** |
| AvgPrec | 88.45 | **88.50** | 88.75 | **89.19** | 90.78 | **90.81** |
| AvgROC | 94.94 | **96.73** | 95.12 | **96.66** | 97.66 | **98.35** |
| Fmax | **93.85** | 93.50 | 93.85 | **93.89** | 94.62 | **94.57** |
| Smin ↓ | 8.69 | **7.96** | 2.09 | **2.09** | 2.40 | **2.32** |

The data in **boldface** denote the better result. 'Original' directly uses annotations in the historical GOA file to predict gene function; 'NoGOA' removes predicted noisy annotations from the historical GOA file and then predicts gene function. ↓ means the lower the value, the better the performance is

only direct annotations can obtain the sources and evidences in archived GOA files, we only count the numbers of direct noisy annotations, predicted noisy annotations and correctly predicted direct ones by NoGOA. These numbers are shown in Table S20-S25 of the supplementary file. Then, we take the first 4 genes ('AAC1', 'AAC3', 'AAD14', 'AAP1'), which have removed annotations in the recently archived (date: September 2016) GOA file of *S. cerevisiae* for illustrative study, and list the correctly (wrongly) predicted direct noisy annotations by NoGOA. The results of *S. cerevisiae* in CC branch are listed in Table 6. Other experimental results of *S. cerevisiae* in other branches are revealed in Additional file 1: Tables S26-S27.

From Additional file 1: Tables S20–S25, we can find that the distribution of predicted noisy annotations for different evidence codes is often approximately consistent with the distribution of noisy annotations. This fact shows the effectiveness of NoGOA in identifying noisy annotations. The number of predicted noisy annotations is often larger than that of direct noisy annotations. That is because if an annotation is predicted as a noisy one of a gene, then its descendant annotations (if any) are also deemed as noisy annotations of that gene. Since the annotations expanded from GO hierarchy and direct annotations maybe supported by different evidence codes, we just report the correctly predicted direct noisy annotations here. In practice, by expanding these direct noisy annotations via the true path rule of GO, the number of correctly predicted noisy annotations can be sharply increased.

In most cases, IEA generally has much more noisy annotations than other evidence codes. That is mainly because the number of IEA annotations is the largest, and it does not mean that IEA annotations are the most unreliable. Similar to IEA, IBA also has many noisy annotations. TAS, IMP or IGI have more noisy annotations in BP than in MF and CC branches. EXP, ISA, ISO, ISM, RCA, IGC, IBD, IKR, IRD and IC annotations are relatively stable and have much fewer noisy annotations. The possible reason is that the number of annotations attached with

these evidence codes is smaller than that of other evidence codes. These statistic numbers show that most evidence codes have no clear pattern of noisy annotations across all the GO branches. These numbers also support our motivation to adaptively set weights to annotations based on the estimated ratio of noisy annotations per evidence code, instead of presetting weights solely based on the categorization (i.e., Experimental and Computational) of evidence codes.

The selected 4 proteins have 16 direct noisy annotations in three branches. NoGOA predicts 20 noisy annotations, and 13 of them are correct. In actual fact, we rechecked the subsequent GOA files (till to February, 2017) of S. cerevisiae, and also found these 13 correctly predicted noisy annotations were always removed in these GOA files. It is anticipated that these correctly predicted noisy annotations could be confirmed by biological experiments. From Table 6 and Additional file 1: Tables S26-S27, we can find that these noisy annotations are attached with different evidence codes (IBA, IPI, IDA, IMP and TAS). In fact, these annotations are reviewed by curators, but they are not always more reliable than IEA [6, 8]. Another interesting observation is that, NoGOA only makes incorrect predictions on 'AAP1'. The reason may be that compared with other genes, 'AAP1' contains more noisy annotations, which heavily mislead the semantic similarity between 'AAP1' and other genes.

## Conclusion

Current efforts toward computational gene function prediction are more focused on predicting GO annotations of un-annotated genes or replenishing missing annotations of partially annotated genes. Given the increasing application of GO annotations in various domains and misleading effect of noisy annotations, it is necessary to identify noisy annotations, which is a rarely studied but important open problem.

In this paper, we investigated whether noisy annotations are predictable or not, and how to predict noisy annotations. For this purpose, we introduced a method

**Table 6** Examples of correctly (√) and wrongly(×) predicted direct noisy annotations by NoGOA in CC branch of *S. cerevisiae*

| Protein | | GO term | Evidence codes | Details |
|---|---|---|---|---|
| AAC1(ADP/ATP carrier) | √ | GO:0005758 (mitochondrial intermembrane space) | TAS | Reactome:R-SCE-1252255 |
| | | GO:0005829 (cytosol) | TAS | Reactome:R-SCE-1252255 |
| AAP1 (Alanine/arginine aminopeptidase) | √ | GO:0005886 (plasma membrane) | IBA | GO_REF:0000033 |
| | | GO:0005664 (nuclear origin of replication recognition complex) | IDA | PMID:9372948 |
| | × | GO:0000276 (mitochondrial proton-transporting ATP synthase complex, coupling factor F(o)) | IDA | PMID:9224714 |

called NoGOA. NoGOA takes advantage of evidence codes attached with annotations and sparse representation to predict noisy annotations. Experimental results on six model species (H. sapiens, A. thaliana, S. cerevisiae, G. gallus, B. Taurus and M. musculus) show that noisy annotations are predictable and NoGOA can more accurately predict noisy annotations than other comparing algorithms. We believe our work will prompt more research toward removing noisy GO annotations.

## Abbreviations
BP: Biological process; CAFA: Critical assessment of protein function annotation; CC: Cellular component; EXP: Inferred from experiment; GO: Gene ontology; GOA: Gene ontology annotations; GOC: Gene ontology consortium; IBA: Inferred from biological aspect of ancestor; IBD: Inferred from biological aspect of descendant; IC: Inferred by curator; IDA: Inferred from direct assay; IEA: Inferred from electronic annotation; IEP: Inferred from expression pattern; IGC: Inferred from genomic context; IGI: Inferred from genetic interaction; IKR: Inferred from key residues; IMP: Inferred from mutant phenotype; IPI: Inferred from physical interaction; IRD: Inferred from rapid divergence; ISA: Inferred from sequence alignment; ISM: Inferred from sequence model; ISO: Inferred from sequence orthology; ISS: Inferred from sequence or structural similarity; MF: Molecular function; NAS: Non-traceable author statement; ND: No biological data available; PPI: Protein-Protein interactions; RCA: Inferred from reviewed computational analysis; TAS: Traceable author statement

## Acknowledgements
We thank the reviewers for insightful and constructive comments on improving our work.

## Funding
This work is supported by Natural Science Foundation of China (No. 61402378), Natural Science Foundation of CQ CSTC (cstc2014jcyjA40031 and cstc2016jcyjA0351), Fundamental Research Funds for the Central Universities of China (2362015XK07 and XDJK2016B009), Science and Technology Development of Jilin Province of China (20150101051JC and 20160520099JH). None of the funding bodies have played any part in the design of the study, in the collection, analysis, and interpretation of the data, or in the writing of the manuscript.

## Authors' contributions
GY initialized the project and solution, conceived the whole process and revised the manuscript. CL performed the experiments, analyzed the results and drafted the manuscript. JW analyzed the results and revised the manuscript. All the authors read and approved the final manuscript.

## Competing interests
The authors declare that they have no competing interests.

## References
1. Ashburner M, Ball CA, Blake JA, Botstein D, Butler H, Cherry JM, Davis AP, Dolinski K, Dwight SS, Eppig JT, et al. Gene ontology: tool for the unification of biology. Nat Genet. 2000;25(1):25–9.
2. Gaudet P, Chisholm R, Berardini T, Dimmer E, FeydictyBase Pt. The gene ontology's reference genome project: a unified framework for functional annotation across species. PLoS Comput Biol. 2009;5(7):e1000431.
3. Schnoes AM, Ream DC, Thorman AW, Babbitt PC, Friedberg I. Biases in the experimental annotations of protein function and their effect on our understanding of protein function space. PLoS Comput Biol. 2013;9(5): e1003063.
4. Radivojac P, Clark WT, Oron TR, Schnoes AM, Wittkop T, Sokolov A, Graim K, Funk C, Verspoor K, Ben-Hur A. A large-scale evaluation of computational protein function prediction. Nat Methods. 2013;10(3): 221–7.
5. Jiang Y, Oron TR, Clark WT, Bankapur AR, D'Andrea D, Lepore R, Funk CS, Kahanda I, Verspoor KM, Ben-Hur A, Koo DCE, Penfold-Brown D, Shasha D, Youngs N, Bonneau R, Lin A. An expanded evaluation of protein function prediction methods shows an improvement in accuracy. Genome Biol. 2016;17(1):184.
6. Škunca N. Quality of computationally inferred gene ontology annotations. PLoS Comput Biol. 2012;8(5):e1002533.
7. Huntley RP, Sawford T, Martin MJ, ODonovan C. Understanding how and why the gene ontology and its annotations evolve: the go within uniprot. GigaScience. 2014;3(1):4.
8. Rhee SY, Wood V, Dolinski K, Draghici S. Use and misuse of the gene ontology annotations. Nat Rev Genet. 2008;9(7):509–15.
9. Koskinen P, Noksokoivisto J, Holm L. Pannzer: high-throughput functional annotation of uncharacterized proteins in an error-prone environment. Bioinformatics. 2015;31(10):1544–52.
10. Binns D, Dimmer E, Huntley R, Barrell D, ODonovan C, Apweiler R. Quickgo: a web-based tool for gene ontology searching. Bioinformatics. 2009;25(22):3045–6.
11. Gross A, Hartung M, Kirsten T, Rahm E. Estimating the quality of ontology-based annotations by considering evolutionary changes. In: International Workshop on Data Integration in the Life Sciences. Berlin: Springer; 2009. p. 71–87.
12. Gross A, Hartung M, Prüfer K, Kelso J, Rahm E. Impact of ontology evolution on functional analyses. Bioinformatics. 2012;28(20):2671–7.
13. Gillis J, Pavlidis P. Assessing identity, redundancy and confounds in gene ontology annotations over time. Bioinformatics. 2013;29(4):476–82.
14. Clarke EL, Loguercio S, Good BM, Su AI. A task-based approach for gene ontology evaluation. J Biomed Semant. 2013;4(S1):4.
15. Mi H, Muruganujan A, Casagrande JT, Thomas PD. Large-scale gene function analysis with the panther classification system. Nat Protoc. 2013;8(8):1551–66.
16. Schlicker A, Lengauer T, Albrecht M. Improving disease gene prioritization using the semantic similarity of gene ontology terms. Bioinformatics. 2010;26(18):561–7.
17. Kissa M, Tsatsaronis G, Schroeder M. Prediction of drug gene associations via ontological profile similarity with application to drug repositioning. Methods. 2015;74:71–82.
18. Faria D, Schlicker A, Pesquita C, Bastos H, Ferreira AEN, Albrecht M, O FA. Mining go annotations for improving annotation consistency. PLoS ONE. 2012;7(7):e40519.
19. Agapito G, Milano M, Guzzi PH, Cannataro M. Improving annotation quality in gene ontology by mining cross-ontology weighted association rules. In: IEEE International Conference on Bioinformatics and Biomedicine. Piscataway: IEEE Press; 2014. p. 1–8.
20. Agapito G, Cannataro M, Guzzi P, Milano M. Extracting cross-ontology weighted association rules from gene ontology annotations. IEEE/ACM Trans Comput Biol Bioinforma. 2016;13(2):197–208.
21. Thomas PD, Mi H, Lewis S. Ontology annotation: mapping genomic regions to biological function. Curr Opin Chem Biol. 2007;11(1):4–11.
22. Clark WT, Radivojac P. Analysis of protein function and its prediction from amino acid sequence. Proteins Struct Funct Bioinforma. 2011;79(7): 2086–96.
23. Buza TJ. Gene ontology annotation quality analysis in model eukaryotes. Nucleic Acids Res. 2008;36(2):12.
24. Jones CE, Brown AL, Baumann AU. Estimating the annotation error rate of curated go database sequence annotations. BMC Bioinforma. 2007;8(1):170.
25. Benabderrahmane S, Smailtabbone M, Poch O, Napoli A, Devignes MD. Intelligo: a new vector-based semantic similarity measure including annotation origin. BMC Bioinforma. 2010;11:588.

26. Caniza H, Romero AE, Heron S, Yang H, Devoto A, Frasca M, Mesiti M, Valentini G, Paccanaro A. Gossto: a user-friendly stand-alone and web tool for calculating semantic similarities on the gene ontology. Bioinformatics. 2014;30(15):2235–6.

27. Pesquita C, Faria D, Falcão AO, Lord P, Couto FM. Semantic similarity in biomedical ontologies. PLoS Comput Biol. 2009;5(7):e1000443.

28. Guzzi PH, Mina M, Guerra C, Cannataro M. Semantic similarity analysis of protein data: assessment with biological features and issues. Brief Bioinform. 2011;13(5):569–85.

29. Tao Y, Li J, Friedman C, Lussier YA. Information theory applied to the sparse gene ontology annotation network to predict novel gene function. Bioinformatics. 2007;23(13):529–38.

30. Wu X, Zhu L, Guo J, Zhang D, Lin K. Prediction of yeast protein-protein interaction network: insights from the gene ontology and annotations. Nucleic Acids Res. 2006;34(7):2137–50.

31. Yu G, Zhu H, Domeniconi C, Liu J. Predicting protein function via downward random walks on a gene ontology. BMC Bioinforma. 2015;15: 271.

32. Lu C, Wang J, Zhang Z, Yang P, Yu G. Noisygoa: noisy go annotations prediction using taxonomic and semantic similarity. Comput Biol Chem. 2016;65:203–11.

33. Donoho DL, Elad M, Temlyakov VN. Stable recovery of sparse overcomplete representations in the presence of noise. IEEE Trans Inf Theory. 2006;52(1):6–18.

34. Wright J, Ma Y, Mairal J, Sapiro G, Huang TS, Yan S. Sparse representation for computer vision and pattern recognition. Proc IEEE. 2010;98(6):1031–44.

35. Noah Y, Duncan PB, Kevin D, Dennis S, Richard B. Parametric bayesian priors and better choice of negative examples improve protein function prediction. Bioinformatics. 2013;29(9):1190–8.

36. Fu G, Wang J, Yang B, Yu G. Neggoa: negative go annotations selection using ontology structure. Bioinformatics. 2016;32(19):2996–3004.

37. Yang H, Nepusz T, Paccanaro A. Improving go semantic similarity measures by exploring the ontology beneath the terms and modelling uncertainty. Bioinformatics. 2012;28(10):1383–9.

38. Teng Z, Guo M, Liu X, Dai Q, Wang C, Xuan P. Measuring gene functional similarity based on group-wise comparison of go terms. Bioinformatics. 2013;29(11):1424–32.

39. Liu J, Ji S, Ye J. Slep: Sparse learning with efficient projections: Arizona State University; 2009. http://yelab.net/software/SLEP/. Accessed 24 Sept 2016.

40. Good BM, Clarke EL, Alfaro LD, Su AI. The gene wiki in 2011: Community intelligence applied to human gene annotation. Nucleic Acids Res. 2011;40(1):1255–61.

41. Good BM, Su AI. Crowdsourcing for bioinformatics. Bioinformatics. 2013;29(16):1925–33.

42. Cover T, Hart P. Nearest neighbor pattern classification. IEEE Trans Inf Theory. 1967;14(1):21–7.

43. Valentini G. True path rule hierarchical ensembles for genome-wide gene function prediction. IEEE/ACM Trans Comput Biol Bioinforma. 2011;8(3): 832–47.

44. The gene ontology annotation files. http://geneontology.org/page/download-annotations. Accessed 24 Sept 2016.

45. The gene ontology database. http://geneontology.org/page/download-ontology. Accessed 24 Sept 2016.

46. Done B, Khatri P, Done A, Drăghici S. Predicting novel human gene ontology annotations using semantic analysis. IEEE/ACM Trans Comput Biol Bioinforma. 2010;7(1):91–9.

47. Salton G. A vector space model for automatic indexing. Commun ACM. 1975;18(11):613–20.

48. Wilcoxon F. Individual comparisons by ranking methods. Biom Bull. 1945;1(6):80–3.

49. Demsar J. Statistical comparisons of classifiers over multiple data sets. J Mach Learn Res. 2006;7(1):1–30.

50. Protein-protein interactions network from biogrid. http://thebiogrid.org/download.php. Accessed 24 Sept 2016.

51. Schwikowski B, Uetz P, Fields S. A network of protein-protein interactions in yeast. Bioinformatics. 2000;18(12):1257–61.

# 3

# SparkBLAST: scalable BLAST processing using in-memory operations

Marcelo Rodrigo de Castro[1], Catherine dos Santos Tostes[2], Alberto M. R. Dávila[2], Hermes Senger[1] and Fabricio A. B. da Silva[3]*

## Abstract

**Background:** The demand for processing ever increasing amounts of genomic data has raised new challenges for the implementation of highly scalable and efficient computational systems. In this paper we propose SparkBLAST, a parallelization of a sequence alignment application (BLAST) that employs cloud computing for the provisioning of computational resources and Apache Spark as the coordination framework. As a proof of concept, some radionuclide-resistant bacterial genomes were selected for similarity analysis.

**Results:** Experiments in Google and Microsoft Azure clouds demonstrated that SparkBLAST outperforms an equivalent system implemented on Hadoop in terms of speedup and execution times.

**Conclusions:** The superior performance of SparkBLAST is mainly due to the in-memory operations available through the Spark framework, consequently reducing the number of local I/O operations required for distributed BLAST processing.

**Keywords:** Cloud computing, Comparative genomics, Scalability, Spark

## Background

Sequence alignment algorithms are a key component of many bioinformatics applications. The NCBI BLAST [1, 2] is a widely used tool that implements algorithms for sequence comparison. These algorithms are the basis for many other types of BLAST searches such as BLASTX, TBLASTN, and BLASTP [3]. The demand for processing large amounts of genomic data that gushes from NGS devices has grown faster than the rate which industry can increase the power of computers (known as Moore's Law). This fact has raised new challenges for the implementation of scalable and efficient computational systems. In this scenario, MapReduce (and its Hadoop implementation) emerged as a paramount framework that supports design patterns which represent general reusable solutions to commonly occurring problems across a variety of problem domains including analysis and assembly of biological sequences [4]. MapReduce has delivered outstanding performance and scalability for a myriad of applications running over hundreds to thousands of processing nodes [5]. On the other hand, over the last decade, cloud computing has emerged as a powerful platform for the agile and dynamic provisioning of computational resources for computational and data intensive problems.

Several tools have been proposed, which combine Hadoop and cloud technologies. Regarding NGS we can cite Crossbow [6] and for sequence analysis: Biodoop [7] and CloudBLAST [8]. Further tools based on Hadoop and related technologies are surveyed in [4].

Despite of its popularity, MapReduce requires algorithms to be adapted according to such design patterns [9]. Although this adaptation may result in efficient implementations for many applications, this is not necessarily true for many other algorithms, which limits the applicability of MapReduce. Moreover, because MapReduce is designed to handle extremely large data sets, its implementation frameworks (e.g. Hadoop and the Amazon's Elastic MapReduce service) constrains the program's ability to process smaller data.

More recently, Apache Spark has emerged as a promising and more flexible framework for the implementation of highly scalable parallel applications [10, 11]. Spark does not oblige programmers to write their algorithms in

*Correspondence: fabricio.silva@fiocruz.br
[3]PROCC, Oswaldo Cruz Foundation, Av. Brasil 4365, 21040-900 Rio de Janeiro, Brazil
Full list of author information is available at the end of the article

terms of the map and reduce parallelism pattern. Spark implements in-memory operations, based on the Resilient Distribution Datasets (RDDs) abstraction [11]. RDD is a collection of objects partitioned across nodes in the Spark cluster so that all partitions can be computed in parallel. We may think of RDDs as a collection of data objects which are transformed into new RDDs as the computation evolves. Spark maintains lists of dependencies among RDDs which are called "lineage". It means RDDs can be recomputed in case of lost data (e.g. in the event of failure or simply when some data has been previously discarded from memory).

In this paper we propose SparkBLAST, which uses the support of Apache Spark to parallelize and manage the execution of BLAST either on dedicated clusters or cloud environments. Spark's *pipe* operator is used to invoke BLAST as an external library on partitioned data of a query. All the input data (the query file and the database) and output data of a query are treated as Spark's RDDs. SparkBLAST was evaluated on both Google and Microsoft Azure Clouds, for several configurations and dataset sizes. Experimental results show that SparkBLAST improves scalability when compared to CloudBLAST in all scenarios presented in this paper.

## Implementation

A design goal is to offer a tool which can be easily operated by users of the unmodified BLAST. Thus, SparkBLAST implements a driver application written in Scala, which receives user commands and orchestrates the whole application execution, including data distribution, tasks execution, and the gathering of results in a transparent way for the user.

Two input files must be provided for a typical operation: (*i*) the target database of bacterial genomic sequences, which will be referred to as *target database* from now on, for short; and (*ii*) the *query file*, which contains a set of query genomic sequences that will be compared to the target's database sequences for matching. As depicted in Fig. 1, SparkBLAST replicates the entire target database on every computing node. The query file is evenly partitioned into data *splits* which are distributed over the nodes for the execution. Thus, each computing node has a local deployment of the BLAST application, and it receives a copy of the entire target database and a set of fragments of the query file (splits).

Note that it is possible to apply different techniques for task and data partitioning. Each data split (i.e., fragment of the query file) can be replicated by the distributed file system (DFS) on a number of nodes, for fault tolerance purposes. Spark's scheduler then partitions the whole computation into tasks, which are assigned to computing nodes based on data locality using delay scheduling [12]. For the execution of each task, the target database and one fragment of the query file are loaded in memory (as RDDs). The target database (RDD) can be reused by other local tasks that execute in the same machine, thus reducing disk access [11].

SparkBLAST uses Spark Pipe to invoke the local installation of the NCBI BLAST2 on each node, and execute multiple parallel and distributed tasks in the cluster.

Spark can execute on top of different resource managers, including Standalone, YARN, and Mesos [13]. We chose YARN because it can be uniformly used by Spark and Hadoop. It is important to avoid the influence of resource scheduling in the performance tests presented in this paper. In fact, YARN was originally developed for Hadoop version 2. With YARN, resources (e.g., cpu, memory) can be allocated and provisioned as *containers* for tasks execution on a distributed computing environment. It plays better the role of managing the cluster configuration, and dynamically shares available resources, providing support for fault tolerance, inter-, and intra-node parallelism. Other applications which have been written or ported to run on top of YARN include Apache HAMA, Apache Giraph, Open MPI, and HBASE[1].

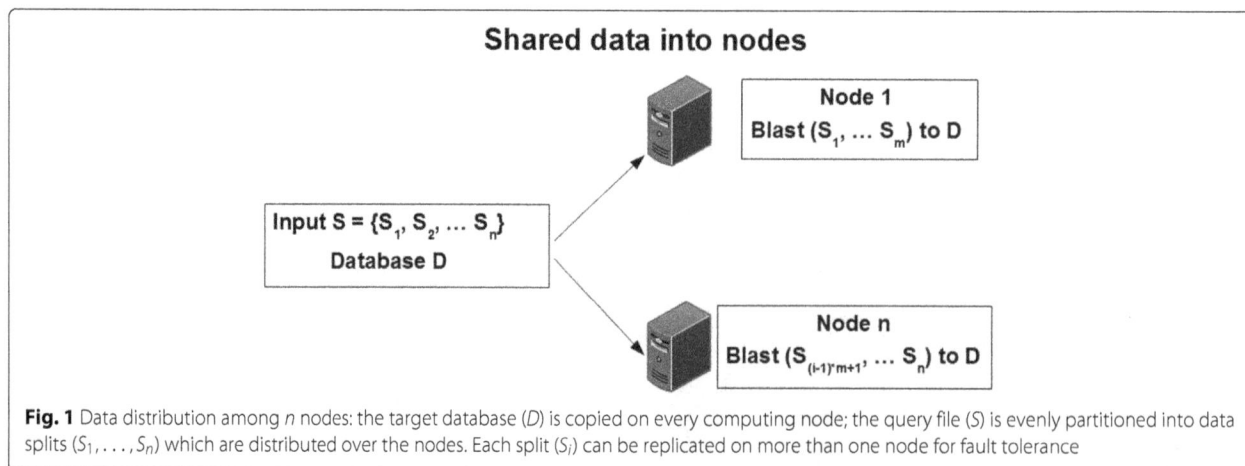

## Shared data into nodes

**Input S = {S$_1$, S$_2$, ... S$_n$}**
**Database D**

**Node 1**
**Blast (S$_1$, ... S$_m$) to D**

**Node n**
**Blast (S$_{(i-1)\cdot m+1}$, ... S$_n$) to D**

**Fig. 1** Data distribution among *n* nodes: the target database (*D*) is copied on every computing node; the query file (*S*) is evenly partitioned into data splits (*S$_1$*,...,*S$_n$*) which are distributed over the nodes. Each split (*S$_i$*) can be replicated on more than one node for fault tolerance

Data processing in SparkBLAST can be divided into three main stages (as depicted in Fig. 2): pre-processing, main processing and post-processing. Such stages are described in the following subsections.

### Execution environment

In order to evaluate the performance and the benefits of SparkBLAST, we present two experiments. The first experiment was executed in the Google Cloud, and the second experiment executed in the Microsoft Azure platform. Both experiments executed with 1, 2, 4, 8, 16, 32, and 64 virtual machines as computing nodes for scalability measurement. For the sake of comparison, each experiment was executed on SparkBLAST and on CloudBLAST. The later is a Hadoop based tool designed to support high scalability on clusters and cloud environments. For the experiments, we used Spark 1.6.1 to execute SparkBLAST on both cloud environments. To execute CloudBLAST, we used Hadoop 2.4.1 on the Google Cloud, and Hadoop 2.5.2 on Azure Cloud. In any case, we configures YARN as the resource scheduler, since our experiments focus on performance. Further details on the experimental setup will be provided in the results section.

### Input data generation

This work was originally inspired and applied in a radionuclides resistance study. Genome sequences of several radiation-resistant microorganisms can be used for comparative genomics to infer the similarities and differences among those species. Homology inference is important to identify genes shared by different species and, as a consequence, species-specific genes can be inferred. Two experiments are considered in this work. The input data for Experiment 1 was composed of 11 bacterial genome protein sequences, 10 of these are radiation-resistant (*Kineococcus radiotolerans* - Accession Number NC_009660.1, *Desulfovibrio desulfuricans* - NC_011883.1, *Desulfovibrio vulgaris* - NC_002937.3, *Rhodobacter sphaeroides* - NC_009429.1, *Escherichia coli* - NC_000913.3, *Deinococcus radiodurans* - NC_001263.1, *Desulfovibrio fructosivorans* - NZ_AECZ01000069.1, *Shewanella oneidensis* - NC_004349.1, *Geobacter sulfurreducens* - NC_002939.5, *Deinococcs geothermalis* - NC_008010.2, *Geobacter metallireducens* - NC_007517.1) for Reciprocal-Best-Hit (RBH) processing.

For Experiment 2, the input query is composed of 10 radiation-resistant bacteria. (i.e., all species listed above but *E. coli*). This similarity-based experiment consisted on the search of potential protein homologs of 10 radiation-resistant genomes in 2 marine metagenomics datasets.

Each input dataset was concatenated into a single multifasta input file named query1.fa (Experiment 1) and query2.fa (Experiment 2). The files query1.fa and query2.fa had 91,108 and 86,968 sequences and a total size of 36.7 MB and 35 MB, respectively. Two target metagenomic datasets obtained from MG-RAST database[2] were used in Experiment 2: (i) Sargaso Sea (Bermuda), coordinates: 32.17,-64.5, 11 GB, 61255,260 proteins (Ber.fasta) and (ii) João Fernandinho (Buzios, Brazil), coordinates: -22.738705, -41874604, 805 MB, 4795,626 proteins (Buz.fasta):

```
$ makeblastdb –dbtype prot –in database.fa –parse_seqids
```

### Pre-processing

In this stage, implemented by SparkBLAST, the query file is evenly partitioned into splits which are written to the DFS. The splits are then distributed among the computing

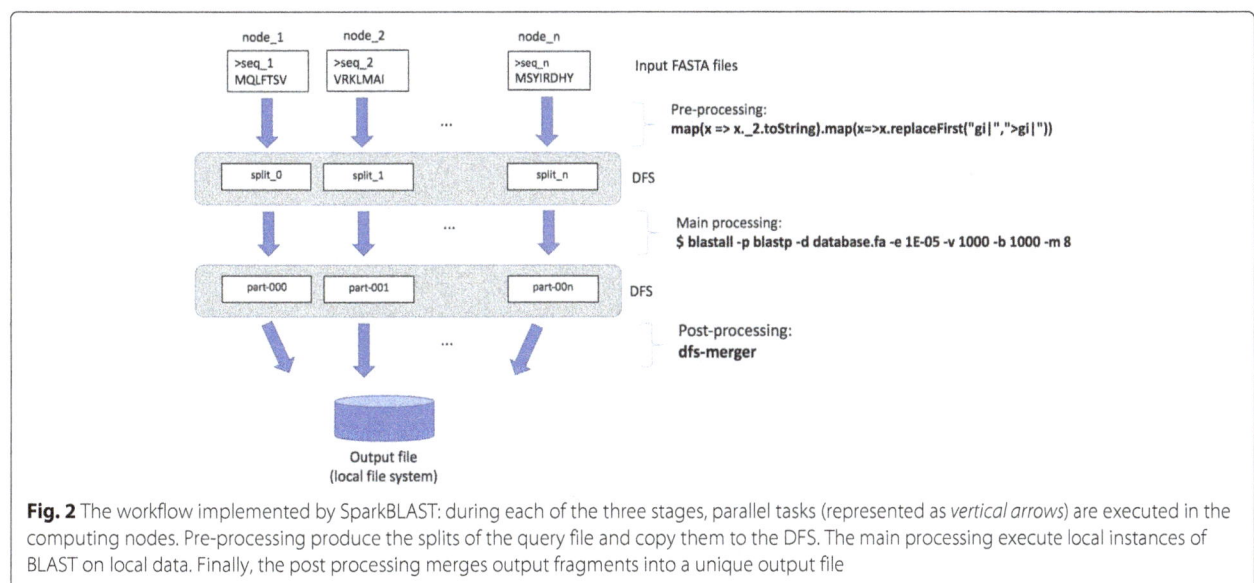

**Fig. 2** The workflow implemented by SparkBLAST: during each of the three stages, parallel tasks (represented as *vertical arrows*) are executed in the computing nodes. Pre-processing produce the splits of the query file and copy them to the DFS. The main processing execute local instances of BLAST on local data. Finally, the post processing merges output fragments into a unique output file

nodes by the DFS, according to some replication policy for fault tolerance. Each split containing a set of (e.g., thousands of) genome sequences can be processed by a different task. Thus, the query file should be partitioned to enable parallelism. Since the input file can be potentially large, the partitioning operation can be also parallelized as illustrated in the following commands:

```
conf.set("textinputformat.record.delimiter", ">")
map(x => x._2.toString).map(x=>x.replaceFirst("gi|",">gi|"))
```

### Main processing
This stage starts after all the input data (i.e., the target database and query file splits) are properly transferred to each processing node. Tasks are then scheduled to execute on each node according to data locality. The amount of tasks executed concurrently on each computing node depends on the number of processing cores available. As soon as a computing core completes the execution of a task, it will be assigned another task. This process repeats until the available cores execute all tasks of the job.

During this stage, each individual task uses Spark pipe to invoke a local execution of BLASTP as illustrated by the following command line:

```
$ blastall -p blastp -d database.fa -e 1E-05 -v 1000 -b 1000 -m 8
```

Note that the query input file to be processed has been omitted because it varies for each task.

In order to measure the scalability and *speedup* of Spark-BLAST we carried out experiments on both the Google Cloud and Microsoft Azure, increasing the platform size from 1 to 64 computing nodes. For the sake of comparison, the same genome searches have been executed with both SparkBLAST and CloudBLAST for each platform size. Every experiment was repeated six times and and the average execution time was considered in results.

For the sake of reproducibility, both experiments with SparkBLAST and CloudBLAST were executed with the following configuration parameters:

Therefore, each node will act as a *mapper*, producing outputs similar to the unmodified BLAST.

### Post-processing
During the previous stage each individual task produces a small output file. During the post-processing stage, Spark-BLAST merges all these small files into a single final output file. For instance, experiment 1 produces a final output file of 610 MB. All output data is written to the DFS, i.e., the Google Cloud Storage or Microsoft Azure's Blob storage service.

Added-value to SparkBLAST, similarity results were obtained by (i) performing a Reciprocal Best Hit analysis [14, 15] among pairs of species, or orthology inference (Experiment 1) and (ii) searching for potential radiation-resistant homologous proteins in 2 marine metagenome datasets (Experiment 2), as described in the following section.

### Results
In order to assess the performance and benefits of Spark-BLAST, we carried out experiments on two cloud platforms: Google Cloud and Microsoft Azure. The same executions were carried out on both SparkBLAST and CloudBLAST.

### Results for experiment 1 - executed on the Google Cloud
In Experiment 1, BLASTP was used to execute queries on a 36 MB database composed of 88,355 sequences from 11 bacterial genomes, in order to identify genes shared by different species. Ten bacteria described in literature as being resistant to ionizing radiation [16] and one species susceptible to radiation were obtained from Refseq database. The same dataset is provided as query and target database, so that an all-to-all bacteria comparison is executed, producing a 610 MB output. BLASTP results were processed to identify RBH among pairs of species.

Experiment 1 was executed on a platform with up to 64 computing nodes plus one master node. Each node is

```
$ spark-submit --executor-memory $memory_per_node
   --driver-memory $memory_node_master$ --num-executors $num_executors
   --executor-cores $qtd_executor_core  --driver-cores $qtd_driver_core
   --class sparkBLAST target/scala-2.10/simple-project_2.10-1.0.jar
   $qtd_splits "blastall -p blastp -d /targetToDB/database.fa -e
   1E-05 -v 1000 -b 1000 -m 8" $input $output

$ hadoop jar targetToHadoop/hadoop-streaming-X.X.X.jar
 -libjars ./StreamPatternRecordReader.jar -input $input
 -output $output -mapper "blastall -p blastp -d /targetToDB/database.fa
 -e 1E-05 -v 1000 -b 1000 -m 8" -numReduceTasks 0
 -inputreader "org.apache.hadoop.streaming.StreamPatternRecordReader, begin=>"
 -cmdenv BLASTDB=/targetToDB/db -jobconf mapreduce.job.maps= $qtd_maps
```

a virtual machine configured as *n1-standard-2* instance (2 vCPUs, 7.5 GB memory, CPU Intel Ivy Bridge). The virtual machines were allocated from 13 different availability zones in the Google Cloud: Asia East (3 zones), Europe West (3 zones), US Central (4 zones) e US East (3 zones). For this scalability test, both SparkBLAST and CloudBLAST were executed on platforms with 1, 2, 4, 8, 16, 32, and 64 nodes. The experiment was repeated six times for each platform size. Thus, Experiment 1 encompasses $2 \times 7 \times 6 = 84$ executions in total, which demanded more than 350 h (wall clock) to execute. As an estimate on the amount of the required computational resources, this experiment consumed 2.420 vCPU-hours to execute on the Google Cloud.

The average execution times are presented in Fig. 3. SparkBLAST achieved a maximum speedup (which is the ratio between execution time of the one node baseline over the run time for the parallel execution) of 41.78, reducing the execution time from 28,983 s in a single node, to 693 s in 64 nodes. In the same scenario, CloudBLAST achieved speedup of 37, reducing the execution time from 30,547 to 825 s on 64 nodes. For this set of executions, both SparkBLAST and CloudBLAST used 2 vCPUs per node for tasks execution. The speedup is presented in Fig. 4. As shown, SparkBLAST presents better scalability than CloudBLAST.

The average execution times and standard deviations are presented in Table 1. Table 2 presents the execution times for SparkBLAST when only one vCPU (core) of each node is used for processing. Table 3 presents the total execution times for SparkBLAST when both cores of each node are used for processing.

Table 4 consolidates results from previous tables and presents mean execution times along with speedup and parallel efficiency figures for the CloudBLAST and SparkBLAST (1 and 2 cores) systems.

Figure 3 compares total execution times of CloudBLAST and SparkBLAST (one and two cores configurations), for platforms composed of 1 up to 64 computing nodes. Execution times presented in correspond to the average for six executions. Parallel efficiency is presented in Fig. 5.

### Results for experiment 2 - executed on the Microsoft Azure

Experiment 2 was executed on a total of 66 nodes allocated on the Microsoft Azure Platform, being all nodes from the same location (East-North US). Two A4 instances (8 cores and 14 GB memory) were configured as master nodes, and 64 A3 (4 cores and 7 GB memory) instances were configured as computing nodes. Both SparkBLAST and CloudBLAST executed queries on two datasets (Buz.fasta, and Ber.fasta), varying the number of cores allocated as 1 (BLAST sequential execution), 4, 12, 28, 60, 124 and 252. Every execution was repeated 6 times for CloudBLAST and six times for SparkBLAST. Thus, Experiment 2 encompasses $2 \times 2 \times 7 \times 6 = 168$ executions in total, which demanded more than 8,118 h (wall clock) to execute. An estimate on the amount of computational resources, this experiment consumed more than 139,595 vCPU-hours to execute on the Azure Cloud.

For the Microsoft Azure platform, SparkBLAST outperforms CloudBlast on all scenarios. Both datasets (Buz.fasta and Ber.fasta) were processed, and results are presented in Fig. 6 (speedup), Fig. 7 (total execution time), Fig. 8 (Efficiency), Table 5 (Buz.fasta), and Table 6 (Ber.fasta). It is worth noting that the largest dataset (Ber.fasta - 11 GB) was larger than the available memory in the computing nodes. For this reason, CloudBLAST could not process the Ber.fasta dataset, while SparkBLAST does

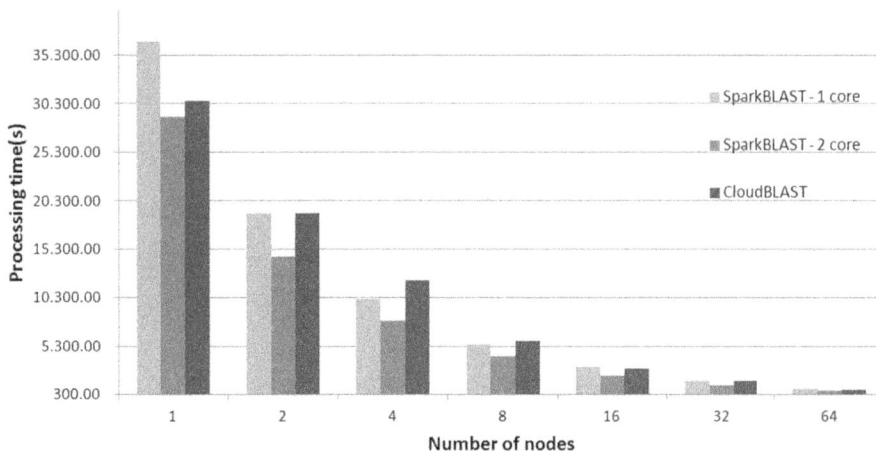

**Fig. 3** Total execution time for CloudBLAST *vs.* SparkBLAST running on the Google Cloud. Values represent the average of six executions for each experiment

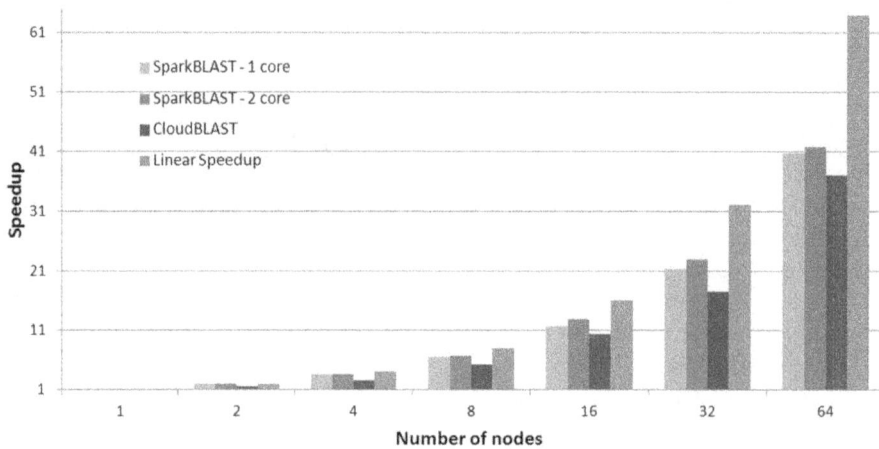

**Fig. 4** Speedup for 1 to 64 nodes in the Google Cloud. SparkBlast was executed on virtual machines with one and two cores. CloudBlast was executed on nodes with two cores

not have this limitation. It is also worth mentioning that larger speedups were achieved on Microsoft Azure when compared to the Google Cloud. This can be partially explained by the fact that all computing nodes allocated on the Microsoft Azure are placed in the same location, while computing nodes on Google Cloud were distributed among 4 different locations.

### Similarity-based inferences
In order to obtain added-value from the SparkBLAST similarity results on the cloud, the output from Spark-BLAST processing of Experiment 1 was used to infer orthology relationships with the RBH approach. In Table 7, numbers represent (RBH) orthologs found between 2 species. Numbers in bold represent (RBH) paralogs found in the same species. The higher number of RBH shared by two species was 264 between *Desulfovibrio vulgaris* and *Desulfovibrio desulfuricans*, and the lower was 15 between *Desulfovibrio fructosivorans* and *Deinococcus radiodurans*. Among the same species,

the higher number of RBH was 572 in *Rhodobacter sphaeroides* and the lower 34 in *Deinococcus geothermalis*. Regarding experiment 2: 1.27% (778,349/61255,260) of the Bermuda metagenomics proteins and 1.4% (68,748/4795,626) of the Búzios metagenomic proteins represent hits or potential homologs to the 10 radiation-resistant bacteria.

### Discussion
In this paper we investigate the parallelization of sequence alignment algorithms through an approach that employs cloud computing for the dynamic provisioning of large amounts of computational resources and Apache Spark as the coordination framework for the parallelization of the application. SparkBLAST, a scalable parallelization of sequence alignment algorithms is presented and assessed. Apache Spark's *pipe* operator and its main abstraction RDD (*resilient distribution dataset*) are used to perform scalable protein alignment searches by invoking BLASTP as an external application library. Experiments

**Table 1** Execution times for CloudBLAST - Google Cloud

| # nodes | 1 | 2 | 4 | 8 | 16 | 32 | 64 |
|---|---|---|---|---|---|---|---|
| Exec. time 1 | 29,921.40 | 19,018.00 | 11,324.00 | 6,204.00 | 2,866.00 | 1,680.00 | 794.00 |
| Exec. time 2 | 30,256.23 | 18,550.25 | 13,799.23 | 5,779.21 | 2,959.65 | 1,828.23 | 900.00 |
| Exec. time 3 | 31.016.85 | 19,221.81 | 12,580.32 | 5,700.52 | 3,004.52 | 1,597.00 | 815.21 |
| Exec. time 4 | 31.350.25 | 19,102.68 | 10,489.53 | 5,850.02 | 2,961.23 | 1,806.25 | 842.30 |
| Exec. time 5 | 30.726.89 | 18,981.32 | 12,721.23 | 5,780.34 | 2,990.81 | 1,780.32 | 799.21 |
| Exec. time 6 | 30.012.14 | 19,118.72 | 11,820.85 | 5,900.64 | 3,008.15 | 1,753.23 | 802.98 |
| Mean | 30,547.29 | 18,998.80 | 12,122.53 | 5,869.12 | 2,965.06 | 1,740.84 | 825.62 |
| Std. Dev. | 576.25 | 235.28 | 1.164.02 | 177.70 | 52.79 | 87.20 | 40.32 |
| Std.Dev./Mean | 1.89% | 1.24% | 9.60% | 3.03% | 1.78% | 5.01% | 4.88% |

**Table 2** Execution times - SparkBLAST 1 core - Google Cloud

| # nodes | 1 | 2 | 4 | 8 | 16 | 32 | 64 |
|---|---|---|---|---|---|---|---|
| Exec. time 1 | 36,106.86 | 18,845.23 | 10,189.11 | 5,556.22 | 3,129.20 | 1,716.10 | 905.21 |
| Exec. time 2 | 36,510.12 | 19,120.32 | 10,199.85 | 5,540.15 | 3,115.12 | 1,730.58 | 899.84 |
| Exec. time 3 | 36,720.86 | 18,952.15 | 10,170.23 | 5,560.88 | 3,140.01 | 1,790.96 | 894.76 |
| Exec. time 4 | 38,120.25 | 18,998.06 | 10,200.01 | 5,543.62 | 3,120.58 | 1,694.69 | 900.42 |
| Exec. time 5 | 36,230.56 | 19,112.23 | 10,178.76 | 5,552.10 | 3,122.15 | 1,701.55 | 897.65 |
| Exec. time 6 | 36,452.53 | 18,880.11 | 10,183.61 | 5,565.11 | 3,127.58 | 1,710.68 | 890.25 |
| Mean | 36,690.20 | 18,984.68 | 10,186.93 | 5,553.01 | 3,125.77 | 1,724.09 | 898.02 |
| Std.Dev | 733.00 | 115.14 | 11.83 | 9.73 | 8.62 | 35.01 | 5.14 |
| Std.Dev/Mean | 2.00% | 0.61% | 0.12% | 0.18% | 0.28% | 2.03% | 0.57% |

**Table 3** Execution times - SparkBLAST 2 cores - Google Cloud

| # nodes | 1 | 2 | 4 | 8 | 16 | 32 | 64 |
|---|---|---|---|---|---|---|---|
| Exec. time 1 | 28,915.52 | 14,500.86 | 7,935.45 | 4,287.85 | 2,249.94 | 1,260.12 | 695.23 |
| Exec. time 2 | 29,002.21 | 14,520.23 | 7,945.10 | 4,290.12 | 2,230.26 | 1,259.28 | 690.04 |
| Exec. time 3 | 29,001.89 | 14,515.35 | 7,950.01 | 4,283.56 | 2,255.04 | 1,260.10 | 701.50 |
| Exec. time 4 | 28,989.52 | 14,557.51 | 7,942.20 | 4,282.21 | 2,242.63 | 1,259.52 | 710.11 |
| Exec. time 5 | 28,990.32 | 14,580.01 | 7,940.80 | 4,310.12 | 2,249.26 | 1,259.82 | 680.80 |
| Exec. time 6 | 29,001.15 | 14,520.23 | 7,950.12 | 4,295.56 | 2,251.08 | 1,262.15 | 682.10 |
| Mean | 28,983.44 | 14,532.37 | 7,943.95 | 4,291.57 | 2,246.37 | 1,260.17 | 693.30 |
| Std.Dev | 33.78 | 29.93 | 5.68 | 10.27 | 8.85 | 1.03 | 11.37 |
| Std.Dev/Mean | 0.12% | 0.21% | 0.07% | 0.24% | 0.39% | 0.08% | 1.64% |

**Table 4** Mean execution times, speedups and parallel efficiency (Experiment 1 - query.fasta - 36 MB) - SparkBLAST vs CloudBLAST - Google Cloud

| # nodes | 1 | 2 | 4 | 8 | 16 | 32 | 64 |
|---|---|---|---|---|---|---|---|
| SparkBLAST | | | | | | | |
| 1 core | | | | | | | |
| Exec. time | 36,690.20 | 18,984.68 | 10,186.93 | 5,553.01 | 3,125.77 | 1,724.09 | 898.02 |
| Speedup | 1 | 1.93 | 3.60 | 6.61 | 11.74 | 21.28 | 40.86 |
| Efficiency | 1 | 0.97 | 0.90 | 0.83 | 0.73 | 0.67 | 0.64 |
| SparkBLAST | | | | | | | |
| 2 cores | | | | | | | |
| Exec. time | 28,983.44 | 14,532.37 | 7,943.95 | 4,291.57 | 2,246.37 | 1,260.17 | 693.30 |
| Speedup | 1.00 | 1.99 | 3.65 | 6.75 | 12.90 | 23.00 | 41,81 |
| Efficiency | 1.00 | 1.00 | 0.91 | 0.84 | 0.81 | 0.72 | 0.65 |
| CloudBLAST | | | | | | | |
| Exec. time | 30,547.29 | 18,998.80 | 12,122.53 | 5,869.12 | 2,965.06 | 1,740.84 | 825.62 |
| Speedup | 1.00 | 1.61 | 2.52 | 5.20 | 10.30 | 17.55 | 37.00 |
| Efficiency | 1.00 | 0.80 | 0.63 | 0.65 | 0.64 | 0.55 | 0.58 |

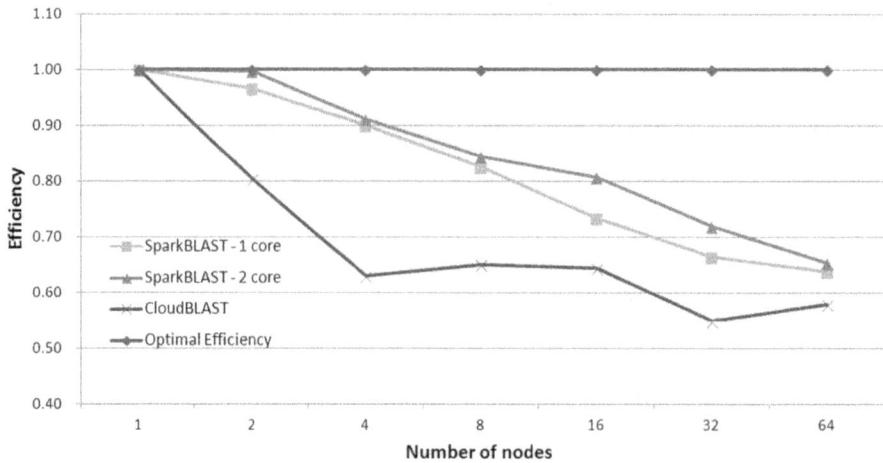

**Fig. 5** Efficiency for CloudBLAST x SparkBLAST running on Google Cloud

on the Google Cloud and Microsoft Azure have demonstrated that the Spark-based system outperforms a state-of-the-art system implemented on Hadoop in terms of speedup and execution times. It is worth noting that SparkBLAST can outperform CloudBlast even when just one of the vCPUs available per node is used by SparkBLAST, as demostrated by results obtained on the Google Cloud. In the experiments presented in this paper, the Hadoop-based system always used all vCPUs available per node.

From Table 4 it is possible to verify that both Speedup and Parallel Efficiency are better for SparkBLAST when compared to CloudBLAST for experiments executed on both the Google Cloud and Microsoft Azure. This is true for both scenarios of SparkBLAST on the Google Cloud (1 and 2 cores per node). It is worth noting that even when total execution time for CloudBLAST is smaller than the 1-core SparkBLAST configuration, Speedup and Parallel Efficiency is always worse for CloudBLAST. When SparkBLAST allocates two cores per node (as Cloud-BLAST does) execution times are always smaller when compared to CloudBLAST.

For the Microsoft Azure platform, all measures (processing time, efficiency and speedup) are better on Spark-BLAST when compared to the corresponding execution of CloudBLAST for the Buz.fasta (805 MB) dataset. It is worth noting that the speedup difference in favor of SparkBLAST increases with the number of computing nodes, which highlights the improved scalability of SparkBLAST over CloudBLAST. As mentioned in the "Results" section, it was not possible to process the larger Ber.fasta (11 GB) dataset using CloudBLAST due to computing node's main memory limitation. This constraint does not affect SparkBLAST, which can process datasets

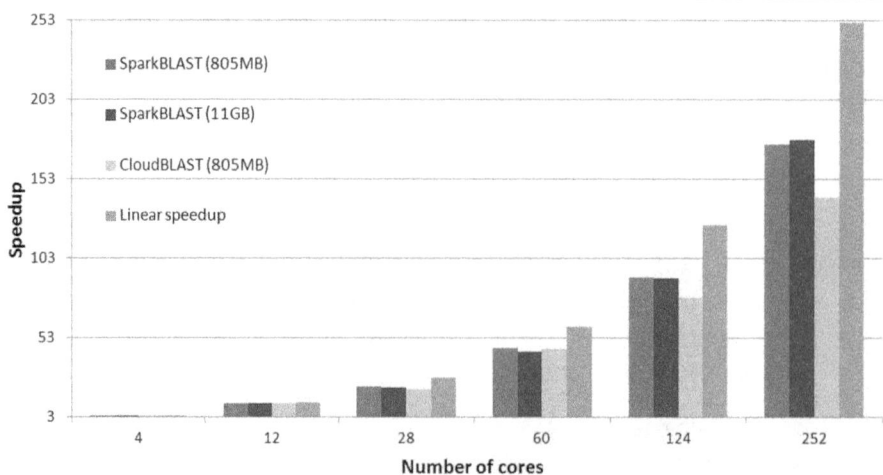

**Fig. 6** Speedup - Microsoft Azure

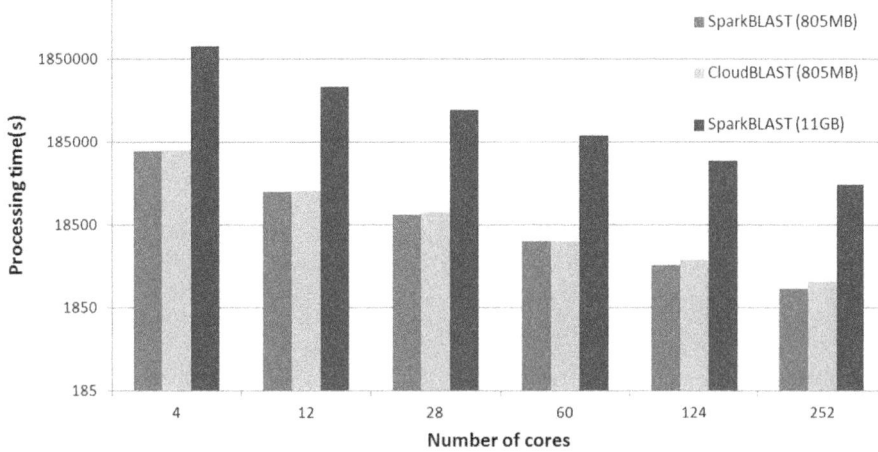

**Fig. 7** Total execution time for CloudBLAST x SparkBLAST on Microsoft Azure

even when they are larger than the main memory available on computing nodes. In the case of Spark, every process invoked by a task (each core is associated to a task) can use RDD even when database do not fit in memory, due memory content reuse and the implementation of circular memory [17]. It is worth noting that RDDs are stored as deserialized Java objects in the JVM. If the RDD does not fit in memory, some partitions will not be cached and will be recomputed on the fly each time they are needed [10]. Indeed, one very important loophole of existing methods that we address in SparkBLAST is the capability of processing large files on the Cloud. As described in this paragraph, SparkBLAST can process much larger files when compared to CloudBLAST, due to a more efficient memory management.

The main reason behind the performance of Spark-BLAST when compared to Hadoop-based systems are the in-memory operations and its related RDD abstraction. The reduced number of Disk IO operations by SparkBLAST results in a significant improvement on overall performance when compared to the Hadoop implementation.

It is clear that in-memory operations available in Spark-BLAST plays a major role both in Speedup and Parallel Efficiency improvements and, as a consequence, also in the scalability of the system. Indeed, the main reason behind the fact that SparkBLAST, even when it allocates only half of nodes processing capacity, achieves performance figures that are superior of those of CloudBLAST is the reduced number of local I/O operations.

Another point to be highlighted is the scalability of SparkBLAST on a worldwide distributed platform such as Google Cloud. For the executions presented in this work, 64 nodes were deployed in 13 zones and it was achieved a speedup of 41.78 in this highly distributed platform. Once

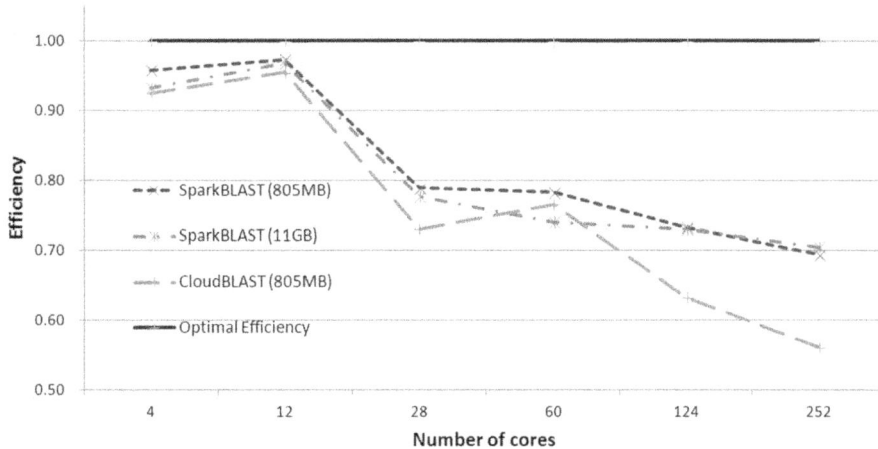

**Fig. 8** Efficiency - CloudBLAST x SparkBLAST - Microsoft Azure

**Table 5** Mean execution times, speedups and parallel efficiency (Experiment 2 - Buz.fasta - 805 MB) - SparkBLAST vs CloudBLAST - Microsoft Azure

| # cores | 4 | 12 | 28 | 60 | 124 | 252 |
|---|---|---|---|---|---|---|
| SparkBLAST | 143,228.95 | 47,031.62 | 24,850.51 | 11,692.45 | 6,041.64 | 3,138.64 |
| Speedup | 3.83 | 11.67 | 22.09 | 46.95 | 90.86 | 174.89 |
| Efficiency | 0.96 | 0.97 | 0.79 | 0.78 | 0.73 | 0.69 |
| CloudBLAST | 148,512.95 | 47,950.05 | 26,858.71 | 11,951.11 | 6,993.52 | 3,879.06 |
| Speedup | 3.7 | 11.45 | 20.44 | 45.93 | 78.49 | 141.51 |
| Efficiency | 0.92 | 0.95 | 0.73 | 0.77 | 0.63 | 0.56 |

again, in-memory operations is a major factor related to this performance.

For applications where the Reduce stage is not a bottleneck, which is the case for SparkBLAST, it is demonstrated in the literature that Spark is much faster than Hadoop. In [18], those authors state that, for this class of application, MapReduce Hadoop is much slower than Spark in task initialization and is less efficient in memory management. Indeed, the supplementary document "Execution Measurements of SparkBLAST and Cloud-BLAST", available in the online version of this paper, presents several measurements performed during Spark-BLAST and CloudBLAST executions on the Microsoft Azure Cloud. These measurements show that task initialization in SparkBLAST is considerably faster than CloudBLAST. It is also shown that SparkBLAST is more efficient in memory management than CloudBLAST. The effect of SparkBLAST's more efficient memory management can be observed in Additional file 1: Figures S5 and S6 of the supplementary information document. These figures show that Hadoop needs to use more memory than Spark, while Spark can maintain a larger cache and less swap to execute. Both factors - task initialization and memory management - are determinant for the improved scalability of SparkBLAST.

Furthermore, CloudBLAST makes use of Hadoop Streaming. In [19], authors shown that the Hadoop Streaming mechanism used in CloudBLAST can decrease application performance because it makes use of OS pipes to transfer input data to the applications' (in this case

BLAST) standard input and from BLAST standard output to disk storage. Data input to BLAST is done by the option: "-inputreader org.apache.hadoop.streaming. StreamPatternRecordReader", which send lines from the input file to BLAST one-by-one, which further degrades performance.

Regarding extended scalability over larger platforms than the ones considered in this paper, it should be highlighted that two authors of this paper have proposed a formal scalability analysis of MapReduce applications [5]. In this analysis the authors prove that the most scalable MapReduce applications are reduceless applications, which is exactly the case of SparkBLAST. Indeed, Theorem 5.2 of [5] states that the increase of amount of computation necessary for a reduceless Scalable MapReduce Computation (SMC) application to maintain a given isoefficiency level is proportional to the number of processors (nodes). This is the most scalable configuration over all scenarios analyzed in [5]. Simulation results that goes up to 10000 nodes corroborate the limits stated in this and other theorems of [5].

Regarding Experiment 1 and RBH inference, we showed that our SparkBLAST results can be post-processed to infer shared genes, then generating added-value to the similarity analysis. That also means that RBH experiments using SparkBLAST are potentially scalable to many more genomes, and can be even used as part of other Blast-based homology inference software such as OrthoMCL [20]. Considering Experiment 2, results indicate that 1.27% of the Bermuda metagenomics proteins and 1.4%

**Table 6** Mean execution times, speedups and parallel efficiency (Experiment 2 - Ber.fasta - 11 GB) - SparkBLAST vs CloudBLAST - Microsoft Azure

| # cores | 4 | 12 | 28 | 60 | 124 | 252 |
|---|---|---|---|---|---|---|
| SparkBLAST | 2,678,902.06 | 859,687.13 | 458,759.75 | 224,869.12 | 110,222.98 | 56,200.21 |
| Speedup | 3.73 | 11.61 | 21.76 | 44.4 | 90.57 | 177.64 |
| Efficiency | 0.93 | 0.97 | 0.78 | 0.74 | 0.73 | 0.7 |
| CloudBLAST | - | - | - | - | - | - |
| Speedup | - | - | - | - | - | - |
| Efficiency | - | - | - | - | - | - |

**Table 7** Numbers of RBH found using data from SparkBLAST cloud processing

| Kineococcus radiotolerans | Desulfovibrio desulfuricans | Desulfovibrio vulgaris | Rhodobacter sphaeroides | Escherichia coli | Deinococcus radiodurans | Desulfovibrio fructosivorans | Shewanella oneidensis | Geobacter sulfurreducens | Deinococcus geothermalis | Geobacter metallireducens | Name | Accession Number | Number of proteins |
|---|---|---|---|---|---|---|---|---|---|---|---|---|---|
| **224** | 43 | 25 | 63 | 35 | 53 | 22 | 21 | 18 | 52 | 20 | Kineococcus radiotolerans | NC_009660.1 | 4,632 |
| | **380** | 264 | 121 | 79 | 38 | 163 | 71 | 88 | 27 | 60 | Desulfovibrio desulfuricans | NC_011883.1 | 10,443 |
| | | **362** | 62 | 46 | 17 | 98 | 53 | 47 | 24 | 37 | Desulfovibrio vulgaris | NC_002937.3 | 12,349 |
| | | | **572** | 114 | 46 | 46 | 98 | 77 | 50 | 44 | Rhodobacter sphaeroides | NC_009429.1 | 20,954 |
| | | | | **98** | 24 | 26 | 155 | 34 | 28 | 26 | Escherichia coli | NC_000913.3 | 4,140 |
| | | | | | **122** | 15 | 21 | 27 | 102 | 17 | Deinococcus radiodurans | NC_001263.1 | 7,671 |
| | | | | | | **84** | 20 | 40 | 17 | 38 | Desulfovibrio fructosivorans | NZ_AECZ01000069.1 | 4,028 |
| | | | | | | | **90** | 36 | 21 | 32 | Shewanella oneidensis | NC_004349.1 | 8,271 |
| | | | | | | | | **146** | 20 | 120 | Geobacter sulfurreducens | NC_002939.5 | 9,340 |
| | | | | | | | | | **34** | 16 | Deinococcs geothermalis | NC_008010.2 | 2,935 |
| | | | | | | | | | | **50** | Geobacter metallireducens | NC_007517.1 | 3,592 |

Numbers in bold represent (RBH) 354 paralogs found in the same species of the Búzios metagenomic proteins represent potential homologs to the 10 radiation-resistant bacteria, and as far as we know no related studies have been published to date. Those potential homologs will be further investigated in another study.

## Conclusion

In this paper we propose SparkBLAST, a parallelization of BLAST that employs cloud computing for the provisioning of computational resources and Apache Spark as the coordination framework. SparkBLAST outperforms CloudBLAST, a Hadoop-based implementation, in speedup, efficiency and scalability in a highly distributed cloud platform. The superior performance of SparkBLAST is mainly due to the in-memory operations available through the Spark framework, consequently reducing the number of local I/O operations required for distributed BLAST processing.

## Endnotes

[1] https://wiki.apache.org/hadoop/PoweredByYarn

[2] http://metagenomics.anl.gov/

## Additional file

**Additional file 1:** Execution Measurements of SparkBLAST and CloudBLAST. In this supplementary document we present performance data collected during the execution of Experiment 2 on the Microsoft Azure Platform. **Figure S1.** CPU utilization for the SparkBLAST execution. **Figure S2.** CPU utilization for the CloudBLAST execution. **Figure S3.** CPU utilization for one worker node running SparkBLAST. **Figure S4.** CPU utilization for one worker node running CloudBLAST. **Figure S5.** Memory utilization for SparkBLAST. **Figure S6.** Memory utilization for CloudBLAST. **Figure S7.** Network traffic produced by SparkBLAST during its execution. **Figure S8.** Network traffic produced by CloudBLAST during its execution.

## Abbreviations

DFS: Distributed file system; NGS: Next generation sequencing; RDD: Resilient distribution datasets; RBH: Reciprocal best hits; SMC: Scalable MapReduce computation; vCPU: Virtual CPU

## Acknowledgements

The authors would like to thank Thais Martins for help with input data preparation, and Rodrigo Jardim for data preparation and preliminary RBH analysis.

## Funding

Authors thank Google, Microsoft Research, CAPES, CNPq, FAPERJ and IFSULDEMINAS for supporting this research project. Hermes Senger thanks CNPq (Contract Number 305032/2015-1) for their support. The authors declare that no funding body played any role in the design or conclusion of this study.

## Authors' contributions

MRC, HS and FABS have implemented SparkBLAST and run execution tests on Google and Microsoft Azure clouds. AMRD and CST have performed RBH and metagenomics analysis. All authors have contributed equally to the writing of this paper. All authors have read and approved the final manuscript.

## Competing interests

The authors declare that they have no competing interests.

## Author details

[1] Computer Science Department, Federal University of São Carlos, Rod. Washington Luís, Km 235, 21040-900 São Carlos, Brazil. [2] LBCS-IOC, Oswaldo Cruz Foundation, Av Brasil 4365, 21040-900 Rio de Janeiro, Brazil. [3] PROCC, Oswaldo Cruz Foundation, Av. Brasil 4365, 21040-900 Rio de Janeiro, Brazil.

## References

1. Altschul SF, Gish W, Miller W, Myers EW, Lipman DJ. Basic local alignment search tool. J Mol Biol. 1990;215(3):403–10.
2. Camacho C, Coulouris G, Avagyan V, Ma N, Papadopoulos J, Bealer K, Madden TL. Blast+: architecture and applications. BMC Bioinforma. 2009;10(1):421.
3. Altschul SF, Madden TL, Schäffer AA, Zhang J, Zhang Z, Miller W, Lipman DJ. Gapped blast and psi-blast: a new generation of protein database search programs. Nucleic Acids Res. 1997;25(17):3389–402.
4. ODriscoll A, Daugelaite J, Sleator R. Big data, hadoop and cloud computing in genomics. J Biomed Inform. 2013;46(5):774–81.
5. Senger H, Gil-Costa V, Arantes L, Marcondes CAC, Marin M, Sato LM, da Silva FAB. BSP Cost and Scalability Analysis for MapReduce Operations. Concurr Comput: Pract Experience. 2016;28(8):2503–27. doi:10.1002/cpe.3628.
6. Langmead B, Hansen KD, Leek JT. Cloud-scale rna-sequencing differential expression analysis with myrna. Genome Biol. 2010;11(8):1–11.
7. Leo S, Santoni F, Zanetti G. Biodoop: bioinformatics on hadoop. In: Intl. Conf. Parallel Processing Workshops ICPPW'09. Los Alamitos: IEEE Computer Society. 2009. p. 415–22.
8. Matsunaga A, Tsugawa M, Fortes J. Cloudblast: Combining mapreduce and virtualization on distributed resources for bioinformatics applications. In: Intl. Conf. on eScience (eScience'08). Los Alamitos: IEEE Computer Society. 2008. p. 222–9.
9. Feng X, Grossman R, Stein L. Peakranger: a cloud-enabled peak caller for chip-seq data. BMC Bioinforma. 2011;12(1):1.
10. Zaharia M, Chowdhury M, Franklin MJ, Shenker S, Stoica I. Spark: Cluster Computing with Working Sets. In: HotCloud'10: Proceedings of the 2Nd USENIX Conference on Hot Topics in Cloud Computing. HotCloud'10. Berkeley: USENIX Association; 2010. p. 10. http://portal.acm.org/citation.cfm?id=1863103.1863113.
11. Zaharia M, Chowdhury M, Das T, Dave A, Ma J, McCauley M, Franklin MJ, Shenker S, Stoica I. Resilient distributed datasets: A fault-tolerant abstraction for in-memory cluster computing. In: Proceedings of the 9th USENIX Conference on Networked Systems Design and Implementation. Berkeley: USENIX Association; 2012. p. 2–2.
12. Zaharia M, Borthakur D, Sen Sarma J, Elmeleegy K, Shenker S, Stoica I. Delay scheduling: a simple technique for achieving locality and fairness in cluster scheduling. In: Proceedings of the 5th European Conference on Computer Systems. New York: ACM. 2010. p. 265–78.
13. Vavilapalli VK, Murthy AC, Douglas C, Agarwal S, Konar M, Evans R, Graves T, Lowe J, Shah H, Seth S, et al. Apache hadoop yarn: Yet another resource negotiator. In: Proceedings of the 4th Annual Symposium on Cloud Computing. New York: ACM. 2013. p. 5.
14. Bork P, Dandekar T, Diaz-Lazcoz Y, Eisenhaber F, Huynen M, Yuan Y. Predicting function: from genes to genomes and back. J Mol Biol. 1998;283(4):707–25.
15. Tatusov RL, Koonin EV, Lipman DJ. A genomic perspective on protein families. Science. 1997;278:631–7.
16. Prakash D, Gabani P, Chandel AK, Ronen Z, Singh OV. Bioremediation: a genuine technology to remediate radionuclides from the environment. Microb Biotechnol. 2013;6(4):349–60.
17. Gopalani S, Arora R. Comparing apache spark and map reduce with performance analysis using k-means. Int J Comput Appl. 2015;113(1):8–11.
18. Shi J, Qiu Y, Minhas UF, Jiao L, Wang C, Reinwald B, Özcan F. Clash of the titans: Mapreduce vs. spark for large scale data analytics. Proc VLDB Endowment. 2015;8(13):2110–21.
19. Ding M, Zheng L, Lu Y, Li L, Guo S, Guo M. More convenient more overhead: the performance evaluation of hadoop streaming. In: Proceedings of the 2011 ACM Symposium on Research in Applied Computation. New York: ACM. 2011. p. 307–13.
20. Li L, Stoeckert CJ, Roos DS. OrthoMCL: identification of ortholog groups for eukaryotic genomes. Genome Res. 2003;13(9):2178–89.

# 4

# Estimating Phred scores of Illumina base calls by logistic regression and sparse modeling

Sheng Zhang[1,2†], Bo Wang[1,2†], Lin Wan[1,2] and Lei M. Li[1,2*]

## Abstract

**Background:** Phred quality scores are essential for downstream DNA analysis such as SNP detection and DNA assembly. Thus a valid model to define them is indispensable for any base-calling software. Recently, we developed the base-caller 3Dec for Illumina sequencing platforms, which reduces base-calling errors by 44-69% compared to the existing ones. However, the model to predict its quality scores has not been fully investigated yet.

**Results:** In this study, we used logistic regression models to evaluate quality scores from predictive features, which include different aspects of the sequencing signals as well as local DNA contents. Sparse models were further obtained by three methods: the backward deletion with either AIC or BIC and the $L_1$ regularization learning method. The $L_1$-regularized one was then compared with the Illumina scoring method.

**Conclusions:** The $L_1$-regularized logistic regression improves the empirical discrimination power by as large as 14 and 25% respectively for two kinds of preprocessed sequencing signals, compared to the Illumina scoring method. Namely, the $L_1$ method identifies more base calls of high fidelity. Computationally, the $L_1$ method can handle large dataset and is efficient enough for daily sequencing. Meanwhile, the logistic model resulted from BIC is more interpretable. The modeling suggested that the most prominent quenching pattern in the current chemistry of Illumina occurred at the dinucleotide "GT". Besides, nucleotides were more likely to be miscalled as the previous bases if the preceding ones were not "G". It suggested that the phasing effect of bases after "G" was somewhat different from those after other nucleotide types.

**Keywords:** Base-calling, Logistic regression, Quality score, $L_1$ regularization, AIC, BIC, Empirical discrimination power

## Background

High-throughput sequencing technology identifies the nucleotide sequences of millions of DNA molecules simultaneously [1]. Its advent in the last decade greatly accelerated biological and medical research and has led to many exciting scientific discoveries. Base calling is the data processing part that reconstructs target DNA sequences from fluorescence intensities or electric signals generated by sequencing machines. Since the influential work of Phred scores [2] in the Sanger sequencing era, it has become an industry standard that base calling software output an error probability, in the form of a quality score, for each base call. The probabilistic interpretation of quality scores allows fair integration of different sequencing reads, possibly from different runs or even from different labs, in the downstream DNA analysis such as SNP detection and DNA assembly [3]. Thus a valid model to define Phred scores is indispensable for any base-calling software.

Many existing base-calling software for high throughput sequencing define quality scores according to the Phred framework [2], which transforms the values of several predictive features of sequencing traces to a probability based on a lookup table. Such a lookup table is obtained by training on data sets of sufficiently large sizes. To keep the size of lookup table in control, the number of predictive

*Correspondence: lilei@amss.ac.cn

†Equal contributors

[1]National Center of Mathematics and Interdisciplinary Sciences, Academy of Mathematics and Systems Science, Chinese Academy of Sciences, 100190 Beijing, China

[2]University of Chinese Academy of Sciences, 100049 Beijing, China

features, also referred to as parameters, in the Phred algorithm is limited.

Thus each complete base-calling software consists of two parts: base-calling and quality score definition. Bustard is the base-caller developed by Illumina/Solexa and is the default method embedded in the Illumina sequencers. Its base-calling module includes image processing, extraction of cluster intensity signals, corrections of phasing and color crosstalk, normalization etc. Its quality scoring module generates error rates using a modification of the Phred algorithm, namely, a lookup table method, on a calibration data set. The Illumina quality scoring system was briefly explained in its manual [4] without details. Recently, we developed a new base-caller 3Dec [5], whose preprocessing further carries out adaptive corrections of spatial crosstalks between neighboring clusters. Compared to other existing methods, it reduces the error rate by 44-69%. However, the model to predict quality scores has not been fully investigated yet.

In this paper, we evaluate the error probabilities of base calls from predictive features of sequencing signals, using logistic regression models [6]. The basic idea of the method is illustrated in Fig. 1. Logistic regression, as one of the most important classes of generalized linear models [7], is widely used in statistics for evaluating success rate of binary data from dependent variables and in machine learning for classification problems. The training of logistic regression models can be implemented by the well-developed maximum likelihood method, and is computed by Newton-Raphson algorithm [8]. Instead of restricting to a limited number of experimental features, we include a large number of candidate features in our model and select predictive features via the sparse modeling. From previous research [9, 10] and our recent work (3Dec [5]), the candidate features for Illumina sequencing platforms should include: signals after correction for color-, cyclic- and spatial-crosstalk, the cycle number of the current positions, the two most likely nucleotide bases of the current positions and the called bases of the neighbor positions. In this article, we select 74 features derived from these factors as the predictive variables in the initial model.

Next, we reduce the initial model by imposing sparsity constraints. That is, we impose a $L_0$ or $L_1$ penalty on the log-likelihood function of the logistic models, and optimize the penalized function. The $L_0$ penalty includes the Akaike information criterion (AIC) and Bayesian information criterion (BIC). However, the exhaustive search of minimum AIC or BIC in all sub-models is a NP-hard problem [11]. An approximate solution can be achieved by the backward deletion strategy, whose computational complexity is polynomial. We note that this strategy coupled with BIC leads to the consistent model estimates in the case of linear regression [12]. Thus it is hypothesized

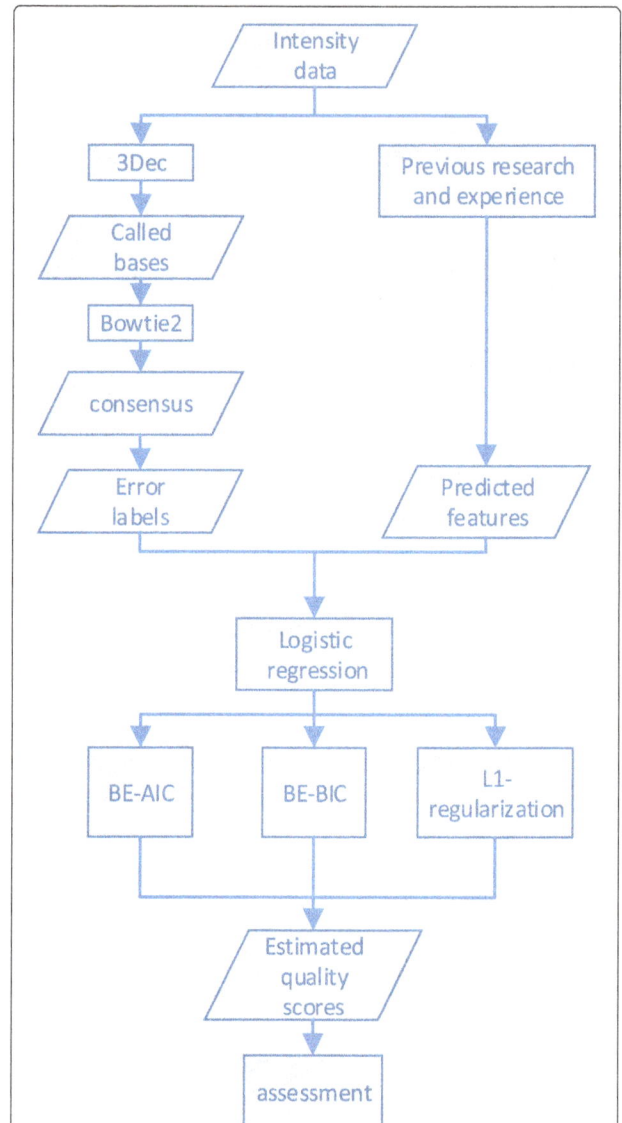

**Fig. 1** The flowchart of the method. The input is the raw intensities from sequencing. Then the called sequences are obtained using 3Dec. Next we used Bowtie2 to map the reads to the reference and defined a consensus sequence. Thus bases that are called different from those in the consensus reference are regarded as base-calling errors. Meanwhile a group of predictive features are calculated from the intensity data followed previous research and experience. Afterwards, three sparse constrained logistic regressions are carried out, and they are backward deletion either with BIC(BE-BIC) and AIC(BE-AIC), and $L_1$-regularization respectively. Finally, we use several measures to assess the predicted quality scores of the above three methods

that the same strategy would lead to a similar consistent asymptotics in the case of logistic regressions. Compared to BIC, AIC is more appropriate in finding the best model for predicting future observations [13]. The $L_1$ regularization, also known as LASSO [14], has recently become a popular tool for feature selection. Its solution can be solved by fast convex optimization algorithms [15]. In this article, we use these three methods to select the most relevant features from the initial ones.

In fact, a logistic model was already used to calibrate the quality values of training data sets that may come from different experiment conditions [16]. The covariates in the logistic model are simple spline functions of original quality scores. The backward deletion strategy coupled with BIC was used to pick up the relevant knots. In the same article, the accuracy of quality scores was examined by the consistency between empirical (aka. observed) error rates and the predicted ones. Besides, the scoring method could be measured by the discrimination power, namely, the ability to discriminate the more accurate base-calls from the less accurate ones. Ewing et al. [2] demonstrated that the bases of high quality scores are more important in the downstream analysis such as deriving the consensus sequence. Technically, they defined the discrimination power as the largest proportion of bases whose expected error rate is less than a given threshold. However, this definition is not perfect if bias exists, to some extent, in the predicted quality scores of a specific data set. Thus, in this article, we propose an empirical version of discrimination power, which is used for comparing the proposed scoring method with that of Illumina.

The sparse modeling using logistic regressions not only defines valid Phred scores, but also provides insights into the error mechanism of the sequencing technology by variable selection. Like the AIC and BIC method, the solution to $L_1$-regularized method is sparse and thereby embeds variable selection. The features identified by the model selection are good explanatory variables that may even lead to the discoveries of causal factors. For example, quenching effect [17] is a factor leading to uneven fluorescence signals, due to short-range interactions between the fluorophore and the nearby molecules. Using the logistic regression methods, we further demonstrated the detailed pattern of G-quenching effect in the Illumina sequencing technology, including G-specific phasing and the reduction of the T-signal following a G.

## Methods

### Data

The source data used in this article were from [18], and were downloaded at [19]. This dataset includes three tiles of raw sequence intensities from Illumina HiSeq 2000 sequencer. Each tile contains about 1,900,000 single-end reads of 101 sequencing cycles, whose intensities are from four channels, namely A, C, G and T. Then we carried out the base calling using 3Dec [5] and obtained the error labels of the called bases by mapping the reads to the consensus sequence. The more than 400X depths of sequencing reads make it possible to define a reliable consensus sequence, and the procedure is the same as [5]. That is, first, Bowtie2 (version 2.2.5, using the default

option of "−sensitive") was used to map the reads to the reference (Bacteriophage PhiX174). Second, in the resulting layout of reads, a new consensus was defined as the most frequent nucleotide at each base position. Finally, this consensus sequence was taken as the updated reference. According to this scheme, the bases that were called different from those in the consensus reference were regarded as the base-calling errors. In this way, we obtained the error labels of the called bases. We selected approximately three million bases of 30 thousand sequences from the first tile as the training set, and tested our methods on a set of bases from the third tile.

Throughout the article, we represent random variables by capital letters, their observations by lowercase ones, and vectors by bold ones. We denote the series of target bases in the training set by $S = S_1 S_2 \cdots S_n$, where $S_i$ is the called base taking any value from the nucleotides A, C, G or T. Let $Y_i$ be the error label of base $S_i$ ($i = 1, 2, \cdots, n$). Therefore,

$$Y_i = \begin{cases} 1 & \text{if base } S_i \text{ is called correctly ,} \\ 0 & \text{otherwise .} \end{cases}$$

### Phred scores

Many existing base-calling software output a quality score $q$ for each base call to measure the error probability after the influential work of Phred scores [2]. Mathematically, let $q_i$ be the quality score of the base $S_i$, then

$$\begin{cases} q_i = -10 \log_{10} \varepsilon_i, \\ \varepsilon_i = \mathbf{Pr}(Y_i = 0 | X_i = \boldsymbol{x}_i), \end{cases} \tag{1}$$

where $\varepsilon_i$ is the error probability of base-calling and $X_i$ is the feature vector described below. For example, if the Phred quality score of a base is 30, the probability that this base is called incorrectly is 0.001. This also indicates that the base call accuracy is 99.9%. The estimation of Phred scores is equivalent to the estimation of the error probabilities.

### Logistic regression model

Ewing et al. proposed the lookup table stratified by four features to predict quality scores [2]. Here, we adopt a different stratification strategy using the logistic regression [6].

Mathematically, the logistic regression model here estimates the probability that a base is called correctly. We denote this probability for the base $S_i$ as

$$p(\boldsymbol{x}_i; \boldsymbol{\beta}) = 1 - \varepsilon_i = \mathbf{Pr}(Y_i = 1 | X_i = \boldsymbol{x}_i; \boldsymbol{\beta}), \tag{2}$$

where $\boldsymbol{\beta}$ is the parameter to be estimated. We assume that $p(\boldsymbol{x}_i; \boldsymbol{\beta})$ follows a logistic form:

$$\log \left( \frac{p(\boldsymbol{x}_i; \boldsymbol{\beta})}{1 - p(\boldsymbol{x}_i; \boldsymbol{\beta})} \right) = \boldsymbol{x}_i^T \boldsymbol{\beta}, \tag{3}$$

where the first element in $\boldsymbol{x}_i$ is a constant, representing the intercept term. Equivalently, the accuracy of base-calling can be represented as:

$$p(\boldsymbol{x}_i; \boldsymbol{\beta}) = \frac{1}{1 + \exp\left(-\boldsymbol{x}_i^T \boldsymbol{\beta}\right)}. \quad (4)$$

The above parameterization leads to the following form of log-likelihood function for the data of base calls:

$$L(\boldsymbol{\beta}; \boldsymbol{x}_1, \cdots, \boldsymbol{x}_n) = \sum_{i=1}^{n} \left(y_i \log p(\boldsymbol{x}_i; \boldsymbol{\beta}) + (1 - y_i) \quad (5)\right.$$
$$\left. \log(1 - p(\boldsymbol{x}_i; \boldsymbol{\beta}))\right),$$

where $y_i$ is the value of $Y_i$, namely 0 or 1, and $\boldsymbol{\beta}$ represents all the unknown parameters. Then $\boldsymbol{\beta}$ is estimated by maximizing the log-likelihood function, and is computed by the Newton-Raphson algorithm [8].

The computation of logistic regression is implemented by the "glm" package provided in the R software [20], in which we take the parameter "family" as binomial and take the "link function" as the logit function.

### Predictive features of Phred scores

Due to the complexity of the lookup table strategy, the number of predictive features in the Phred algorithm is limited for Sanger sequencing reads. Ewing et al. [2] used only four trace features such as peaking spacing, uncalled/called ratio and peak resolution to discriminate errors from correct base-calls [2]. However, these features are specific in the Sanger sequencing technology and are no longer suitable for next generation sequencers. In addition, next generation sequencing methods have their own error mechanism leading to incorrect base-calls such as the phasing effect. From previous research [9, 10] and our recent work [5], it should be noted that the error rates of the base calls in the Illumina platforms are related to the factors such as the signals after correction for color-, cyclic- and spatial crosstalk, the cycle number of current positions, the two most likely nucleotide bases of current positions and the called bases of the neighbor positions. Therefore, a total of 74 candidate features are included as the predictive variables in the initial model. Let $X_i = (X_{i,0}, X_{i,1}, \cdots, X_{i,74})$ be the vector of the predictive features for the base $S_i$, and we explain them in groups as follows. Notice that some features are trimmed off to reduce their statistical influence of outliers.

- $X_{i,0}$ equals 1, representing the intercept term.
- $X_{i,1}, X_{i,2}$ are the largest and second largest intensities in the $i^{th}$ cycle, respectively. Because 3Dec [5] assigns the called base of $i^{th}$ cycle as the type with the largest intensity, the signal intensities such as $X_{i,1}$ and $X_{i,2}$ are crucial to the estimation of error probability. It makes sense that the called base is more accurate if

$X_{i,1}$ is larger. On the contrary, the called base $S_i$ has a tendency to be miscalled if $X_{i,2}$ is large as well, because the base calling software may be confused to determine the base with two similar intensities.

- $X_{i,3}, X_{i,4}$ and $X_{i,5}$ are the average of $X_{i,1}$, the average and standard error of $|X_{i,1} - X_{i,2}|$ in all the cycles in that sequence, respectively. The average signals outside [0.02, 3] and the standard error outside [0.02, 1] were trimmed off. $X_{i,3}$ to $X_{i,5}$ are common statistics that describe the intensities over the whole sequence.
- $X_{i,6}, X_{i,7}, X_{i,8}$ are $1/X_{i,3}$, $\sqrt{X_{i,5}}$ and $\log(X_{i,5})$, respectively. $X_{i,9}$ to $X_{i,17}$ are nine different piecewise linear functions of $|X_{i,1} - X_{i,2}|$, which are similar to [16]. $X_{i,6}$ to $X_{i,17}$ are used to approximate the potential non-linear effects of the former features.
- $X_{i,18}$ equals the current cycle number $i$, and $X_{i,19}$ is the inverse of the distance between the current and last cycle. These two features are derived from the position in the sequence due to the facts that bases close to both ends of sequences are more likely to be miscalled [18].
- $X_{i,20}$ to $X_{i,26}$ are seven dummy variables [21], each representing whether the current cycle $i$ is the first, the second, ..., the seventh, respectively. We add these seven features because the error rates in the first seven cycles of this dataset are fairly high [18].
- $X_{i,27}$ to $X_{i,74}$ are 48 dummy variables, each representing a 3-letter-sequence. The first letter indicates the called base in the previous cycle; the second and third letter respectively correspond to the nucleotide type with the largest and the second largest intensity in the current cycle. It is worth noting that these 3-letter-seuqnces involve only two DNA neighbor positions, instead of three. Take "A(AC)" as an example, the first letter "A" indicates the called base of the previous cycle, namely $S_{i-1}$; the second letter "A" in the parenthesis represents the called base in the current cycle, namely $S_i$; and the third letter "C" in the parenthesis is corresponding to the nucleotide type with the second largest intensity in the current cycle. All the 48 possible combinations of such 3-letter sequences are sorted in lexicographical order, which are "A(CA)", "A(CG)", "A(CT)", ..., "T(GT)", respectively.

The 48 features are chosen based on the facts that the error rate of a base varies when preceded by different bases [10]. These 3-letter sequences derived from the two neighboring bases can help us understand the differences among the error rates of the bases preceded by "A", "C", "G" and "T". Back to the example mentioned earlier, if the coefficient of "A(AC)" was positive, in other word, the presence of "A(AC)" led to a higher quality score, we would consider that an "A" after another "A" was more likely to be called correctly. On the contrary, if the coefficient of "A(AC)" was negative, the presence of "A(AC)" would reduce the quality score. In this case, there would be a high probability that the second "A" was an error while the correct one was "C". Thus it would indicate a substitution error pattern between "A" and "C" proceeded by base "A".

## Sparse modeling and model selection

To avoid overfitting and to select a subset of significant features, we reduce the initial logistic regression model by imposing sparsity constraints. That is, we impose a $L_0$ or $L_1$ penalty to the log-likelihood function of the logistic models, and optimize the penalized function.

The $L_0$ penalty includes AIC and BIC, which are respectively defined as

$$AIC = 2k - 2\hat{L}, \tag{6}$$

$$BIC = k\log(n) - 2\hat{L}, \tag{7}$$

where $k$ is the number of non-zero parameters in the trained model referred to as $||\beta||_0$, $n$ is the number of samples, and $\hat{L}$ is the maximum of the log-likelihood function defined in Eq. (5). AIC and BIC look for a tradeoff between the goodness of fit (the log-likelihood function) and the model complexity (the number of parameters). The smaller the AIC/BIC score is, the better the model is. The exhaustive search of minimum of AIC or BIC among all sub-models is a NP-hard problem [11], thus approximate approaches such as backward deletion are usually used in practice. The computational complexity of the backward deletion strategy is only polynomial. In fact, we note that this strategy coupled with BIC leads to the consistent model estimates in the case of linear regression [12]. Thus it is hypothesized that the same strategy would lead to a similar consistent asymptotics in the case of logistic regressions. Compared to BIC, AIC is more appropriate in finding the best model for predicting future observations [13].

The details of backward deletion are as follows. First, we implement the logistic regression with all features and calculate the AIC and BIC scores. Second, we remove each feature, recalculate the logistic regression models as well as their AIC and BIC scores, then delete the feature resulting in the lowest AIC or BIC score if it was removed. Last, we repeat the second step in the remaining features until AIC or BIC score no longer decreases. We note that this heuristic algorithm is still very time consuming due to the repetitive calculation of the logistic regression.

An alternative approach for sparse modeling is $L_1$ regularization. It imposes a $L_1$ norm penalty on the objective function, rather than the hard constraint on the number of nonzero parameters. Specifically, $L_1$-regularized logistic regression is to minimize the log-likelihood function penalized by the $L_1$ norm penalty of the parameters as follows:

$$\min_{\beta} -L(\beta; x_1, \cdots, x_n) + \lambda||\beta||_1, \tag{8}$$

where $||\beta||_1$ is the sum of the absolute value of each element in $\beta$, and $\lambda$ is specified based on a certain cross-validation procedure. The $L_1$ regularization, also known as LASSO [14], is applied here due to its two merits: first, it leads to a convex optimization problem which is well studied and can be solved very fast; second, it often produces a sparse solution which embeds feature selection and enables model interpretation. We further extended LASSO to the elastic net model [22], and the details were described in Additional file 1.

All these three methods seek for a tradeoff between the goodness of fit and model complexity. They also extract underlying sparse patterns from high dimensional features to enhance the model interpretability. However, they may result in different sparse solutions. If the data size $n$ is large enough, $\log(n)$ is much larger than 2, then backward deletion with BIC results in a sparser result than the AIC procedure does. Similarly, the sparsity of $L_1$ regularization depends on $\lambda$. The larger $\lambda$ is, the sparser the solution is.

The backward deletion with either AIC or BIC is implemented by the "stepAIC" function in "MASS" package provided in R [20], and $L_1$-regularized logistic regression is implemented in C++ using the liblinear library [23].

## Model assessment

### Consistency between predictive and empirical error rates

First, we follow Ewing et al. [2] and Li et. al. [16] to calculate the observed score stratified by the predicted ones. The observed score for the predicted quality score $q$ is calculated by

$$q_{obs}(q) = -10 \cdot \log_{10}\left(\frac{Err_q}{Err_q + Corr_q}\right), \tag{9}$$

where $Err_q$ and $Corr_q$ are, respectively, the number of incorrect and correct base-calls at quality score $q$. The consistency between the empirical scores with the predicted ones indicates the accuracy of the model.

### Empirical discrimination power

Second, Ewing et al. [2] proposed that the quality scores could be evaluated by the discrimination power, which is the ability to discriminate the more accurate base-calls from the less accurate ones.

Let $B$ be a set of base-calls and $e(b)$ be the error probability assigned by a valid method for each called base $b$. For any given error rate $r$, there exists a unique largest set of base-calls, $B_r$, satisfying two properties: (1) the expected error rate of $B_r$, i.e. the average assigned error probabilities of $B_r$ is less than $r$; (2) whenever $B_r$ includes a base-call $b$, it includes all other base-calls whose error probabilities are less than $e(b)$. The discrimination power at the error rate $r$ is defined as

$$P_r = \frac{|B_r|}{|B|}.$$

However, if bias exists, to some extent, in the predicted quality scores of a specific data set, the above definition is not perfectly fair. For example, if an inconsistent method assigns each base call a fairly large score, then $P_r$ reaches 1 at any $r$. Therefore, its discrimination power is much larger than any consistent method, which is obviously unfair. Thus we proposed an empirical version of discrimination power, defined as:

$$\widetilde{P}_r = \frac{|\widetilde{B}_r|}{|B|}, \tag{10}$$

where the above $B_r$ is replaced by $\widetilde{B}_r$ having the properties that: $(\widetilde{1})$ the empirical error rate of $\widetilde{B}_r$, i.e. the number of errors divided by the number of base-calls in $\widetilde{B}_r$, is less than $r$. (2) the same as that in Ewing et al's definition, see above. When little bias exists in the estimated error rates, the empirical discrimination power converges to the one proposed in Ewing et al. [2].

We note that the calculation of empirical discrimination power requires the information of base call errors, which could be obtained by mapping reads to a reference. Then $\widetilde{P}_r$ is calculated as follows: (1) sort the bases in descending order by their predicted quality scores; (2) for each base, generate a set containing the bases from the beginning to the current one, and calculate its empirical error rate; (3) for a given error rate $r$, select the largest set whose empirical error rate is less than $r$; (4) $\widetilde{P}_r$ equals the number of base calls in the selected set divided by the number of total bases.

We can take the quality scores as reliability measures of base-calls, assuming that the bases with higher scores are more accurate. Therefore, a higher $\widetilde{P}_r$ indicates that a method could identify more reliable bases for a given empirical error rate. By plotting the empirical discrimination power versus the empirical error rate $r$, we can compare the performance of different methods.

### ROC curve
Last, we plot the ROC and Precision-Recall curve to compare the methods. That is, by adjusting various quality score thresholds, we can classify the bases as correct and incorrect calls based on their estimated scores, and calculate the true positive rate against false positive rate and plot the ROC curve [24]. The area under the ROC curve (AUC) represents the probability that the quality score of a randomly chosen correctly called base is larger than the score of a randomly chosen incorrectly one.

## Results and discussion
### Model training
First we trained the model by the AIC, BIC and $L_1$ regularization method using a data set of about 3 million bases from a tile. The computation was implemented on a Dell T7500 workstation that has an Intel Xeon E5645 CPU and 192 GB RAM. It took about 50 h to train the model using the backward deletion coupled with either AIC or BIC. In comparison, the $L_1$ regularization training took about 2 min only. As we increased the size of training data set to 5- and 50-folds, the workstation could no longer finish the training by the AIC or BIC method in a reasonable period of time while it respectively took 5 and 15 min for the $L_1$ regularization training.

The coefficients of the trained model using a data set of 3 million bases are shown in Table 1. If we compare the models trained on the 3 million base dataset, the backward deletion with BIC deleted 53 variables, and the backward deletion with AIC eliminated 14 ones. Besides, the latter is a subset of the former. Unlike AIC/BIC, the sparsity of $L_1$-regularized logistic regression depends on the parameter $\lambda$. Here $\lambda$ was chosen by a cross-validation method that maximizes AUC as described in Methods. When we took $\lambda$ to be 1.0, the $L_1$ regularization removed 11 variables, two of them were not removed by BIC. Overall, The BIC method selected the least number of features, thus was most helpful for model interpretation.

We also calculated the contribution of each feature, defined as the t-score, namely, the coefficient divided by its standard error. As shown in Table 2, we listed the contribution of each feature, and classified the features into different groups by the method it was selected. The features contributing the most to all three methods were $x_{10}$ to $x_{14}$, which were the transformations of $x_1 - x_2$, namely the difference between the largest and second largest intensities. It makes sense that the model could discriminate called-bases more accurate if the largest intensity is much larger than the second largest intensity.

We have defined a consensus sequence described in Methods. This strategy may not eliminate the influence of polymorphism. Polymorphisms do occur in this data set of sequencing reads of Bacteriophage PhiX174, but they are very rare. Generally, we could use variant calling methods, such as GATK-HC [25] and Samtools [26], to identify variants and then remove those bases mapped to the variants. This could be achieved by replacing the corresponding bases in the reference by "N"s before the second mapping (the first mapping is for variant calling). In addition, this proposal has been implemented in the updated training module of 3Dec, which was published in the accompany paper [5].

**Table 1** The coefficients of the 74 predictive variables in the three methods

| x | Description | L1LR | BE-AIC | BE-BIC |
|---|---|---|---|---|
| x0 | intercept | 1.09 | 11.47 | 7.63 |
| x1 | largest intensity | 1.48 | - | - |
| x2 | second largest intensity | -1.73 | -4.84 | -4.42 |
| x3 | average of x1 | -1.18 | - | - |
| x4 | average of (x1-x2) | -4.65 | -6.2 | -5.65 |
| x5 | standard error of (x1-x2) | 3.19 | -10.03 | - |
| x6 | 1/x3 | -2.37 | -3.22 | -2.88 |
| x7 | $\sqrt{x5}$ | 0.54 | 1.42 | 0.77 |
| x8 | log(x5) | -0.93 | 2.69 | - |
| x9 | piecewise function of \|x1-x2\| | 0.59 | - | - |
| x10 | | 3.53 | 4.94 | 4.71 |
| x11 | | 3.45 | 6.62 | 6.3 |
| x12 | | 2.42 | 9.32 | 8.74 |
| x13 | | 1.44 | 12.35 | 11.43 |
| x14 | | 0.34 | 15.41 | 14.21 |
| x15 | | - | 23.06 | 21.45 |
| x16 | | - | 118.87 | 46.79 |
| x17 | | - | - | - |
| x18 | current cycle number | -0.016 | -0.019 | -0.018 |
| x19 | inverse distance | -0.24 | - | - |
| x20 | indicators of the first 7th cycles | -0.3 | -2.99 | - |
| x21 | | -0.15 | - | - |
| x22 | | - | - | - |
| x23 | | - | - | - |
| x24 | | -0.25 | - | - |
| x25 | | -0.54 | -1.22 | - |
| x26 | | 0.32 | 12.49 | - |
| x27 | A(AC) | -0.11 | - | - |
| x28 | A(AG) | -0.91 | -2.21 | -1.32 |
| x29 | A(AT) | -0.67 | -3.39 | -1.15 |
| x30 | A(CA) | 1.29 | - | - |
| x31 | A(CG) | 0.86 | -2.89 | - |
| x32 | A(CT) | 0.25 | -5.31 | - |
| x33 | A(GA) | 1.44 | -3.23 | - |
| x34 | A(GC) | 0.21 | -5.8 | - |
| x35 | A(GT) | 1.66 | -6.51 | - |
| x36 | A(TA) | 0.89 | -6.96 | - |
| x37 | A(TC) | 0.44 | -8.77 | - |
| x38 | A(TG) | - | -10.79 | - |
| x39 | C(AC) | 2.27 | 2.88 | 2.29 |
| x40 | C(AG) | - | -1.34 | - |
| x41 | C(AT) | - | -2.77 | - |
| x42 | C(CA) | -0.95 | -2.65 | -1.4 |
| x43 | C(CG) | -0.7 | -5.29 | - |
| x44 | C(CT) | -0.7 | -5.29 | - |
| x45 | C(GA) | -1.29 | -7.09 | -1.68 |
| x46 | C(GC) | 0.89 | -3.51 | - |
| x47 | C(GT) | 0.63 | -5.31 | - |
| x48 | C(TA) | 0.68 | -7.14 | - |
| x49 | C(TC) | - | -9.25 | - |
| x50 | C(TG) | -0.54 | -11.32 | - |
| x51 | G(AC) | 0.58 | -1.09 | - |
| x52 | G(AG) | 0.05 | -1.09 | - |
| x53 | G(AT) | -0.45 | -3.32 | -1.1 |
| x54 | G(CA) | 0.18 | -1.4 | - |
| x55 | G(CG) | -0.18 | -4.54 | - |
| x56 | G(CT) | -1.02 | -6.89 | -1.52 |
| x57 | G(GA) | 1.6 | -2.78 | - |
| x58 | G(GC) | 0.24 | -5.76 | - |
| x59 | G(GT) | -0.75 | -9.81 | -1.28 |
| x60 | G(TA) | 0.93 | -7.26 | - |
| x61 | G(TC) | 0.24 | -9.18 | - |
| x62 | G(TG) | 0.7 | -10.12 | - |
| x63 | T(AC) | - | - | - |
| x64 | T(AG) | -0.23 | -1.68 | - |
| x65 | T(AT) | 2.03 | - | - |
| x66 | T(CA) | 0.21 | -1.28 | - |
| x67 | T(CG) | -0.74 | -5.27 | - |
| x68 | T(CT) | -0.1 | -5.72 | - |
| x69 | T(GA) | 0.16 | -4.64 | - |
| x70 | T(GC) | 0.73 | -5.15 | - |
| x71 | T(GT) | 1.94 | -6.55 | - |
| x72 | T(TA) | - | -8.09 | - |
| x73 | T(TC) | -0.29 | -9.76 | - |
| x74 | T(TG) | -0.99 | -11.72 | - |

We denote these 74 variables by $x = (x_0, x_1, \cdots, x_{74})$. In the first row of the table, 'L1LR' means the $L_1$-regularized logistic regression, 'BE-AIC' indicates the backward deletion with AIC, and 'BE-BIC' represents the backward deletion with BIC. The details of the variables in each row are described in Methods. $x_{27}$ to $x_{74}$ are corresponding to the 3-letter sequences, which indicate the type of the base in the previous cycle, type of the base with the largest and the second largest intensity in current cycle. Meanwhile, '-' implies that the method has removed the feature

### Consistency between predictive and empirical error rates

We assessed the quality-scoring methods in several aspects. First, following Ewing et al. [2] and Li et al. [16], we assess the consistency of error rates predicted by each model. That is, we plotted the observed scores against the predicted ones obtained from each method. The results from the 3 million base training dataset are shown in Fig. 2. Little bias was observed when the score is below 20. All the three methods slightly overestimate the error

**Table 2** The contribution of each feature in the three methods: the backward deletion with either AIC or BIC and the $L_1$ regularization method

|   | Selected methods | | Contribution | | |
|---|---|---|---|---|---|
|   | L1 & AIC & BIC | Description | L1 | AIC | BIC |
| 1 | x2 | second largest intensity | -7.2243 | -20.211 | -18.458 |
| 2 | x4 | average of (x1-x2) | -21.93 | -29.24 | -26.646 |
| 3 | x6 | 1/x3 | -8.696 | -11.815 | -10.567 |
| 4 | x7 | $\sqrt{x5}$ | 0.47013 | 1.2363 | 0.67037 |
| 5 | x10 | piecewise function of \|x1-x2\| | 35.389 | 49.525 | 47.219 |
| 6 | x11 | | 14.579 | 27.974 | 26.622 |
| 7 | x12 | | 7.4602 | 28.731 | 26.943 |
| 8 | x13 | | 4.3013 | 36.89 | 34.142 |
| 9 | x14 | | 2.3397 | 106.05 | 97.787 |
| 10 | x18 | current cycle number | -0.00054878 | -0.00065167 | -0.00061738 |
| 11 | x28 | A(AG) | -5.3348 | -12.956 | -7.7384 |
| 12 | x29 | A(AT) | -4.2171 | -21.337 | -7.2382 |
| 13 | x39 | C(AC) | 14.771 | 18.74 | 14.901 |
| 14 | x42 | C(CA) | -8.0916 | -22.571 | -11.925 |
| 15 | x45 | C(GA) | -10.411 | -57.22 | -13.558 |
| 16 | x53 | G(AT) | -3.3127 | -24.44 | -8.0976 |
| 17 | x56 | G(CT) | -7.7223 | -52.163 | -11.508 |
| 18 | x59 | G(GT) | -5.893 | -77.08 | -10.057 |
|   | AIC & BIC | Description | L1 | AIC | BIC |
| 1 | x15 | piecewise function of \|x1-x2\| | 0 | 859.11 | 799.13 |
| 2 | x16 | | 0 | 19180 | 7549.6 |
|   | L1 & AIC | Description | L1 | AIC | BIC |
| 1 | x5 | standard error of (x1-x2) | 121.95 | -383.44 | 0 |
| 2 | x8 | log(x5) | -2.8995 | 8.3866 | 0 |
| 3 | x20 | indicators of the first 7th cycles | -3.0296 | -30.195 | 0 |
| 4 | x25 | | -5.4533 | -12.32 | 0 |
| 5 | x26 | | 3.2316 | 126.13 | 0 |
| 6 | x31 | A(CG) | 5.7108 | -19.191 | 0 |
| 7 | x32 | A(CT) | 1.864 | -39.591 | 0 |
| 8 | x33 | A(GA) | 9.9989 | -22.428 | 0 |
| 9 | x34 | A(GC) | 1.4679 | -40.542 | 0 |
| 10 | x35 | A(GT) | 11.712 | -45.93 | 0 |
| 11 | x36 | A(TA) | 5.7597 | -45.042 | 0 |
| 12 | x37 | A(TC) | 2.8418 | -56.643 | 0 |
| 13 | x43 | C(CG) | -5.6759 | -42.894 | 0 |
| 14 | x44 | C(CT) | -5.8779 | -44.42 | 0 |
| 15 | x46 | C(GC) | 7.3038 | -28.805 | 0 |
| 16 | x47 | C(GT) | 5.051 | -42.573 | 0 |
| 17 | x48 | C(TA) | 4.8103 | -50.508 | 0 |
| 18 | x50 | C(TG) | -4.0949 | -85.841 | 0 |
| 19 | x51 | G(AC) | 4.0089 | -7.5339 | 0 |
| 20 | x52 | G(AG) | 0.3575 | -7.7936 | 0 |

**Table 2** The contribution of each feature in the three methods: the backward deletion with either AIC or BIC and the $L_1$ regularization method *(Continued)*

| # | | Description | L1 | AIC | BIC |
|---|---|---|---|---|---|
| 21 | x54 | G(CA) | 1.1807 | -9.1835 | 0 |
| 22 | x55 | G(CG) | -1.2149 | -30.643 | 0 |
| 23 | x57 | G(GA) | 13.802 | -23.981 | 0 |
| 24 | x58 | G(GC) | 1.9919 | -47.805 | 0 |
| 25 | x60 | G(TA) | 6.2969 | -49.157 | 0 |
| 26 | x61 | G(TC) | 1.9621 | -75.049 | 0 |
| 27 | x62 | G(TG) | 5.7278 | -82.807 | 0 |
| 28 | x64 | T(AG) | -1.6306 | -11.911 | 0 |
| 29 | x66 | T(CA) | 1.6158 | -9.8488 | 0 |
| 30 | x67 | T(CG) | -5.0538 | -35.991 | 0 |
| 31 | x68 | T(CT) | -0.72808 | -41.646 | 0 |
| 32 | x69 | T(GA) | 1.0712 | -31.065 | 0 |
| 33 | x70 | T(GC) | 5.0709 | -35.774 | 0 |
| 34 | x71 | T(GT) | 12.661 | -42.749 | 0 |
| 35 | x73 | T(TC) | -1.7016 | -57.268 | 0 |
| 36 | x74 | T(TG) | -6.1194 | -72.443 | 0 |
| | **AIC** | Description | L1 | AIC | BIC |
| 1 | x38 | A(TG) | 0 | -72.981 | 0 |
| 2 | x40 | C(AG) | 0 | -8.7049 | 0 |
| 3 | x41 | C(AT) | 0 | -19.167 | 0 |
| 4 | x49 | C(TC) | 0 | -63.57 | 0 |
| 5 | x72 | T(TA) | 0 | -48.82 | 0 |
| | **L1** | Description | L1 | AIC | BIC |
| 1 | x1 | largest intensity | 1.0896 | 0 | 0 |
| 2 | x3 | average of x1 | -6.0558 | 0 | 0 |
| 3 | x9 | piecewise function of \|x1-x2\| | 13.791 | 0 | 0 |
| 4 | x19 | inverse distance | -2.0626 | 0 | 0 |
| 5 | x21 | indicators of the first 7th cycles | -1.5148 | 0 | 0 |
| 6 | x24 | | -2.5247 | 0 | 0 |
| 7 | x27 | A(AC) | -0.59405 | 0 | 0 |
| 8 | x30 | A(CA) | 11.7 | 0 | 0 |
| 9 | x65 | T(AT) | 15.749 | 0 | 0 |
| | **None** | Description | L1 | AIC | BIC |
| 1 | x17 | piecewise function of \|x1-x2\| | 0 | 0 | 0 |
| 2 | x22 | indicators of the first 7th cycles | 0 | 0 | 0 |
| 3 | x23 | | 0 | 0 | 0 |
| 4 | x63 | T(AC) | 0 | 0 | 0 |

The contribution is defined by the t-score, namely the coefficient divides by its standard error. All 74 features are classified into different groups by the method it is selected

rates between 20 and 35, and underestimate the error rates after 35.

The bias decreases as we increased the size of the training dataset to 5- and 50-folds, as shown in Fig. 2. But in these two cases, only $L_1$ regularization results are available due to the computational complexity. Thus if we expect more accurate estimates of error rates, then we need larger training datasets and the $L_1$ regularization training is the computational choice.

### Empirical discrimination power

As described in "Methods" section, a good quality scoring method is expected to have high empirical discrimination power, especially in the high quality score range. We

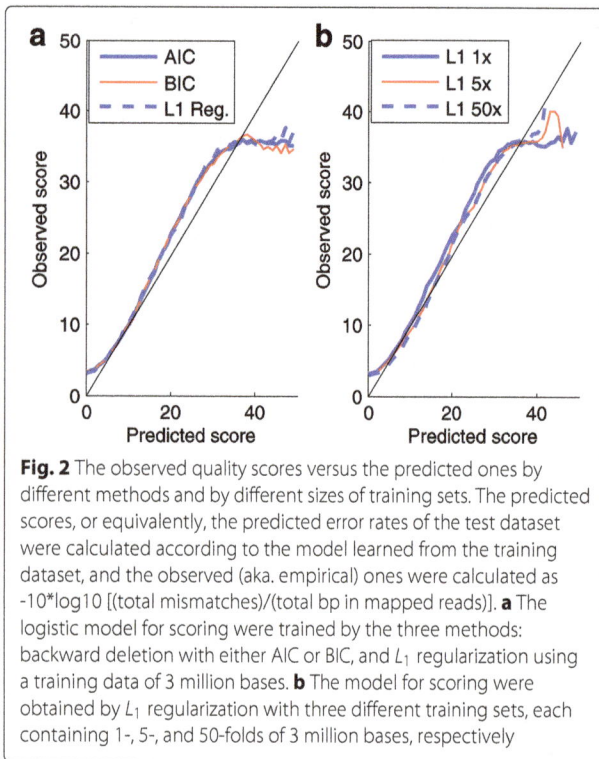

**Fig. 2** The observed quality scores versus the predicted ones by different methods and by different sizes of training sets. The predicted scores, or equivalently, the predicted error rates of the test dataset were calculated according to the model learned from the training dataset, and the observed (aka. empirical) ones were calculated as -10*log10 [(total mismatches)/(total bp in mapped reads)]. **a** The logistic model for scoring were trained by the three methods: backward deletion with either AIC or BIC, and $L_1$ regularization using a training data of 3 million bases. **b** The model for scoring were obtained by $L_1$ regularization with three different training sets, each containing 1-, 5-, and 50-folds of 3 million bases, respectively

calculated the empirical discrimination power for each method, based on the error status of the alignable bases in the test sets.

The results are shown in Fig. 3, where the x-aixs is the -log10(error rate) in the range between 3.3 to 3.7, and the y-axis is the empirical discrimination power. If we

took the 3 million bases training dataset, the BIC and the $L_1$ method show comparable discrimination powers, and both outperform the AIC method by around 60% at the error rate $3.58 \times 10^{-4}$. On average, the empirical discrimination power of the BIC and the $L_1$ method is 6% higher than that of the AIC method.

Moreover, we compared the empirical discrimination power of the $L_1$ regularization method with different training sets. As the size of training data goes up, higher empirical discrimination power is achieved at almost any error rate by the $L_1$ regularization method. The 5-, 50-folds data respectively gains 10 and 14% higher empirical discrimination power than 1-fold data on average. This implies that the $L_1$ method could identify more highly reliable bases with more training data.

We also used the concepts in classification such as the ROC and the Precision-Recall curve to assess the three methods. As shown in Fig. 4, the $L_1$ regularization achieves the highest precision in the range of high-quality scores, and in most other cases the three methods perform similarly. The AUC scores of the ROC curve for AIC, BIC, and $L_1$ regularization were 0.9141, 0.9161, and 0.9175, respectively, which show no significant difference.

The detailed results of the elastic net model were described in Additional file 1.

### Comparison with the Illumina scoring method

To be clear, hereafter we refer to the Illumina base-calling as Bustard, and the Illumina quality scoring method as Lookup. Similarly, we refer to the new base calling method [5] as 3Dec and the new quality scoring scheme as Logistic.

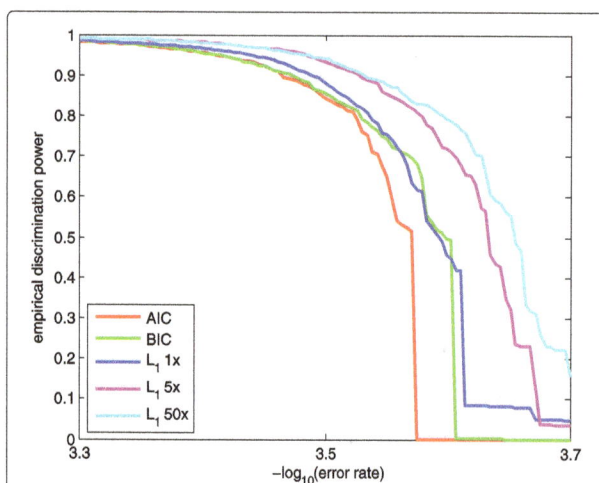

**Fig. 3** Empirical discrimination powers for three methods: backward deletion with either AIC or BIC, and $L_1$ regularization. The x-axis is the -log10 (error rate) in the range between 3.3 and 3.7. The y-axis is the empirical discrimination power defined as the largest proportion of bases whose empirical error rate is less than $10^{-x}$. $L_1$ 1x, 5x, 50x indicates that the $L_1$-regularized model is trained with 1-, 5-, 50-folds of 3 million bases, respectively

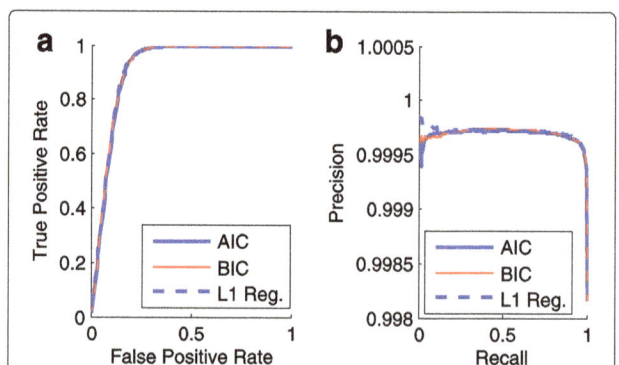

**Fig. 4** The ROC and Precision-Recall curve for the three methods. A logistic model can be considered as a classifier if we set a threshold to the Phred scores. The predicted condition of a base is positive/negative if its Phred score is larger/smaller than the threshold. The true condition of a base is obtained from the mapping of reads to the reference. Consequently, bases will be divided into four categories: true positives, false positive, true negatives, and false negatives. **a** The ROC curve on the test set by the three methods: the backward deletion with either AIC or BIC, and the $L_1$ regularization. **b** The corresponding precision-recall curve

In fact, we could exchange the use of Lookup and Logistic with the two base calling methods Bustard and 3Dec. We abbreviate these four schemes by Bustard+Lookup, Bustard+Logistic, 3Dec+Lookup, 3Dec+Logistic respectively, and the details are shown in Table 3. We note that the training of logistic models here involves only $L_1$ regularization with 100-folds data.

To have a systematic comparison of the scoring methods, we need to implement the four schemes in practice. First, Bustard+Lookup is the default method of Illumina. Second, we notice that the definition of quality scores depends on the cluster intensity files but not on the corresponding base calls. We have successfully extracted cluster intensity files preprocessed by Bustard, and input them into the Logistic scoring model. In this way, we implemented Bustard+Logistic.

As to the remaining two schemes, the implementation of 3Dec+Lookup is challenging because it is very hard to separate the quality scoring module from the Illumina systems. As a good approximation, we input the cluster intensity files preprocessed by 3Dec into Bustard, and consequently obtain the quality scores defined by the Phred algorithm provided by Illumina. A subtle issue needs to be explained here. The 3Dec preprocessing of cluster intensity files in fact corrects the spatial crosstalk as well as the phasing and color crosstalk effects. But the Illumina system routinely estimates the effects of phasing and color crosstalk and remove them even if it is unnecessary. Nevertheless, we found this extra step would make little change on the cluster intensity signals. This is supported by the fact: taking the 3Dec preprocessed cluster intensity files as input, the Illumina system outputs base calls highly identical to those by 3Dec. The resulting quality scores are surrogates for those from the 3Dec+Lookup scheme to a good extent. To make a fair comparison, we also use the same cluster intensity signals that have been preprocessed by both 3Dec and Bustard for the Logistic scoring. The resulting quality scores are surrogates for those from the 3Dec+Logistic scheme to a good extent.

We compare 3Dec+Logistic versus 3Dec+Lookup from the two aspects: consistency and empirical discrimination power, as shown in Fig. 5. In terms of consistency,

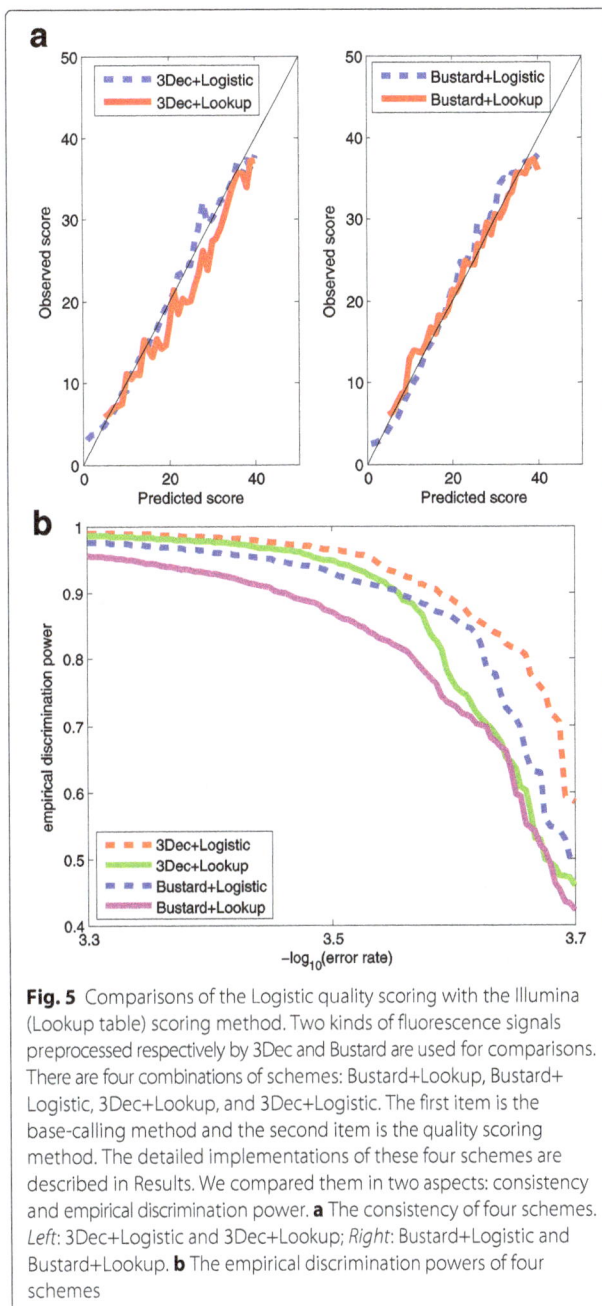

**Fig. 5** Comparisons of the Logistic quality scoring with the Illumina (Lookup table) scoring method. Two kinds of fluorescence signals preprocessed respectively by 3Dec and Bustard are used for comparisons. There are four combinations of schemes: Bustard+Lookup, Bustard+ Logistic, 3Dec+Lookup, and 3Dec+Logistic. The first item is the base-calling method and the second item is the quality scoring method. The detailed implementations of these four schemes are described in Results. We compared them in two aspects: consistency and empirical discrimination power. **a** The consistency of four schemes. *Left*: 3Dec+Logistic and 3Dec+Lookup; *Right*: Bustard+Logistic and Bustard+Lookup. **b** The empirical discrimination powers of four schemes

**Table 3** Four combinations of schemes between two base-calling and two quality scoring methods

| | | Quality scoring | |
| --- | --- | --- | --- |
| | | L1-regularized logistic regression | Lookup table strategy |
| Base-calling | Bustard | Bustard+Logistic | Bustard+Lookup |
| | 3Dec | 3Dec+Logistic | 3Dec+Lookup |

We have two base calling methods: Bustard, the default method embedded in the Illumina sequencers; 3Dec, our newly developed method. We also have two quality scoring methods: Lookup, the lookup table strategy adopted by Illumina; Logistic, the $L_1$-regularized logistic regression model proposed in this study

by and large, Logistic shows less bias than Lookup does, especially when the scores are less than 25 or between 30 and 40. In terms of discrimination power, Logistic outperforms Lookup across the board. Logistic achieves 25% higher empirical discrimination power at the error rate $3.67 \times 10^{-4}$, and 6% higher on average than Lookup does.

By the same token, we compare Bustard+Logistic versus Bustard+Lookup, see Fig. 5. In terms of consistency, Logistic shows some bias at the high score end while Lookup shows some bias at the low score end. In terms of discrimination power, Logistic outperforms Lookup by

14% at the error rate $3.62 \times 10^{-4}$. On average, the empirical discrimination power of Logistic increases by 6% than that of Lookup.

Overall, Logistic defines better quality scores than Lookup does, particularly in the sense that it identifies more base calls of high quality.

## Biological insights

The error patterns identified by the model selection results provide insights into the error mechanism of the sequencing technology. For example, the coefficients of the 3-letter sequences "G(AT)", "G(CT)", and "G(GT)" ($x_{53}$, $x_{56}$, and $x_{59}$) are all negative across the three methods. This implies that a nucleotide "T" after a "G" was more likely to be miscalled. To verify this, we plotted the kernel density of fluorescence intensities of "T" stratified by the types of the preceding nucleotide bases. That is, we read the corrected fluorescence signals and the called sequences of the first tile. Then for each nucleotide type X (X="A", "C", "G", or "T"), we found the sequence fragments "XT" in Cycle 8-12 in all the sequences, and calculated the kernel densities of the signals of "T" in these fragments, respectively. As shown in Fig. 6, the signals of "T" after "G" are lower than those after other types of nucleotide bases. One factor that causes uneven fluorescence signals is the quenching effect [17], due to short-range interactions between the fluorophore and the nearby molecules. The G-quenching factor was included in the quality score definition of the Illumina base-calling [4]. In comparison, our sparse modeling of logistic regression suggested that the most prominent quenching pattern in the current chemistry of Illumina occurred at the dinucleotide "GT".

Phasing is a phenomenon specific to the technique of reversible terminators. In the presence of phasing, a nucleotide has a larger chance to be miscalled as the preceding one. Interestingly, we found that in the BIC model, only 7 coefficients are negative, of which 5 are corresponding to the pattern "X(XY)" ($x_{28}$, $x_{29}$, $x_{42}$, $x_{44}$, and $x_{59}$). Similarly, we noticed that in the AIC and $L_1$ regularization model, most coefficients of the pattern "X(XY)" are non-positive, except those when "X" represents "G". This implies that nucleotides were more likely to be miscalled as the previous bases if the preceding ones were not "G". It suggested that the phasing effect of bases after "G" was somewhat different from those after other nucleotide types.

## Conclusions

In the recent years, next-generation sequencing technology has been greatly developed. However, the errors in the both ends of reads are still very high, and low quality called bases result in missing or wrong alignments that strongly affect downstream analysis [27]. So a valid and accurate method to estimate the quality scores is still essential and indispensable. In this article, we applied logistic regression and sparse modeling to predict the quality scores for Illumina sequencing technology. Both the Phred algorithm and our method belong to the supervised learning, since the labels of base-calling errors are obtained from sequence alignment results. Meanwhile, our method has some distinct merits that we explain as follows.

First, the logistic model can take many relevant features. As shown in Fig. 7, the AUC of the $L_1$ method

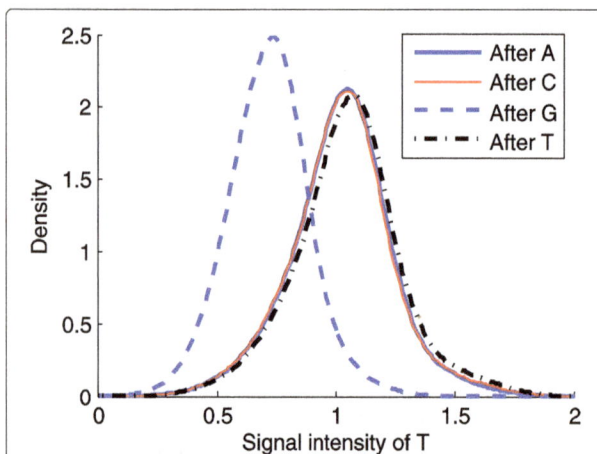

**Fig. 6** The density plots of "T" signals stratified by the preceding nucleotide bases. First we read the corrected fluorescence signals and the called sequences of the first tile. Then for each nucleotide type X (X="A", "C", "G", or "T"), we found the sequence fragments "XT" in Cycle 8-12 in all the sequences, and draw the density curves of the signals of "T" in these fragments, respectively. The curve was calculated using the Gaussian kernel with a fixed width of 0.01. As shown in the figure, the signals of "T" preceded by "G" are lower than those after other nucleotide bases

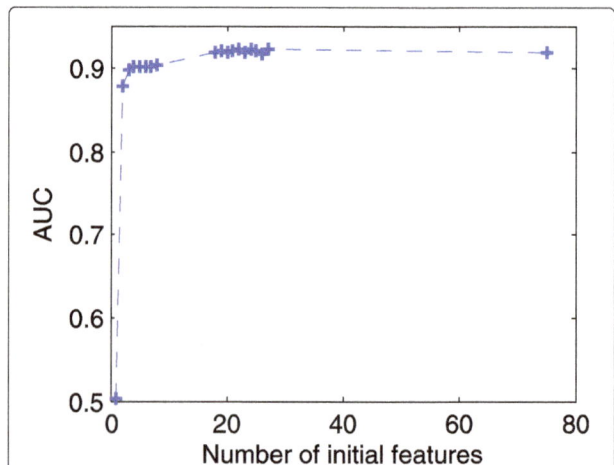

**Fig. 7** The AUC of the ROC curve versus the number of features in the initial model. In the logistic model, we sequentially include one more feature starting from $x_0$ to $x_{74}$ in Table 1 (the number of features is shown by the x-axis), and calculated each AUC (shown by the y-axis) using the $L_1$-regularized method

increases monotonically as we put more features in the model. Therefore, any features that are thought to be associated with the error rates could be included in the initial model. The possible overfitting problem is then overcome by the $L_0$ or $L_1$ regularization.

Second, the $L_1$-regularized logistic regression can be solved in a short period of time, and it has improved performance with more training data. Thus it can handle large dataset and is efficient enough for daily sequencing. Compared to the $L_1$ method, backward deletion with either AIC or BIC takes a long training time, and it fails to complete the training in a reasonable period of time for the 50-folds dataset. However, the BIC method selects the least number of features, which greatly helps for model interpretation.

Third, our method can be easily modified to adjust other base callers. The features we used are not software-specific. As shown in Fig. 5, the $L_1$ scoring method outperforms the Illumina scoring method by a great margin in terms of the empirical discrimination power, based on the fluorescence signals preprocessed either by 3Dec or by Bustard. We note that the Illumina system does not have an option that allows us to train it based on the same dataset used by the Logistic method. In conclusion, we recommend the logistic regression with $L_1$ regularization method to estimate the quality scores.

Fourth, the sparse modeling also helps us discover error patterns that help the downstream analysis. One important application of the sequencing technology is SNP calling. Our results indicate that not only allele frequencies, but also sequencing error patterns can help improve the SNP calling accuracy. Using the logistic regression methods, we further demonstrated the detailed pattern of G-quenching effect including G-specific phasing and the reduction of the T-signal following a G. Therefore, one should take the preceding bases into consideration when performing SNP calling.

Finally, the proposed training method is applicable to sequencing data from any sequencing technique. Meanwhile the resulting model including predictive features and error patterns is specific to the corresponding sequencing technique such as Illumina. Furthermore, the training method is adaptive to the experimental conditions.

## Additional file

**Additional file 1:** Supplementary information about the elastic net model. This file contains the following sections: **S1** - Introduction to the elastic net model and its advantages. **S2** - Results of the elastic net mode include training time, coefficients, consistency and empirical discrimination power. **Table S1** - The coefficients of 74 predicted features of the elastic net model. **Figure S1** - The consistency of the elastic net model with three different training sets. **Figure S2** - The empirical discrimination power of the elastic net model with three different training sets.

## Abbreviations
AIC: Akaike information criterion; AUC: area under the curve; BE-AIC: Backward deletion with AIC; BE-BIC: Backward deletion with BIC; BIC: Bayesian information criterion; L1LR: $L_1$-regularized logistic regression; LASSO: Least absolute shrinkage and selection operator; Logistic: the $L_1$-regularized logistic regression model; Lookup: the lookup table strategy; ROC: Receiver operating characteristic; SNP: Single nucleotide polymorphism

## Acknowledgements
We thank the editor and four anonymous reviewers for their insightful and critical comments, which have led to significant improvements of our manuscript.

## Funding
This work was supported by the National Natural Science Foundation of China (Grant No. 91130008 and No. 91530105), the Strategic Priority Research Program of the Chinese Academy of Sciences (Grant No. XDB13040600), the National Center for Mathematics and Interdisciplinary Sciences of the CAS, and the Key Laboratory of Systems and Control of the CAS. Lei M Li's research was also supported by the Program of One Hundred Talented People, CAS. LW is also supported by the NSFC grants (No. 11571349 and No. 11201460), the Youth Innovation Promotion Association of the CAS.

## Authors' contributions
BW participated in method design and data analysis, wrote the program, and drafted the manuscript. SZ participated in method design and data analysis, wrote the program, and drafted the manuscript. LW conceived the project, and participated in method design and writing. LML conceived and designed the study, and participated in method design and writing. All authors read and approved the manuscript.

## Competing interests
The authors declare that they have no competing interests.

## References
1. Mardis ER. Next-generation dna sequencing methods. Annu Rev Genomics Hum Genet. 2008;9:387–402.
2. Ewing B, Green P. Base-calling of automated sequencer traces using Phred. ii. error probabilities. Genome Res. 1998;8(3):186–94.
3. Bokulich NA, Subramanian S, Faith JJ, Gevers D, Gordon JI, Mills DA, Caporaso JG. Quality-filtering vastly improves diversity estimates from Illumina amplicon sequencing. Nat Methods. 2013;10(1):57–9.
4. HCS 1.4/RTA 1.12 Theory of Operation. Illumina Inc. http://www.illumina.com/Documents/products/technotes/technote_rta_theory_operations.pdf. Accessed 20 July 2016.
5. Wang B, Wan L, Wang A, Li LM. An adaptive decorrelation method removes Illumina DNA base-calling errors caused by crosstalk between adjacent clusters. Sci Rep. 2017;7:41348.
6. Hosmer Jr DW, Lemeshow S. Applied Logistic Regression. Hoboken: Wiley; 2004.
7. Mccullagh P, Nelder JA. Generalized Linear Models. vol. 37. 2nd ed. London: Chapman and Hall; 1989.

8.  Ypma TJ. Historical development of the Newton-Raphson method. SIAM Rev. 1995;37(4):531–51.

9.  Dohm JC, Lottaz C, Borodina T, Himmelbauer H. Substantial biases in ultra-short read data sets from high-throughput dna sequencing. Nucleic Acids Res. 2008;36(16):105.

10. Minoche AE, Dohm JC, Himmelbauer H. Evaluation of genomic high-throughput sequencing data generated on Illumina hiseq and genome analyzer systems. Genome Biol. 2011;12(11):1–15.

11. Rish I, Grabarnik G. Sparse Modeling: Theory, Algorithms, and Applications. Beaverton: CRC Press, Inc; 2014.

12. An H, Gu L. On the selection of regression variables. Acta Math Applicatae Sin. 1985;2(1):27–36.

13. Chakrabarti A, Ghosh JK. AIC, BIC, and recent advances in model selection. Handbook of the philosophy of science. 2011;7:583–605.

14. Tibshirani RJ. Regression shrinkage and selection via the lasso. J R Stat Soc. 1996;58:267–88.

15. Friedman J, Hastie T, Tibshirani R. Regularization paths for generalized linear models via coordinate descent. J Stat Softw. 2010;33(1):1–22.

16. Li M, Nordborg M, Li LM. Adjust quality scores from alignment and improve sequencing accuracy. Nucleic Acids Res. 2004;32(17):5183–91.

17. Seidel CAM, And AS, Sauer MHM. Nucleobase-specific quenching of fluorescent dyes. 1. nucleobase one-electron redox potentials and their correlation with static and dynamic quenching efficiencies. J Phys Chem. 1996;100(13):5541–53.

18. Ye C, Hsiao C, Corrada BH. Blindcall: ultra-fast base-calling of high-throughput sequencing data by blind deconvolution. Bioinformatics. 2014;30(9):1214–9.

19. Bravo HC. Research Webpage. http://www.cbcb.umd.edu/%7Ehcorrada/secgen. Accessed 20 July 2016.

20. R Core Team. R: A Language and Environment for Statistical Computing. Vienna: R Foundation for Statistical Computing; 2017. R Foundation for Statistical Computing. https://www.R-project.org/.

21. Mcclave JT, Sincich T. Statistics, 8th, annotat instructor's edn. Upper Saddler River: Prentice Hall; 2000.

22. Zou H, Hastie T. Regularization and variable selection via the elastic net. J R Stat Soc Ser B Stat Methodol. 2005;67(2):301–20.

23. Fan RE, Chang KW, Hsieh CJ, Wang XR, Lin CJ. Liblinear: A library for large linear classification. J Mach Learn Res. 2010;9(12):1871–4.

24. Hanley JA, Mcneil BJ. The meaning and use of the area under a receiver operating characteristic (roc) curve. Radiology. 1982;143(1):29–36.

25. McKenna A, Hanna M, Banks E, Sivachenko A, Cibulskis K, Kernytsky A, Garimella K, Altshuler D, Gabriel S, Daly M, DePristo MA. The Genome Analysis Toolkit: A MapReduce framework for analyzing next-generation DNA sequencing data. Genome Res. 2010;20(9):1297–303.

26. Li H, Handsaker B, Wysoker A, Fennell T, Ruan J, Homer N, Marth G, Abecasis G, Durbin R. The Sequence Alignment/Map format and SAMtools. Bioinformatics. 2009;25(16):2078–79.

27. Del Fabbro C, Scalabrin S, Morgante M, Giorgi FM. An extensive evaluation of read trimming effects on illumina NGS data analysis. PLoS ONE. 2013;8(12):1–13.

# 5

# NucDiff: in-depth characterization and annotation of differences between two sets of DNA sequences

Ksenia Khelik[1], Karin Lagesen[1,2], Geir Kjetil Sandve[1], Torbjørn Rognes[1,3] and Alexander Johan Nederbragt[1,4*]

## Abstract

**Background:** Comparing sets of sequences is a situation frequently encountered in bioinformatics, examples being comparing an assembly to a reference genome, or two genomes to each other. The purpose of the comparison is usually to find where the two sets differ, e.g. to find where a subsequence is repeated or deleted, or where insertions have been introduced. Such comparisons can be done using whole-genome alignments. Several tools for making such alignments exist, but none of them 1) provides detailed information about the types and locations of all differences between the two sets of sequences, 2) enables visualisation of alignment results at different levels of detail, and 3) carefully takes genomic repeats into consideration.

**Results:** We here present NucDiff, a tool aimed at locating and categorizing differences between two sets of closely related DNA sequences. NucDiff is able to deal with very fragmented genomes, repeated sequences, and various local differences and structural rearrangements. NucDiff determines differences by a rigorous analysis of alignment results obtained by the NUCmer, delta-filter and show-snps programs in the MUMmer sequence alignment package. All differences found are categorized according to a carefully defined classification scheme covering all possible differences between two sequences. Information about the differences is made available as GFF3 files, thus enabling visualisation using genome browsers as well as usage of the results as a component in an analysis pipeline. NucDiff was tested with varying parameters for the alignment step and compared with existing alternatives, called QUAST and dnadiff.

**Conclusions:** We have developed a whole genome alignment difference classification scheme together with the program NucDiff for finding such differences. The proposed classification scheme is comprehensive and can be used by other tools. NucDiff performs comparably to QUAST and dnadiff but gives much more detailed results that can easily be visualized. NucDiff is freely available on https://github.com/uio-cels/NucDiff under the MPL license.

**Keywords:** Whole-genome alignment, Comparative analysis, Whole-genome assembly, Annotation of differences

## Background

Advances in whole genome sequencing strategies and assembly approaches have brought on a need for methods for comparing sets of sequences to each other. Common questions asked are how assemblies of the same read set obtained with different assembly programs differ from each other, or how genomes from different strains of the same bacterial species differ from each other. Whole genome alignment (WGA) methods are often used for performing such analyses and have long been studied in bioinformatics. WGA "is, in general, the prediction of homologous pairs of positions between two or more sequences" [1]. WGA is mainly used for identifying conserved sequences between genomes, e.g. genes, regulatory regions, non-coding RNA sequences, and other functional elements [2, 3], thus aiding, for instance, genome (functional) annotation, detecting large scale evolutionary changes between genomes, and phylogenetic inference [1, 2]. This field has been under continuous development since the 1970s, and many

* Correspondence: lex.nederbragt@ibv.uio.no
[1]Biomedical Informatics Research Group, Department of Informatics, University of Oslo, PO Box 1080, 0316 Oslo, Norway
[4]Centre for Ecological and Evolutionary Synthesis, Department of Biosciences, University of Oslo, PO Box 1066 Blindern, 0316 Oslo, Norway
Full list of author information is available at the end of the article

methods and tools for WGA have been created. Reviews of existing methods and tools can be found in [1, 4, 5].

For the purpose of detecting differences between sequence sets, tools that can be used to perform WGA analysis should come with certain features. First, they should be able to deal with very fragmented genomes, structural rearrangements, genome sequence duplications, and various differences that are often related to repeated regions. Second, the comparative analysis results should provide information about the types of differences and their locations. This information should be stored in ways suitable for further analysis. Such comparison information may, for example, be used for scaffolding purposes, for reference-assisted genome assembly, assembly error detection, and comparison of different assemblies. Third, they should enable visualisations of alignment results at different levels of detail. Global scale visualisation can be used for examining duplications, structural rearrangements, and uncovered regions, while local scale visualisation can provide information about small differences, such as substitutions, insertions and deletions (collectively called 'indels').

Three different tools are available today that partially satisfy these criteria: MAUVE [6], QUAST [7] and dnadiff [8]. MAUVE performs multiple genome alignment, identifies conserved genomic regions, rearrangements and inversions in these regions, and the exact sequence breakpoints of such rearrangements across multiple genomes as well as nucleotide substitutions and small indels [6]. It also enables analysis of results through interactive visualisation and stores information in separate files. However, only information about small differences (substitutions, indels) is easily accessible without running accessory programs.

QUAST is a tool for quality assessment of genome assemblies, which outputs different metrics on assembly quality in the presence of a reference genome. It gives information about the locations of structural and long local differences, specifying the types of structural differences only. QUAST enables visualisation in an accompanying genome browser called Icarus. However, QUAST lacks visualisation of small local differences, only providing summary statistics for them.

Dnadiff is a wrapper for the NUCmer alignment program from MUMmer [9] that quantifies the differences and provides alignment statistics and other high-level metrics [8]. Similar to QUAST, dnadiff can be used for quality assessment of assemblies and comparison of genomes, but it does not provide any visualization of the detected differences.

Here we present the tool NucDiff, which uses the NUCmer, delta-filter and show-snps programs from MUMmer for sequence comparison. NUCmer aligns sequences and outputs information about aligned sequence

regions. Rigorous analysis of the relative positions of these regions enables detection of various types of differences, including rearrangements and inversions, and in some cases also to ascertain their connection with repeated regions. NucDiff identifies the differences between two sets of closely related sequences and classifies the differences into several subtypes. The precise locations of all differences using coordinates systems with respect to both input sequences are output as GFF3 (Generic Feature Format version 3, [10]) files. These precise locations enables both visualisation and further analysis. The information provided by NucDiff can thus significantly help clarify how two sets of sequences differ.

## Implementation

NucDiff determines the various types of differences between two sets of sequences, usually referred to as a reference genome and a query, by parsing alignment results produced by the NUCmer, delta-filter and show-snps programs from the MUMmer sequence alignment package [9]. NUCmer performs DNA sequence alignment, while delta-filter filters the alignment results according to specified criteria. With the settings used by NucDiff by default, delta-filter also selects the longest consistent alignments for the query sequences. NUCmer alignment results contain information about fragments of sequences that match, which we here refer to as query and reference fragments. NUCmer output contains the exact coordinates of all fragments in relation to their source sequences, directions of query fragments relative to corresponding reference fragments, and percent similarity of the alignment. The show-snps results contain information about all inserted, deleted and substituted bases in the query fragments compared to the corresponding reference fragments.

If we represent the output fragments as blocks on the query and reference sequences, then a possible NUCmer alignment result may look as illustrated in Fig. 1.

During the alignment process, NUCmer searches for maximal exact matches of a given minimum length, then

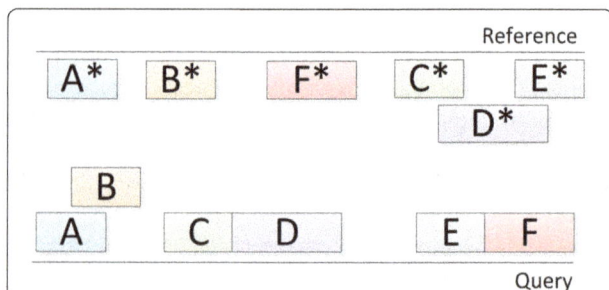

**Fig. 1** NUCmer alignment. A,...,F represent query fragments, while A*,..., F* represent reference fragments. A*-A, ..., F*-F are matches according to NUCmer

clusters these matches to form larger inexact alignment regions, and finally extends alignments outwards from each of the matches to join the clusters into a single high scoring pairwise alignment [11]. If the query sequences contain long (by default, more than 200 bp) insertions, deletions, substitutions, or any structural rearrangements, the alignment will be broken and subsequently consist of separate fragments with the ends coinciding with the locations of these differences. NucDiff classifies the alignment fragments by analysing the placement of all pairs of neighbouring query fragments (A-B, B-C, etc. in Fig. 1), their placement on the reference sequences (A*-B*, B*-C*, etc. in Fig. 1), and their orientations (5′ to 3′, or 3′ to 5′). The obtained differences together with the differences from show-snps form the set of all differences between query and reference sequences.

The NucDiff workflow is shown in Fig. 2. An overview of all types of differences that NucDiff is able to detect is presented in the Types of differences section. A description of the steps involved in their detection is given in the Stepwise detection of differences section.

## Types of differences

We classify all types of differences into 3 main groups: global, local and structural (Fig. 3). These differences are here denoted as changes in the query when compared to the reference.

### Global differences

Global differences affect the whole query sequence. This group consists of only one type, called unaligned sequence.

- unaligned sequence - a query sequence that has no matches of length equal to or longer than a given number of bases (65 by default) with the reference genome.

### Local differences

Local differences involve various types of insertions, deletions and substitutions. NucDiff distinguishes between six types of insertions (the insertion subgroup in Fig. 3):

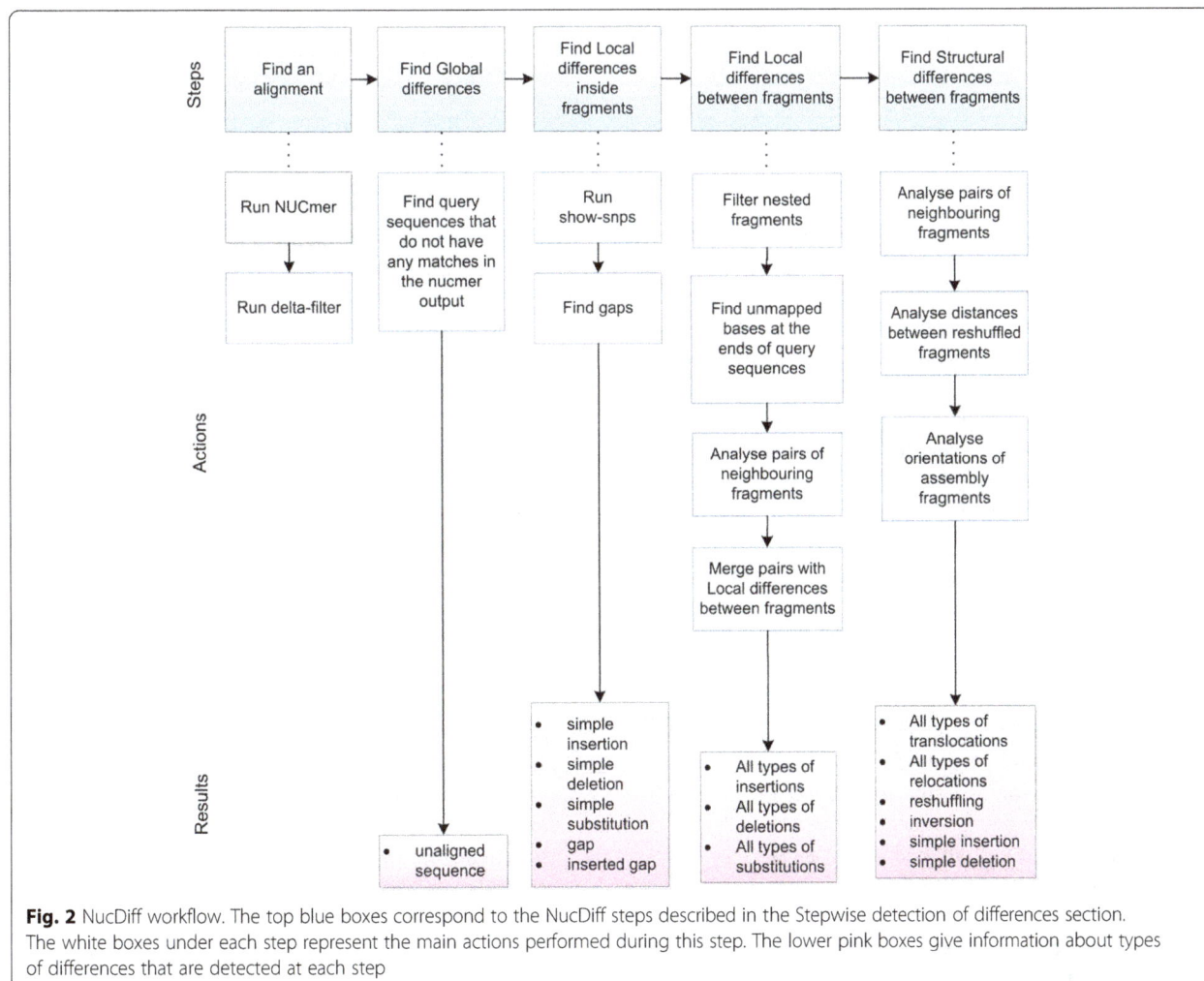

**Fig. 2** NucDiff workflow. The top blue boxes correspond to the NucDiff steps described in the Stepwise detection of differences section. The white boxes under each step represent the main actions performed during this step. The lower pink boxes give information about types of differences that are detected at each step

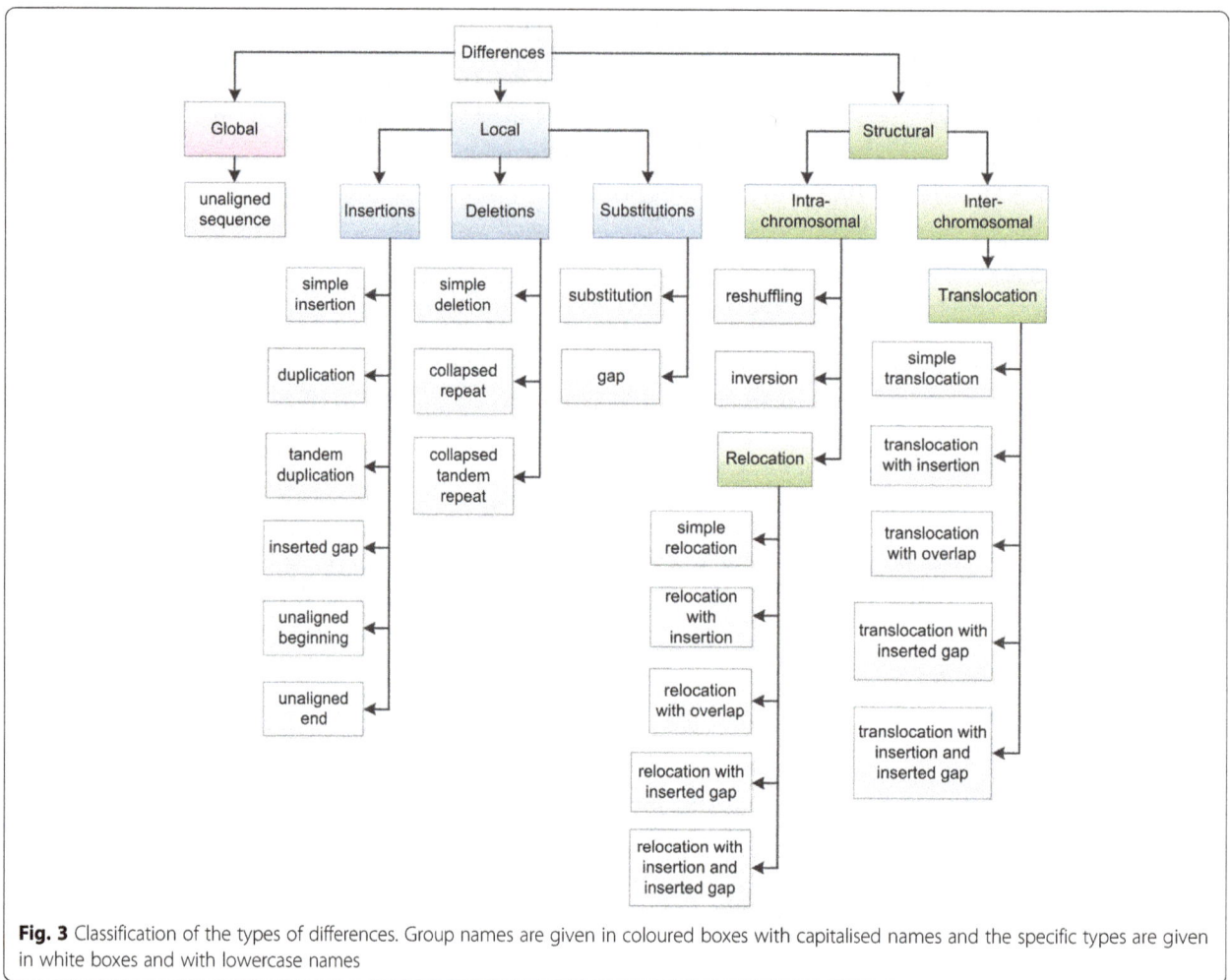

**Fig. 3** Classification of the types of differences. Group names are given in coloured boxes with capitalised names and the specific types are given in white boxes and with lowercase names

- simple insertion - an insertion of bases in the query sequence that were not present anywhere on the reference genome.
- duplication - an insertion in the query sequence of an extra copy of some reference sequence not adjacent to this region, creating an interspersed repeat, or increasing the copy number of an interspersed repeat
- tandem duplication - an insertion of an extra copy of some reference sequence region adjacent to this region in the query sequence
- inserted gap - an insertion of unknown bases (N's) in the query sequence in a region which is continuous (without a gap) in the reference, or which results in an elongation of a region of unknown bases in the reference.
- unaligned beginning - unaligned bases in the beginning of a query sequence
- unaligned end - unaligned bases at the end of query sequence

There are several types of deletions (the deletion subgroup in Fig. 3):

- simple deletion - a deletion of some bases, present in the reference sequence, from a query sequence
- collapsed repeat - a deletion of one copy of an interspersed repeat from the reference sequence in a query sequence
- collapsed tandem repeat - a deletion of one or more tandem repeat units from the reference sequence in a query sequence

And, last, there are two types of substitutions (the substitution subgroup in Fig. 3):

- substitution - a substitution of some reference sequence region with another sequence of the exact same length not present anywhere in the reference genome (note that this sequence is not categorised as unaligned sequence because it is within a fragment that overlaps between query and reference). SNPs can be considered as a subcategory of substitutions.
- gap - a substitution where a reference subsequence is replaced by an unknown sequence (N's) of the

same length. If the query has an enlarged gap, it will be classified as a combination of a gap and an inserted gap, while a shortened gap is classified as a gap and a simple deletion.

### Structural differences

NucDiff detects several structural differences. These can be grouped into intra- and inter-chromosomal differences, and some of these contain groups of types:

- translocation - a group of different types of inter-chromosomal structural rearrangements which occur when two regions located on different reference sequences are placed nearby in the same query sequence. The detailed description of all translocation types is given in the Structural difference detection between aligned fragments section.
- relocation - a group of different types of intra-chromosomal structural rearrangements which occur when two regions located in different parts of the same reference sequence are placed nearby in the same query sequence. The detailed description of all relocation types is given in the Structural difference detection between aligned fragments section.

- reshuffling - an intra-chromosomal structural rearrangement which occurs when several neighbouring reference sequence regions are placed in a different order in a query sequence.
- inversion - an intra-chromosomal structural rearrangement which occurs when a query sequence region is the reverse complement of a reference sequence region.

The translocation type belongs to the inter-chromosomal subgroup, while relocation, reshuffling and inversion types belong to the intra-chromosomal subgroup (see Fig. 3). Examples of structural differences are given in Fig. 4.

### Stepwise detection of differences

The steps in this section refer to Fig. 2.

### Global difference detection

NucDiff starts the detection of differences by finding unaligned sequence differences. NUCmer does not output any information about sequences without mapped subsequences longer or equal to a predefined length. Therefore, to find unaligned sequences, NucDiff looks for query sequences with names not mentioned in the NUCmer

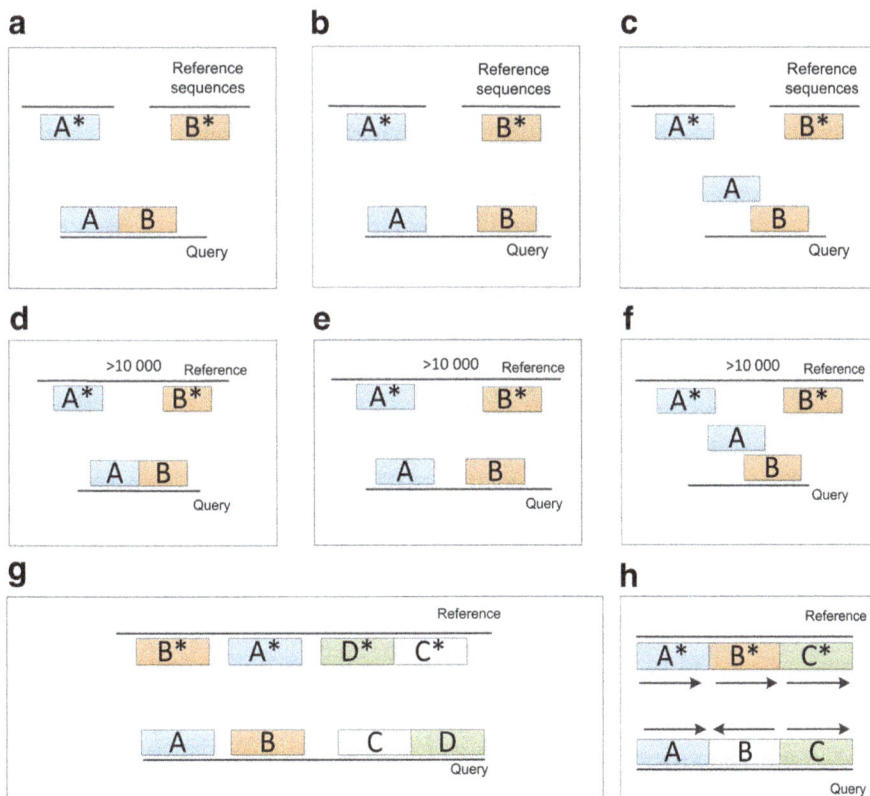

**Fig. 4** Examples of structural differences. **a** Simple translocation. **b** Translocation with insertion/with inserted gap/with insertion and inserted gap. **c** Translocation with overlap. **d** Simple relocation. **e** Translocation with insertion/with inserted gap/with insertion and inserted gap. **f** Relocation with overlap. **g** Reshuffling. **h** Inversion

output. By default, all query sequences shorter than 65 bp will be treated as unaligned sequences. This threshold may be changed using the NUCmer minimum cluster length option.

### Local difference detection inside aligned fragments

Four types of simple differences may be detected inside the query fragments: simple insertion, simple deletion, simple substitution and gap. The lengths of the differences of these types are limited by how far NUCmer will attempt to extend poorly scoring regions before giving up and are up to 200 bases by default (this threshold may be changed using the NUCmer minimum length of a maximal exact match parameter). Information about the positions of all local differences, except gaps, is found in the show-snps output file. NucDiff parses this file to find simple insertions, simple deletions, and substitutions. To find gaps, NucDiff searches for N's in the query fragment sequences and outputs their locations.

### Local difference detection between aligned fragments

NucDiff starts with examining the reason for alignment fragmentation by looking at fragmentation caused by local differences. First, it filters nested fragments in the query and reference sequences. A query nested fragment occurs when two (nearly) identical reference sequence regions have been merged together into one fragment in the query sequence. A reference nested fragment occurs when one reference sequence region is duplicated in the query sequence. Nested fragments provide important information about duplications and collapsed repeats. However, they can cause rather complicated interactions between aligned fragments, which can be difficult to resolve programmatically. Thus, the nested fragments are discarded, and all duplications and collapsed repeats are detected as simple insertions and deletions at later stages of the analysis. Then, NucDiff identifies bases in both ends of the query sequences that were not mapped to the reference sequences. Such bases will be output as unaligned beginning and unaligned end differences.

NucDiff next searches for pairs of neighbouring fragments that were not joined together by NUCmer during the alignment process due to the presence of simple differences, rather than structural differences. Such pairs of fragments should satisfy the following criteria:

- The pair of query fragments as well as the corresponding pair of reference fragments may overlap, be adjacent to each other, or be separated by an inserted region not mapped anywhere on the reference genome.
- The two query fragments should have the same direction. Their two corresponding reference fragments should also have the same direction, but it may be opposite to the direction of the query fragments.

- If the query fragments have the same direction as their corresponding reference fragments, then the reference fragments should be placed in the same order as the query fragments ([Additional file 1: Figure S1a]).
- If the query fragments have the reverse direction of their corresponding reference fragments, then the reference fragments should be in reverse order ([Additional file 1: Figure S1a]).
- The distance between corresponding reference fragments should not be more than a user-defined distance, by default 10,000 bases.

If all these criteria are fulfilled, NucDiff determines the differences based on the placement of the query and reference fragments relative to each other. Examples of all possible placement cases and the corresponding differences are shown in [Additional file 1: Table S1].

After detecting differences between the current pair of neighbouring fragments, NucDiff merges the pair of reference fragments as well as the pair of query fragments together, creating new continuous reference and query fragments, and then searches for the next pair.

### Structural difference detection between aligned fragments

Fragments not merged during the previous step were kept separate by NUCmer due to structural rearrangements between the query and reference sequences. First, NucDiff searches for translocations, which is one type of inter-chromosomal differences, by searching for a pair of neighbouring query fragments that correspond to fragments located on different reference sequences. We distinguish between 5 types of translocations depending on the placement of the query fragments relative to each other (see also examples in Fig. 4a-c):

- simple translocation - a translocation where two query fragments are placed adjacent to each other.
- translocation with insertion - a translocation where two query fragments have a stretch of bases (not N's) inserted between them, not mapped anywhere on the reference genome. The inserted region is treated as a simple insertion difference.
- translocation with inserted gap - a translocation where two query fragments have a stretch of unknown bases (N's) inserted between them. The inserted region is treated as an inserted gap difference.
- translocation with insertion and inserted gap - a translocation where two query fragments have a stretch of bases (A, C, G, T or N's) inserted between them, not mapped anywhere on the reference

genome. The inserted region is treated as both a simple insertion and an inserted gap.

- translocation with overlap - a translocation with a partial overlap between the two query fragments.

In the next step, NucDiff searches for relocations, which is one type of intra-chromosomal differences, by looking for pairs of neighbouring query fragments that were mapped to fragments located on the same reference sequence (e.g. the same chromosome) but separated from each other by at least 10,000 bases, by default. In addition, these fragments should not belong to the group of query fragments placed nearby each other (with the distance between each pair less than 10,000 bases) on the reference sequence in the wrong order, as that would be considered as a reshuffling (see further down). If these two conditions are fulfilled, then there is a relocation. There are 5 types of relocations (see also examples in Fig. 4d-f):

- simple relocation - a relocation where two query fragments are placed adjacent to each other.
- relocation with insertion - a relocation where two query fragments have a stretch of bases (not N's) inserted between them, not mapped anywhere on the reference genome. The inserted region is treated as a simple insertion difference.
- relocation with inserted gap - a relocation where two query fragments have a stretch of unknown bases (N's) inserted between them. The inserted region is treated as an inserted gap difference.
- relocation with insertion and inserted gap - a relocation where two query fragments have a stretch of bases (both ATGC's and N's) inserted between them, not mapped anywhere on the reference genome. The inserted region is treated as both a simple insertion and an inserted gap.
- relocation with overlap - a relocation with a partial overlap between the two query fragments.

For circular genomes, there is one special case that causes alignment fragmentation: when the start of the query sequence does not coincide with the start of the reference sequence ([Additional file 1: Figure S2]). It satisfies all the criteria for relocations but is not treated as a difference, although it is included in the output.

In the case of translocations and relocations, the query and the corresponding reference fragments may be placed in any direction and order relative to each other. The translocated fragment may contain none, two or more relocated fragments inside. Before the detection of the types of relocations and translocations, NucDiff searches for the pairs of relocated or translocated query fragments that have an overlap between corresponding

reference fragments. If such a pair is found, NucDiff truncates the rightmost fragment, so the overlap disappears. In this case information about the repeated nature of the insertion events will be lost.

Third, NucDiff searches for a group of nearby query fragments whose corresponding reference fragments are located on the same reference sequence (chromosome) but in a different order. The distance between two neighbouring reference fragments should not be more than 10,000 bases. If a group satisfying these conditions is found, then there is a reshuffling difference in the query. There may be simple insertion and simple deletion differences between reshuffled fragments. To find them, NucDiff first truncates fragments so that all overlaps between query or reference fragments are removed. It then searches for unmapped bases between neighbouring query fragments to find simple insertions and then searches for unmapped bases between neighbouring reference fragments to find simple deletions.

Finally, NucDiff searches for the last type of intra-chromosomal structural difference, inversions. If a query sequence has several mapped fragments and one or more of them, but not all, have directions opposite to the directions of the corresponding reference fragments, then such fragments are inversions. Some examples of possible alignments of query sequences in cases with reshuffling and inversion are shown in Fig. 4g-h.

Reshufflings and inversions may be present inside translocated and relocated fragments. During reshuffling detection, the directions of reshuffled fragments are not taken into account. Their directions are checked during the inversion detection step. Simple insertions and simple deletions found during this step may be connected to repeated regions, but this connection will not be detected.

### Datasets

We created ten simulated reference and query DNA sequences. The genomes were constructed from random DNA sequences, and different types of controlled genome modifications were subsequently applied to these sequences (e.g. relocation of different fragments, or deletions, or duplications of fragments). The detailed description of implemented genome modifications can be found in [Additional file 1: Table S2].

In addition, we used data produced for the GAGE-B article [12] for the demonstrations of the comparison of several assemblies. The assemblies from the ABySS [13], CABOG [14], MaSuRCA [15], SGA [16], SOAPdenovo [17] (shown as SOAP in the figures), SPAdes [18] and Velvet [19] assemblers for *Vibrio cholerae* based on HiSeq reads were used. These assemblies together with the *V. cholerae* reference genome were downloaded from the GAGE-B website [20].

For the demonstration of the comparison of genomes from different strains of the same species, 22 *Escherichia coli* K12 reference genomes were downloaded from the NCBI database [21]. Their accession numbers can be found in [Additional file 1: Table S3]. In the sections with the demonstrations, we also used annotations for the *V. cholerae* reference genome and *E. coli* K12 MG1655. They were downloaded from the NCBI database [22, 23], respectively.

## Results
### The NucDiff tool
We have created a tool, called NucDiff, which is primarily aimed at locating and categorizing differences between any two sets of closely related nucleotide sequences. It is able to handle very fragmented genomes and various structural rearrangements. These features make NucDiff suitable for comparing, for instance, different assemblies with each other, or an assembly with a reference genome. NucDiff first runs the NUCmer, delta-filter and show-snps programs from MUMmer and parses the alignment results to detect differences. These differences are subsequently categorized according to a carefully defined classification scheme of all possible differences between two sequences.

A unique feature of NucDiff is that it provides detailed information about the exact genomic locations of the differences in the form of four GFF3 files: two files with information for small and medium local differences that do not cause alignment fragmentation, two others for structural differences and local differences that cause alignment fragmentation. All locations of the differences are output in query - and reference-based coordinates, separately. Each GFF3 entry is additionally annotated with the location of the difference in the opposite coordinate system as well. A detailed description of the format of these GFF files can be found in the GitHub repository of NucDiff. NucDiff also finds the coordinates of mapped blocks (the query sequences split at the points of translocation, relocation, inversions, and/or reshuffling) and then stores them in the GFF3 files, one based on query coordinates and another with reference-based coordinates. Uploading these GFF3 files into a genome browser such as the Integrated Genome Viewer (IGV) [24, 25] enables visualisation of the differences as well as the coverage of a reference genome by query sequences, making it possible to see all uncovered reference bases or if any reference regions are covered multiple times.

In addition, NucDiff generates a summary file containing information about the number of differences of each type. The detailed level of reporting enables users to create their own custom summary from the NucDiff output (e.g. taking into account the length of differences, joining several types of differences together, and so on) if desired.

### Effect of different MUMmer parameters
The alignment results parsed by NucDiff depend on the values of the input parameters for two MUMmer programs, NUCmer and delta-filter. NUCmer performs DNA sequence alignment, while delta-filter filters the alignment results according to specified criteria. Running these programs with different input parameters may result in alternative sets of matches, since the choice of parameters affects the sensitivity of the detection of matching sequence fragments as well as the stringency of the subsequent filtering. To analyse the influence of the different parameters on the alignment and on the subsequent NucDiff results, we compared the results of running NucDiff on the simulated genomes described in the Datasets section with different NUCmer and delta-filter input parameters values. The specific values for each test can be found in [Additional file 1: Table S4]. We also ran one test to enable comparison of QUAST and NucDiff as described in Comparison with QUAST section, since QUAST uses the same underlying tools as NucDiff.

The locations and types of simulated differences were compared with the results obtained from NucDiff, and the number of correctly detected differences was calculated for each test (see [Additional file 1] for details). The results with the total average number of correctly detected expected differences for each type are presented in Table 1. The detailed results for each implemented modification case (see in [Additional file 1: Table S2]) and for each parameter configuration set can be found in [Additional file 2].

We did not expect NucDiff to be able to detect all simulated differences of most types. This is confirmed in the results presented in Table 1, where NucDiff misses many differences of several types, no matter what parameter settings were used. A small deviation from the simulated results was expected since the fixed 30 bp limit for lengths of duplications in reference and query sequences and relocated blocks is much lower than the variable NUCmer and delta-filter thresholds. Another reason for the result deviation is that some difference locations were shifted a few bp due to accidental base similarity at the region borders. In such cases, the differences were considered wrongly resolved in spite of correctly detected types. These reasons are applicable to all difference types with the observed deviation to a greater or lesser extent. All other reasons are related to the chosen NUCmer and delta-filter parameter settings and NucDiff limitations and are discussed below.

The detailed results from [Additional file 2] indicate that increasing the alignment extension distance (–b parameter) led to the loss of information about repeat related local differences and inverted, relocated and substituted fragments. With a greater -b parameter value, NUCmer more successfully expands low scoring regions. It enables detection of

**Table 1** Average number of correctly detected simulated differences by NucDiff with different parameter settings and QUAST

| Difference | Truth | Default | c30 | c120 | l10 | l65 | b80 | b350 | QUAST-like | QUAST |
|---|---|---|---|---|---|---|---|---|---|---|
| insertion | 1650 | 1634 | 1634 | 1634 | 1634 | 1634 | 1632 | 1634 | 1634 | 858 |
| deletion | 1719 | 1678 | 1678 | 1678 | 1677 | 1678 | 1679 | 1676 | 1674 | 465 |
| duplication | 251 | 136 | 122 | 150 | 136 | 136 | 137 | 124 | 136 | 196 |
| tandem_duplication | 60 | 57 | 57 | 57 | 57 | 57 | 60 | 54 | 57 | 57 |
| collapsed_repeat | 58 | 53 | 53 | 53 | 53 | 53 | 54 | 51 | 53 | 53 |
| collapsed_tandem_repeat | 59 | 55 | 55 | 55 | 55 | 55 | 56 | 53 | 55 | 55 |
| relocation | 217 | 127 | 142 | 108 | 127 | 127 | 136 | 112 | 127 | 130 |
| relocation-insertion | 13 | 13 | 13 | 13 | 13 | 13 | 13 | 13 | 13 | 13 |
| relocation-insertion_ATGC | 13 | 13 | 13 | 13 | 13 | 13 | 13 | 13 | 13 | 13 |
| relocation-inserted_gap | 13 | 13 | 13 | 13 | 13 | 13 | 13 | 13 | 13 | 13 |
| relocation-overlap | 13 | 13 | 13 | 13 | 13 | 13 | 13 | 13 | 13 | 12 |
| translocation | 111 | 50 | 57 | 43 | 50 | 50 | 50 | 50 | 50 | 62 |
| translocation-insertion | 13 | 12 | 12 | 12 | 12 | 12 | 12 | 12 | 12 | 13 |
| translocation-insertion_ATGC | 13 | 13 | 13 | 13 | 13 | 13 | 13 | 13 | 13 | 13 |
| translocation-inserted_gap | 13 | 13 | 13 | 13 | 13 | 13 | 13 | 13 | 13 | 13 |
| translocation-overlap | 13 | 13 | 13 | 13 | 13 | 13 | 13 | 13 | 13 | 11 |
| inversion | 534 | 530 | 530 | 528 | 529 | 530 | 531 | 526 | 530 | 528 |
| reshuffling | 2585 | 2585 | 2585 | 2585 | 2585 | 2585 | 2585 | 2585 | 2585 | 2536 |
| substitution | 115 | 81 | 81 | 81 | 80 | 81 | 89 | 68 | 81 | 84 |
| gap | 49 | 46 | 46 | 46 | 46 | 46 | 48 | 46 | 45 | 34 |
| inserted_gap | 21 | 21 | 21 | 21 | 21 | 21 | 21 | 21 | 20 | 16 |
| mapped_seq | 13 | 10 | 11 | 6 | 10 | 10 | 10 | 10 | 10 | 10 |
| unaligned_sequence | 13 | 13 | 13 | 13 | 13 | 13 | 13 | 13 | 13 | 13 |

more differences inside fragments and a reduction of the number of aligned fragments. However, at the same time, it does not allow tracking of possible locations of query regions involved in differences in the reference sequences. This leads to loss of information about the repeated, inverted and substituted nature of the regions. Changing the maximal exact match length (–l parameter) did not influence significantly on the obtained results within the considered simulations. Increasing the parameter value for minimum alignment identity (–i parameter) (see columns l65 and QUAST-like in Table 1) led to an increased number of wrongly discarded valid mapped short fragments as well as query sequences containing even a small number of short and medium length differences.

Increasing the values for the minimum cluster length (–c parameter) increases the number of discarded correct query sequences and discarded valid mapped fragments. This leads to 1) the undesirable loss of information about the inverted, relocated and translocated nature of some fragments and 2) the misrepresentation of correct query sequences as being unaligned.

Additional result deviations can be explained by the specifics and limitations of the approach implemented in NucDiff independent on the parameter values used.

First, due to some simplifications during the NucDiff structural difference detection step, NucDiff does not allow detection of both relocations/translocations and duplications at the same time in cases when simple relocations/translocations are followed by duplications (see [Additional file 1: Table S2], relocation case 2 and translocation case 1). In such cases, the differences are detected either as a combination of a simple relocation/translocation and a simple insertion or as a combination of a simple insertion and a duplication depending on the length of a relocated or translocated fragment.

Second, another problem with duplication detection occurs in situations when reference fragments are duplicated and inserted into query sequences somewhere far away from their original locations (see [Additional file 1: Table S2], insertions, case 2). The duplications are detected by NUCmer but are filtered out by the delta-filter program as being aligned fragments with smaller length*identity weighted LIS [longest increasing subset]. This option is set by the -q parameter and is always used in NucDiff. As a result, NucDiff detects such duplications as simple insertions.

Third, in cases with a combination of a gap and an inserted gap, the order of the gap and the inserted gap varies depending on whether a subsequence of N's caused alignment fragmentation or not. Since in the simulated results a gap is always followed by an inserted gap, the number of correctly detected gaps was slightly lower than the expected number for all parameter settings. However, this behavior influences only the numbers in Table 1 but not the quality of the obtained results.

### Comparison with QUAST

Both NucDiff and QUAST use the NUCmer package in their pipeline. However, QUAST only provides information about the locations of regions where the reference sequences were split during the alignment process and specifies the general reasons for the alignment fragmentations (e.g. local misassembly, relocation and so on). As with NucDiff, we calculated the number of correctly detected simulated differences. Since QUAST only separates the differences into broad categories, it is not possible to make direct one-to-one comparisons. We therefore grouped the simulated differences into types as described in [Additional file 1: Table S5]. A simulated difference is considered correctly detected if it overlaps with a QUAST difference that belongs to the same general category. In cases with repeat related types, a difference is considered correctly detected when one of the repeated fragments involved in the simulated difference overlaps with the QUAST difference. The obtained average total number for each type of difference is shown in Table 1. The detailed results for each simulated case (see in [Additional file 1: Table S2]) can be found in the [Additional file 2].

As expected, the results presented in Table 1 show that QUAST, as well as NucDiff, was not able to detect all simulated differences in most groups. The small deviation of QUAST results in all problematic groups can also be explained by the introduced 30 bp limit for lengths of duplications in reference and query sequences and relocated blocks and shifted locations of some differences. However, there are some additional reasons specific to QUAST.

First, QUAST does not output any information about the locations of small differences obtained after parsing the results given by the show-snps package, only providing information about their total number. This is reflected in a large deviation between the numbers of simulated and detected insertions, deletions, substitutions, gaps, and inserted gaps. Second, QUAST is unable to distinguish differences of several types at the identical locations. For example, duplications and reshufflings were not reported as stand-alone differences when they were located together with relocations or translocations. The same is also true for insertions and deletions when they were introduced between inverted and reshuffled blocks. Third, the comparison of the QUAST results with the NucDiff results obtained with the QUAST-

like parameters settings suggests that QUAST has its own internal length threshold for filtering mapped fragments. This value is somewhat higher than the NUCmer -c parameter value used. This led to a reduced number of correctly detected relocation and translocation events.

During comparison of the QUAST results with the NucDiff results obtained with the QUAST-like settings, we noticed that QUAST was able to detect more duplication and translocation events. This can be explained by less strict requirements for correspondence between the simulated and obtained types for QUAST. For example, in situations where NucDiff detected simple translocations and duplications as translocation with insertions and simple insertions, respectively (see translocation case 1 in [Aditional file 1: Table S2]), the differences were considered wrongly resolved by NucDiff and correctly resolved by QUAST. The same problem is also applicable to simple relocations. However, since fewer relocations were detected by QUAST because of its filtering approach, the significant divergence between numbers is not apparent in Table 1.

### Comparison with dnadiff

The NucDiff, dnadiff and QUAST tools provide a quantification of the differences between two sets of genomes. In this section, we compare the numbers output by these tools. Due to the way these tools report their results, it is very difficult to make a fair comparison between them. All tools were run on the same simulated genome described in Datasets section. NUCmer, whose output was used by NucDiff and dnadiff, was run with the QUAST-like parameter settings (see [Additional file 1: Table S4]). Since dnadiff only provides the number of differences and not their locations, we cannot know for sure whether the differences are actually in the same places as reported by the other tools. To perform the comparison, we created a set of categories suitable for comparison and grouped the differences reported into these categories (see [Additional file 1: Table S6] for grouping). The results are presented in Table 2.

The results showed that the obtained counts for NucDiff and dnadiff are largely similar, while QUAST has a tendency to detect fewer differences than NucDiff and dnadiff in almost all categories. A large deviation between the results from QUAST and the other tools was observed in the nonTandem and Relocations groups. In both cases, it can be explained by how the comparison is performed and not necessarily by the performance of the tool.

### Comparison of several assemblies of the same read set to the same reference genome

We downloaded assemblies of the same *V. cholerae* read set as described in the Datasets section, and compared

**Table 2** Number of simulated differences (Truth) and differences obtained by NucDiff, dnadiff and QUAST

| Group | Truth | NucDiff | dnadiff | QUAST |
|---|---|---|---|---|
| nonTandem | 3448 | 6717 | 7460 | 2814 |
| Tandem | 119 | 116 | 116 | 0 |
| Substitutions | 164 | 423 | 234 | 354 |
| Relocations | 2854 | 2802 | 2211 | 185 |
| Translocations | 163 | 137 | 137 | 117 |
| Inversions | 1068 | 1060 | 1060 | 1053 |
| UnalignedSeq | 13 | 21 | 16 | 21 |

In the nonTandem group, the values shown for the simulated differences (Truth) and QUAST are the number of events, while in the other columns the values are the sum of the number of bases involved in the differences for short and medium local differences (found by the show-snps program) and the number of events of long local differences (those causing alignment fragmentation). In the Inversions group, the numbers of simulated inversions and inversions found by NucDiff were multiplied by two to enable a fair comparison, because QUAST and dnadiff report the number of fragment ends, while NucDiff reports the number of fragments. In the Substitutions group, the values shown are the number of bases, while in the other rows the values are the number of events. The reshuffling differences are contained in the nonTandem group for QUAST, but placed in the Relocations group in all other cases

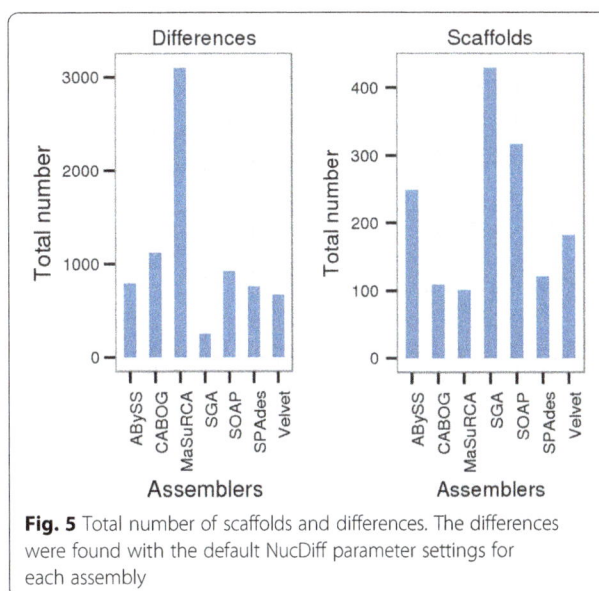

**Fig. 5** Total number of scaffolds and differences. The differences were found with the default NucDiff parameter settings for each assembly

them to a *V. cholerae* reference using NucDiff with default parameter settings (see in [Additional file 1: Table S4]). The number of detected differences is presented in [Additional file 1: Figure S3]. The total number of scaffolds and differences is shown in Fig. 5. The resulting GFF3 files with mapped blocks and differences (shown with reference-based coordinates) were displayed using the IGV genome browser, and an example of assembly comparison is shown in [Additional file 1: Figure S4]. As is evident, we were able not only to compare quantitative metrics (i.e. the number of each type of difference, the number of uncovered reference bases, etc.) but also to analyse the placement of contigs/scaffolds and differences relative to each other and the exact location of the different types of detected differences.

Based on the obtained results, it is possible to conclude for the given examples that SGA gives the most fragmented assembly compared with other assemblers, while MaSuRCA gives the solution with the highest number of errors (differences are considered as errors in this case, since we are comparing to a good quality reference genome), mainly suffering from substitution errors (2839 out of 3106 differences). SOAPdenovo has rather high numbers of errors in all categories, confirming the result from GAGE-B using QUAST, which states that SOAPdenovo has "a larger number of errors than most other methods" [12]. It is also possible to see the large fragmentation in the SGA assembly and the large total number of differences in the MaSuRCA assembly by visualisation in IGV in [Additional file 1: Figure S4a and c].

According to GAGE-B results, MaSuRCA has produced the assembly with the best N50 size. However, in our experience, MaSuRCA did not distinguish itself when compared to other assemblers. All assemblers have managed to resolve some regions where most of the other assemblers failed to get continuous solutions. In addition, we noticed that there are some differences that were produced by all assemblers in the same places. For example, we detected two deletions, one of length 1255 bp, overlapping with two open reading frames of a transposase ([Additional file 1: Figure S5a]) and a second of length 1367 bp, overlapping with two genes of unknown function ([Additional file 1: Figure S5b]), and many short insertions, deletions and substitutions through the genome. We suspect that such errors may actually be true variations between the sequenced genome and the reference genome rather than errors in the assemblies in many cases. For example, in the case of the transposase, this may have inserted itself in the strain sequenced for the reference genome, while it was absent from the DNA of the strain sequenced for GAGE-B.

## Comparison of genomes from different strains of the same species

With NucDiff, it is also possible to compare genomes of different strains of the same species to show genomic differences between them. We have compared the genomes of 21 different strains of *E. coli* K12 available in the NCBI database to the *E. coli* K12 MG1655 as the reference genome. We have calculated the total number of differences of each type at every base of each query reference. The result was saved in the bedGraph format and uploaded into the IGV genome browser together with the *E. coli* K12 MG1655 annotation (see Fig. 6).

The results show that the differences are not distributed randomly, they tend to be clustered in some

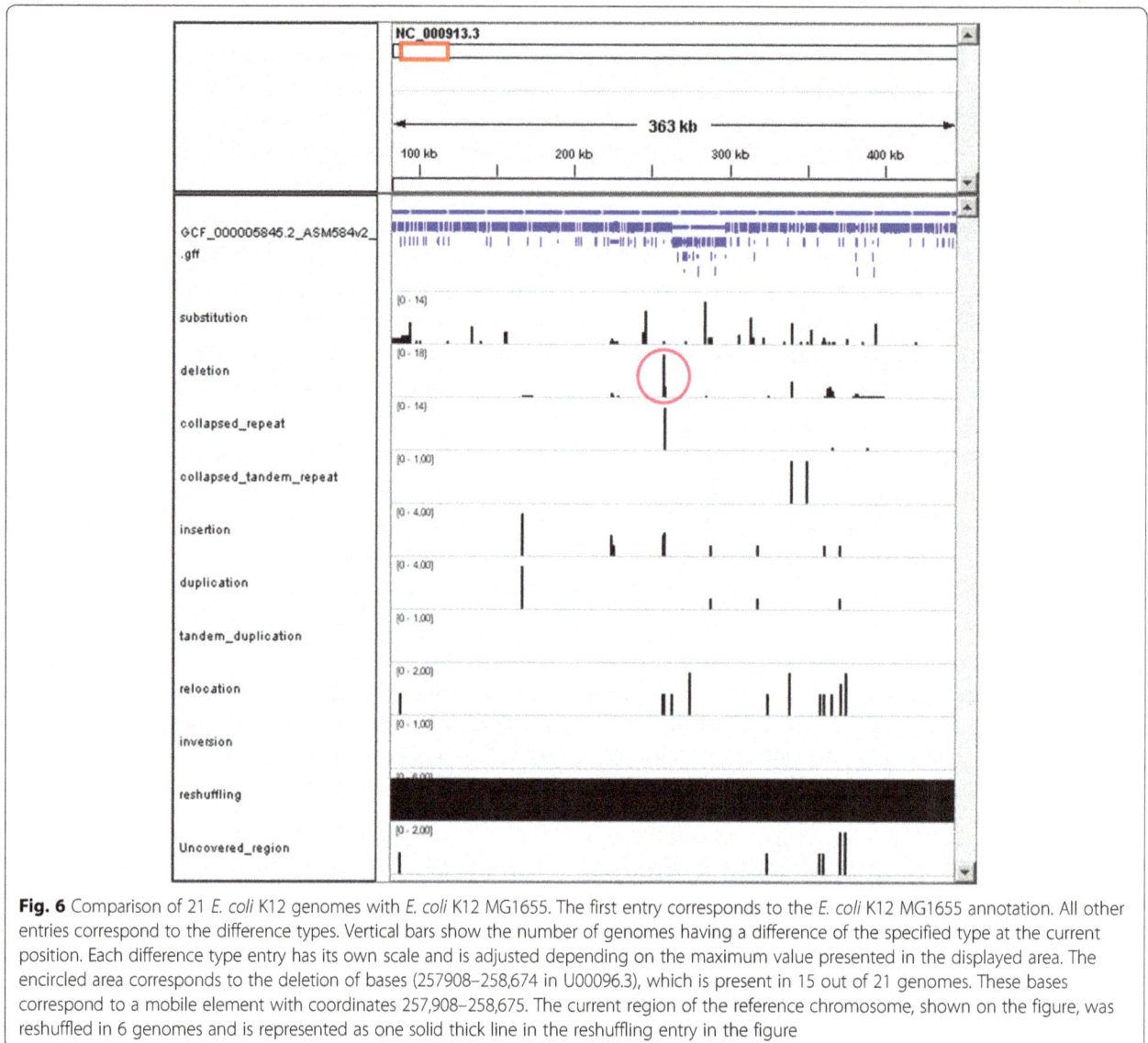

**Fig. 6** Comparison of 21 *E. coli* K12 genomes with *E. coli* K12 MG1655. The first entry corresponds to the *E. coli* K12 MG1655 annotation. All other entries correspond to the difference types. Vertical bars show the number of genomes having a difference of the specified type at the current position. Each difference type entry has its own scale and is adjusted depending on the maximum value presented in the displayed area. The encircled area corresponds to the deletion of bases (257908–258,674 in U00096.3), which is present in 15 out of 21 genomes. These bases correspond to a mobile element with coordinates 257,908–258,675. The current region of the reference chromosome, shown on the figure, was reshuffled in 6 genomes and is represented as one solid thick line in the reshuffling entry in the figure

locations. For example, in 15 out of 21 genomes there is a deletion of bases, starting from base 257,908 and ending with base 258,674 in U00096.3 (a circle in Fig. 6). These bases correspond to a mobile element with almost the same starting coordinate and ending in position 258,675.

## Discussion

In this paper, we have described a tool, called NucDiff, which detects and describes the differences between any two sets of closely related DNA sequences according to our comprehensive classification scheme. The tool has several properties that make it very useful for doing comparative analysis of assemblies and reference genomes. 1) It is able to work with very fragmented genome assemblies and genomes with various structural rearrangements. We have demonstrated this with the *V. cholerae* assemblies and *E. coli* K12 reference genomes.

Moreover, NucDiff is in many cases able to detect differences that are associated with repeated regions (for example, in case of duplication, tandem duplication, collapsed repeat and tandem collapsed repeat differences). However, it is not able to detect such associations for simple insertions and simple deletions found between mapped blocks. 2) The tool gives information about the locations and types of differences. This information is stored in the widely used GFF3 format with both query-based and reference-based coordinates, which can be used with existing genome browsers for visualizing the differences. The NucDiff output also enables users to incorporate the tool in a large variety of applications where detecting differences is either the final goal or as a component in an analysis pipeline. 3) The tool enables visualisation of alignment results, as a result of outputting differences in the GFF3 format. We have shown two different applications of visualisation to further inspect the results. First, by uploading the files with locations of differences and mapped blocks for all compared assemblies at once into a genome browser. This approach enables comparison of all differences at any level of detail across all datasets to look for patterns. Second, by uploading the files with the summarized counts of all difference types in all query genomes for each reference sequence base. This visualisation gives a better overview of the comparison analysis results when the number of compared genomes is high, revealing common and distinct patterns in the structures of genome sequences. However, such an approach leads to the loss of some information about differences (e.g. which query genome(s) have the specific differences and the difference locations in these genome(s)).

The benchmarking results showed that the NucDiff output depends on the NUCmer and delta-filter parameters values. The values mainly influence the types of differences and not the total number, revealing or hiding information about the repeated, inverted, substituted, or relocated nature of the short- and medium-sized differences. The locations of regions containing differences remain the same in most cases. As for NucDiff result quality, we have noticed systematic loss of information about the repeated nature of some differences in specific cases. This was due to the limitations of the approach implemented in NucDiff.

Together with NUCmer, MUMmer provides another alignment program called PROmer. Unlike NUCmer, PROmer can be used for highly divergent sequences that show little DNA sequence conservation. Since both tools output the alignments results in the same format, it is possible to run NucDiff with PROmer output file as an input parameter, thus enabling detection of differences between two highly divergent sequences.

There are similarities between NucDiff and QUAST, a software tool for comparing assemblies to reference genomes. Both use NUCmer as a part of their analysis pipeline to align the input sequences. However, QUAST assesses genome quality mainly based on contiguity and gene complement completeness, producing various reports, plots and tables. QUAST will output quality metrics (e.g. number of misassemblies, indels and so on) only when a reference genome is available. In this case, it reports information about similar reference and query sequences, unmapped query sequences, and the locations of the regions where the reference and query sequences were split during the alignment process, giving general explanations for the fragmentation. It does not output the locations of small indels and substitutions obtained after parsing results given by the show-snps package. It provides only the raw show-snps output and summary statistics for these types of differences. Our experiments showed that QUAST tends to count several differences located at the same position as one difference. Comparing to QUAST, our tool is also able to give more detailed information about the locations of all differences as well as a more detailed classification of them. In addition, NucDiff allows the users to upload the results to different genome browsers, while QUAST output can be directly visualised only in its own genome browser, Icarus, that does not handle uploading of additional tracks.

We have also compared NucDiff with dnadiff. Both tools parse the NUCmer output and produce detailed information about the differences between two sets of sequences. Their results are very similar, but, in contrast to NucDiff, dnadiff does not allow visualization of differences and is not able to quantify them at the same level of detail.

Our results from analyses of different real assemblies have revealed a complication related to assembly comparison. It is not always enough to only use the quality and contiguity summary metrics when choosing the "best" assembly. The ability to visualize results and manually inspect the regions where the differences are located may dramatically influence this choice.

## Conclusions

We present the tool NucDiff for the comparison of two sets of closely related sequences. NucDiff outputs information about the types and locations of the differences between the sequences. Special attention has been paid to detection of differences involving repeated regions. All differences are categorized according to a proposed detailed classification scheme. The output from NucDiff enables the user to visualise the results using a genome browser, and we demonstrate two different applications of such visualisations. The ability to 1) give detailed information about the differences, 2) handle small local differences as well as structural rearrangements, and 3) visualise the comparison results makes NucDiff convenient for whole-genome sequence comparison or as an intermediate step in an analysis pipeline.

## Additional files

**Additional file 1: Figure S1.** Reference fragments placement order depending on query fragment orientations during detection of local differences. **Figure S2.** Circular genome alignment alternatives. **Figure S3.** Number of differences in each category obtained by NucDiff with the default parameter settings for all assemblers. **Figure S4.** Comparison of multiple assemblies against one reference using NucDiff. **Figure S5.** Examples of detection of long deletions located in all assemblies at the same place in the reference sequence. **Table S1.** Alignment fragmentation cases caused by simple differences. **Table S2.** Genome modifications implemented during the simulation process. **Table S3.** List of *E. coli* genomes used in the Comparison of genomes from different strains of the same species section. **Table S4.** Parameter values used for each parameter settings. **Table S5.** Correspondence between the QUAST difference types and the simulated difference types. **Table S6.** Correspondence between the QUAST, dnadiff and NucDiff difference types and the expected difference types.

**Additional file 2:** Detailed results for Table 1.

## Abbreviation
WGA: Whole-genome alignment

## Acknowledgements
The authors wish to thank the Centre for Ecological and Evolutionary Synthesis (CEES) for access to the computational infrastructure ('cod' servers) that enabled the bioinformatics analysis for this project.

## Funding
KK was funded by the Computational Life Science initiative (CLSi) at the University of Oslo. The funding body played no role in the design or conclusions of this study.

## Authors' contributions
KK designed and implemented NucDiff. AJN and KK developed the proposed classification of the differences. TR, AJN, GKS, and KL suggested the demonstration examples and other experiments performed. KK performed all the experiments. KK, TR and AJN wrote the manuscript. KL and GKS revised the manuscript. All authors read and approved the final manuscript.

## Competing interests
The authors declare that they have no competing interests.

## Author details
[1]Biomedical Informatics Research Group, Department of Informatics, University of Oslo, PO Box 1080, 0316 Oslo, Norway. [2]Norwegian Veterinary Institute, PO Box 750 Sentrum, 0106 Oslo, Norway. [3]Department of Microbiology, Oslo University Hospital, Rikshospitalet, PO Box 4950 Nydalen, 0424 Oslo, Norway. [4]Centre for Ecological and Evolutionary Synthesis, Department of Biosciences, University of Oslo, PO Box 1066 Blindern, 0316 Oslo, Norway.

## References
1. Dewey CN. Whole-genome alignment. Methods Mol Biol. 2012;855:237–57. doi:10.1007/978-1-61779-582-4_8.
2. Engels R, Yu T, Burge C, Mesirov JP, DeCaprio D, Galagan JE. Combo: a whole genome comparative browser. Bioinformatics. 2006;22(14):1782–3. doi:10.1093/bioinformatics/btl193.
3. Choi JH, Cho HG, Kim S. GAME: a simple and efficient whole genome alignment method using maximal exact match filtering. Comput Biol Chem. 2005;29(3):244–53. doi:10.1016/j.compbiolchem.2005.04.004.
4. Blanchette M. Computation and analysis of genomic multi-sequence alignments. Annu Rev Genomics Hum Genet. 2007;8:193–213. doi:10.1146/annurev.genom.8.080706.092300.
5. Belal NA, Heath LS. A theoretical model for whole genome alignment. J Comput Biol J Comput Biol. 2011;18(5):705–28. doi:10.1089/cmb.2010.0101.
6. Darling AC, Mau B, Blattner FR, Perna NT. Mauve: multiple alignment of conserved genomic sequence with rearrangements. Genome Res. 2004;14(7):1394–403. doi:10.1101/gr.2289704.
7. Gurevich A, Saveliev V, Vyahhi N, Tesler G. QUAST: quality assessment tool for genome assemblies. Bioinformatics. 2013;29(8):1072–5. doi:10.1093/bioinformatics/btt086.
8. dnadiff. https://github.com/marbl/MUMmer3/blob/master/docs/dnadiff. README. Accessed 8 July 2017.
9. Kurtz S, Phillippy A, Delcher AL, Smoot M, Shumway M, Antonescu C, Salzberg SL. Versatile and open software for comparing large genomes. Genome Biol. 2004;5(2):R12. doi:10.1186/gb-2004-5-2-r12.
10. Stein L. GFF3 format specification. 2013. https://github.com/The-Sequence-Ontology/Specifications/blob/master/gff3.md. Accessed 8 July 2017.
11. The MUMmer manual. http://mummer.sourceforge.net/manual/. Accessed 8 July 2017.
12. Magoc T, Pabinger S, Canzar S, Liu X, Su Q, Puiu D, Tallon LJ, Salzberg SL. GAGE-B: an evaluation of genome assemblers for bacterial organisms. Bioinformatics. 2013;29(14):1718–25. doi:10.1093/bioinformatics/btt273.
13. Simpson JT, Wong K, Jackman SD, Schein JE, Jones SJ, Birol I. ABySS: a parallel assembler for short read sequence data. Genome Res. 2009;19(6): 1117–23. doi:10.1101/gr.089532.108.
14. Miller JR, Delcher AL, Koren S, Venter E, Walenz BP, Brownley A, Johnson J, Li K, Mobarry C, Sutton G. Aggressive assembly of pyrosequencing reads with mates. Bioinformatics. 2008;24(24):2818–24. doi:10.1093/bioinformatics/btn548.
15. Zimin AV, Marçais G, Puiu D, Roberts M, Salzberg SL, Yorke JA. The MaSuRCA genome assembler. Bioinformatics. 2013;29(21):2669–77. doi:10.1093/bioinformatics/btt476.
16. Simpson JT, Durbin R. Efficient de novo assembly of large genomes using compressed data structures. Genome Res. 2012;22(3):549–56. doi:10.1101/gr.126953.111.
17. Luo R, Liu B, Xie Y, Li Z, Huang W, Yuan J, He G, Chen Y, Pan Q, Liu Y, Tang J, Wu G, Zhang H, Shi Y, Liu Y, Yu C, Wang B, Lu Y, Han C, Cheung DW, Yiu SM, Peng S, Xiaoqian Z, Liu G, Liao X, Li Y, Yang H, Wang J, Lam TW, Wang J. SOAPdenovo2: an empirically improved memory-efficient short-read de novo assembler. Gigascience. 2012 Dec 27;1(1):18. doi:10.1186/2047-217X-1-181 comment on PubPeer (by: Comment from PubMed Commons).
18. Bankevich A1, Nurk S, Antipov D, Gurevich AA, Dvorkin M, Kulikov AS, Lesin VM, Nikolenko SI, Pham S, Prjibelski AD, Pyshkin AV, Sirotkin AV, Vyahhi N, Tesler G, Alekseyev MA, Pevzner PA. SPAdes: a new genome assembly algorithm and its applications to single-cell sequencing. J Comput Biol. 2012 May;19(5):455–77. doi:10.1089/cmb.2012.0021.
19. Zerbino DR, Birney E. Velvet: algorithms for de novo short read assembly using de Bruijn graphs. Genome Res. 2008;18(5):821–9. doi:10.1101/gr.074492.107.
20. The GAGE-B website. http://ccb.jhu.edu/gage_b/. Accessed 8 July 2017.
21. The NCBI database: E. coli K12 references. http://www.ncbi.nlm.nih.gov/genome/genomes/167. Accessed 8 July 2017.
22. The NCBI database: V. cholerae annotation. ftp://ftp.ncbi.nlm.nih.gov/genomes/refseq/bacteria/Vibrio_cholerae/reference/GCF_000006745.1_ASM674v1/. Accessed 8 July 2017.
23. The NCBI database: E. coli K12 MG1655 annotation. ftp://ftp.ncbi.nlm.nih.gov/genomes/refseq/bacteria/Escherichia_coli/all_assembly_versions/GCF_000005845.2_ASM584v2/. Accessed 8 July 2017.
24. Robinson JT, Thorvaldsdóttir H, Winckler W, Guttman M, Lander ES, Getz G, Mesirov JP. Integrative genomics viewer. Nat Biotechnol. 2011;29(1):24–6. doi:10.1038/nbt.1754.
25. Thorvaldsdóttir H, Robinson JT, Mesirov JP. Integrative genomics viewer (IGV): high-performance genomics data visualization and exploration. Brief Bioinform. 2013;14(2):178–92. doi:10.1093/bib/bbs017.

# 6

# A systematic evaluation of nucleotide properties for CRISPR sgRNA design

Pei Fen Kuan[1]* ⓘ, Scott Powers[2], Shuyao He[1], Kaiqiao Li[1], Xiaoyu Zhao[2] and Bo Huang[3]

## Abstract

**Background:** CRISPR is a versatile gene editing tool which has revolutionized genetic research in the past few years. Optimizing sgRNA design to improve the efficiency of target/DNA cleavage is critical to ensure the success of CRISPR screens.

**Results:** By borrowing knowledge from oligonucleotide design and nucleosome occupancy models, we systematically evaluated candidate features computed from a number of nucleic acid, thermodynamic and secondary structure models on real CRISPR datasets. Our results showed that taking into account position-dependent dinucleotide features improved the design of effective sgRNAs with area under the receiver operating characteristic curve (AUC) > 0.8, and the inclusion of additional features offered marginal improvement (∼2% increase in AUC).

**Conclusion:** Using a machine-learning approach, we proposed an accurate prediction model for sgRNA design efficiency. An R package `predictSGRNA` implementing the predictive model is available at http://www.ams.sunysb.edu/~pfkuan/softwares.html#predictsgrna.

**Keywords:** CRISPR, Machine learning, Predictive modeling, Thermodynamics

## Background

Clustered Regularly Interspaced Short Palindromic Repeats (CRISPR)/Cas system is a heritable and adaptive prokaryotic immune system that protects cells by destroying foreign genetic elements [1]. Over the past few years, CRISPR has emerged as a powerful gene editing technology [2, 3]. CRISPR consists of a single guide RNA (sgRNA) and an enzyme called Cas9. The sgRNA is composed of a short synthetic RNA (approximately 20 base pairs (bp), known as spacer target) located within a N-bp scaffold. The spacer target is designed to bind to a specific sequence in the genome, whereas the Cas9 protein acts as a biomolecular scissor. This system has proven to be a powerful tool for studying individual gene function and for genome engineering.

The design of sgRNA is an important aspect to ensure the success of CRISPR-Cas9 screens. It is desirable to design sgRNA libraries which have maximum on-target and minimum off-target effects. The binding specificity

of the sgRNA is determined by the 20 bp spacer target and a protospacer adjacent motif (PAM) sequence (generally NGG or NAG) on the genome. Once the sgRNA binds to the target sequence, the Cas9 nuclease cuts 3-bp upstream of the PAM sequence. Different groups have studied the sequence features of spacer target sites that predict sgRNA on-target efficiency [4–7]. In particular, [5] investigated the position-dependent sequence on sgRNA efficiency and whether these features could reproducibly predict sgRNA efficiency in several publicly available CRISPR datasets. They proposed a predictive model using the position-dependent mono-nucleotide composition across a 40 bp sequence encompassing 5' flanking, spacer target and 3' flanking region; and further demonstrated that their model performed better than the model of [4]. On the other hand, [6, 7] proposed a predictive model based on gradient-boosted regression trees using position-dependent and independent sequence properties, location of the sgRNA within the protein and melting temperatures.

Aspects of sgRNA design share similarities to oligonucleotide designs used for microarrays. In both cases, optimal oligonucleotide design aims to increase binding sensitivity and specificity while minimizing off target

*Correspondence: peifen.kuan@stonybrook.edu
[1] Department of Applied Mathematics and Statistics, Stony Brook University, 100 Nicolls Road, 11794 Stony Brook, USA
Full list of author information is available at the end of the article

hybridization. A position dependent sequence bias has been observed in the design of oligonucleotides in Affymetrix microarrays [8], whereas in our earlier work [9] we showed that the thermodynamic and secondary features of the oligonucleotides affect the hybridization intensities in Nimblegen arrays. In addition, [6, 7] investigated position dependent and independent features, position of the guide within the genes, interaction with the PAM sequence and melting temperatures, and showed that these features improved the prediction model in CRISPR/Cas9 screens; whereas microhomology features did not improve the prediction. In this paper, we computed a comprehensive list of features of the target sequence from a number of nucleic acid, thermodynamic, and secondary structure models by adopting some ideas of microarray designs. In a similar manner as [6, 7], we systematically characterized the effect of these features on the efficiency of sgRNA design, and seek to understand if the inclusion of these features improves the design of effective sgRNAs in CRISPR/Cas9 knockout screens.

## Methods

We used the sets of efficient and inefficient sgRNAs from the CRISPR/Cas9 screens of [10] and [11] compiled by [5]. The first dataset consists of 731 efficient and 438 inefficient sgRNAs targeting ribosomal genes [10], the second dataset consists of 671 efficient and 237 inefficient sgRNAs targeting non-ribosomal genes [10] and the third dataset consists of 830 efficient and 234 inefficient sgRNAs targeting essential genes in mouse embryonic stem cell (mESC) line, JM8 [11]. The procedures for identifying efficient and inefficient sgRNAs were used exactly as described in [5]. Spacer lengths in the reported studies were 20 bp [10] and 19 bp [11]. Using these sets of sgRNAs, we computed primary sequence, thermodynamic, and secondary structures as candidate features. Further details are provided below.

### DNA sequence candidate features
#### Position-dependent nucleotide composition

Similar to [5], we created vectors of position-dependent mono-nucleotide composition (PD Mono) for the 40 bp long sequences comprised of the spacer targets, and 5' and 3' flanking regions. In addition, we extracted position-dependent dinucleotide composition (PD Dinuc) for these 40 bp sequences and computed the single and dinucleotide frequencies (Freq) for the spacer target. Since positions 32 and 33 were part of the PAM sequence (GG), they were excluded from the analysis.

### Thermodynamics and secondary structure properties of [9] (Thermo)

Motivated by our earlier work which studied the relationship between oligonucleotide properties and hybridization signal intensities in microarray design [9], we computed the thermodynamic properties: melting temperature $(T_m)$, GC content, entropy change $(\Delta S)$, enthalpy change $(\Delta H)$, free energy change $(\Delta G)$; and secondary structures: longest polyN, repetitive sequence (repeat), length of a potential stem-loop (LSL) and minimum energy folding (MEF). $T_m$ was computed according the formula

$$T_m = 81.5 + 16.6 \left( \log_{10}([Na^+]) \right) + 0.41 * (\%GC) - 600/L$$

where $[Na^+]$ was assumed to be 0.2M [12]. $\Delta G$, $\Delta H$ and $\Delta G$ were calculated by summing the respective entropy, enthalpy and free energy parameters of each dinucleotide, including the initiation parameters and penalty for self complementary duplexes according to the position-dependent nearest neighbor approach as described [13]. These parameters were provided in Tables 1 and 2 of [13]. MEF was computed using the `hybrid-ss-min` program in OligoArrayAux package, whereas LSL was computed using the palindrome function in the EMBOSS package. Longest polyN and repeat were calculated as previously described [9]. These properties were computed for the spacer target sequence.

### DNA secondary structures based on dinucleotide and tetra nucleotide properties of [14] and [15] (Packer)

Following a previously described approach [16], we computed the minimum, maximum and average values of both the tetranucleotide energy and flexibility scores as described [15]. These scores were given in Tables 3 and 4 of [15]. In addition, we computed the minimum, maximum and average values of the dinucleotide roll, twist, slide and shift scores as described [14]. The dinucleotide values of these properties were given in Tables 1, 2 and 3 of [14]. These scores were representations of the three-dimensional DNA structure and anisotropic flexibility [14]. Similar to above, we computed these properties for the spacer target sequence.

### Physiochemical properties of [17] (PhyChem)

We adapted the approach described by [17] which was developed for predicting nucleosome occupancy and computed the 12 physiochemical properties (A-pillicity, base-stacking, B-DNA twist, bendability, DNA bending stiffness, DNA denaturation, duplex disrupt energy, duplex free energy, propeller twist, protein deformation, protein-DNA twist and Z-DNA). For each property, we computed the minimum, maximum and average dinucleotide scores for the spacer target sequence. The dinucleotide values of the 12 physicochemical properties were given in Table 1 of [17].

*Pseudo k-tuple nucleotide composition of [18] (PseKNC)*

The PseKNC model was also originally developed for predicting nucleosome occupancy by taking into account global sequence-order effects. PseKNC represents the DNA sequence as vectors $\left[\frac{f_1}{d}, \ldots, \frac{f_{4^k}}{d}, \frac{w\theta_1}{d}, \ldots, \frac{w\theta_\lambda}{d}\right]^T$ where $d = \sum_{j=1}^{4^k} f_j + w\sum_{j=1}^{\lambda}\theta_j, f_j$'s are the k-tuple nucleotide frequencies and

$$\theta_j = \frac{1}{m(L-j-1)}\sum_{s=1}^{L-j-1}\sum_{t=1}^{m}\left[P_t(r_s r_{s+1}) - P_t\left(r_{s+j}r_{s+j+1}\right)\right]^2$$

$m$ is the number of local DNA properties considered, $P_t(r_s r_{s+1})$ and $P_t\left(r_{s+j}r_{s+j+1}\right)$ are the score of the $t$-th DNA local structural property for dinucleotide $r_s r_{s+1}$ and $r_{s+j}r_{s+j+1}$ at position $s$ and $s + j$, respectively. $\lambda$ is the order of correlations along the DNA sequence and $w$ is the weight factor. Our candidate $k$, $\lambda$ and $w$ took values of $k = 2, 3, \ldots, 6$, $\lambda = 1, 2, \ldots, 15$, and $w = 0, 0.1, 0.2, \ldots, 1$. We considered the following strategy to choose the optimal parameters for the PseKNC model. A three way cross validation was performed on each dataset using elastic net [19]. The parameters corresponding to the PseKNC model with the largest average area under the receiver operating characteristic curve (AUC) were selected for subsequent analysis. Based on this criterion, we set $k = 2$, $\lambda = 1$ and $w = 0.5$. Similar to [18], we considered $m = 6$ DNA local structural properties which were divided into local translational (rise, slide and shift) and angular (twist, roll and tilt).

*Optimal pairwise alignment (Align)*

We computed the optimal global pairwise alignment scores between the seed region and scaffold using the Needleman-Wunsch algorithm [20] which served as a measure of the potential of the $k$ PAM-proximal seed region of the spacer target to interact with the scaffold sequence. The seed region was defined as the immediate $k$ nucleotides next to the PAM sequence. We considered $k = 5, \ldots, L$, where $L$ is the length of spacer target.

## Results and discussion

For each dataset, we computed a score for every feature as a measure of strength of association with sgRNA efficiency. If the feature was a binary variable, a log odds ratio between efficient and inefficient sgRNAs was computed. If the feature was a continuous variable, two-sample t-statistic was computed. We divided the features into 8 classes (1) position-dependent mono-nucleotide (PD Mono), (2) position-dependent dinucleotide (PD Dinuc), (3) frequencies of mono and dinucleotides (Freq) (4) optimal pairwise alignment between spacer target and

scaffold (Align) (5) thermodynamics and secondary structures of [9] (Thermo) (6) secondary structures of [14, 15] (Packer) (7) physiochemical properties (PhyChem) of [17] and (8) pseudo k-tuple nucleotide composition of [18] (PseKNC). We found that most of the features were consistently associated with sgRNA efficiency across datasets (Figs. 1 and 2).

### Candidate feature ranking

To rank the contribution of each feature to the efficiency of sgRNA design, we fitted a logistic regression model within each dataset using the binary sgRNA efficiency indicator as the response and the features as predictors. The Bayesian Information Criterion (BIC) for the fitted model was computed. The features were ranked by the BIC scores and the top 10 most important features were shown in Additional file 1: Figure S1. The top ranked feature based on average BIC scores across the three datasets was the 16-th feature from PseKNC model. This feature is a function of TT dinucleotide frequency. In addition, we computed the area under receiver operating characteristic curves (AUCs) for continuous features. The top 10 features ranked by AUC were shown in Fig. 3, in which the 16-th feature from the PseKNC model was also ranked number one. The third measure we considered for feature ranking was the permutation based variable importance score from the random forest prediction algorithm. Random forest [21] is a non-parametric ensemble approach based on a large number of classification trees trained on bootstrap samples. The permutation based variable importance score of a feature is defined as the difference in prediction accuracy before and after permuting this feature, averaging over all trees. We used the unscaled version of variable importance score as recommended by [22, 23] to avoid bias due to number of trees grown. The top 10 features ranked by variable importance are shown in Additional file 1: Figure S2. Based on these results, the frequencies of T and TT had the strongest association with sgRNA efficiency, in which higher frequencies of T and TT were associated with decreased efficiency.

### Predictive modeling

To assess the contribution of the 8 different feature classes in prediction sgRNA efficiency, we formed all possible combinations of feature classes ($\sum_{i=1}^{8}\binom{8}{i} = 255$ combinations). We adapted the strategy in [5] in constructing and evaluating the predictive model for sgRNA efficiency:

1. To evaluate intra-platform consistency within the same class of genes, we performed 3-way cross validation within dataset 1 (sgRNA targeting ribosomal genes) from [10]. We randomly split

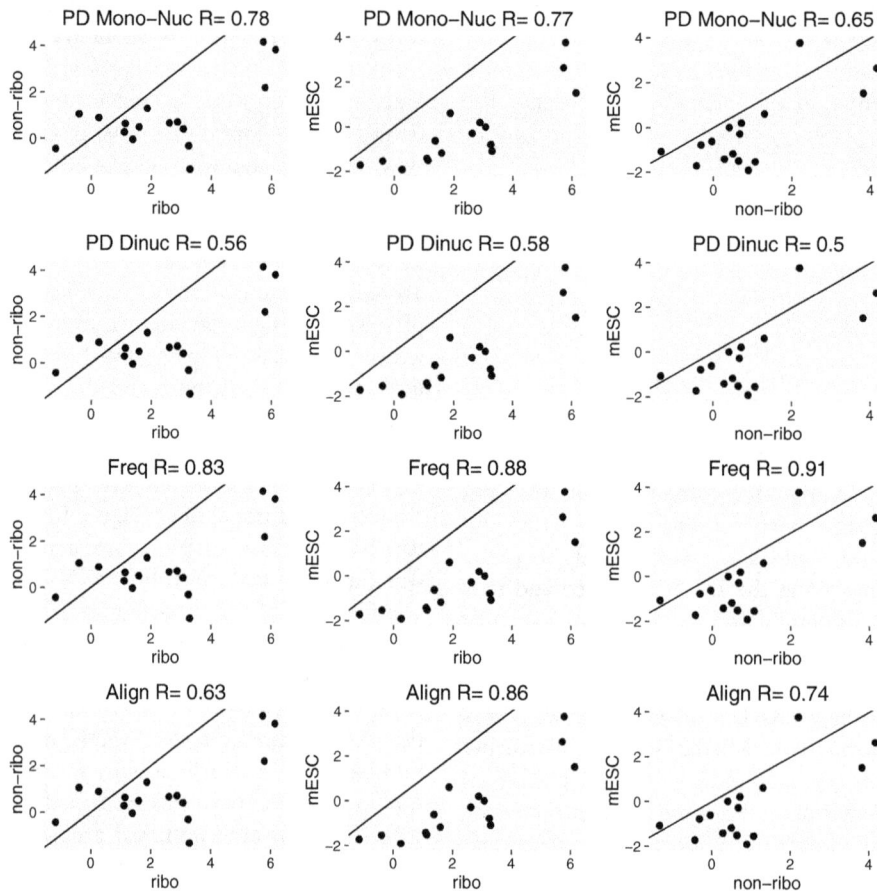

**Fig. 1** Pairwise correlation plot for each class of features. *Left* column is the pairwise correlation plot between ribosomal and non-ribosomal genes from [10]. *Middle* column is the pairwise correlation plots between ribosomal genes from [10] and mESC essential genes from [11]. *Right* column is the pairwise correlation plots between non-ribosomal genes from [10] and mESC essential genes from [11]. Each point is a feature

dataset 1 into 3 parts of equal sample size, trained the model on two parts (training set) and evaluated the performance of the resulting predictive model on the remaining part (test set). This process was repeated 3 times by leaving out a different test set, and results were averaged over 10 iterations of random sampling.

2. To evaluate intra-platform consistency across different classes of genes, the predictive algorithm was trained on dataset 1 (ribosomal genes) and tested on dataset 2 (non-ribosomal genes).

3. To evaluate inter-platform consistency, the predictive algorithm was trained on datasets 1 and 2 (ribosomal+non-ribosomal genes) from [10] and tested on dataset 3 (mESC essential genes) from [11].

The elastic net algorithm [19] was used in constructing the predictive model on the training set based on 10 fold cross-validation. Since the features we considered in this paper were functions of the nucleotide composition, they were correlated and the elastic net algorithm automatically selected non-redundant informative features. The

objective function of elastic net consists of a loss function + penalty:

$$\min_{\beta} ||\mathbf{y} - \mathbf{X}\beta||^2 + \lambda \left\{ \alpha ||\beta||_1 + (1-\alpha)||\beta||^2 \right\}$$

where $||\beta||_1 = \sum_{j=1}^{p} |\beta_j|$ and $||\beta||^2 = \sum_{j=1}^{p} \beta_j^2$.

We evaluated the performance on the test set in terms of AUC. The optimal cutpoints were determined by maximizing the Youden index$(J)$=Se+Sp−1, where Sensitivity(Se)= $\frac{TP}{TP+FN}$ and Specificity(Sp)= $\frac{TN}{TN+FP}$. The results were shown in Tables 1, 2 and 3. For each test set, we reported these performance measures for the predictive models constructed using each of the 8 feature classes, as well as the combinations of feature classes with the maximum AUC (Comb Feature). Across all comparisons, integrating multiple feature classes showed improvements in terms of AUC compared to position-dependent mono-nucleotide models (PD Mono) in [5]. Among the 8 individual feature classes, position-dependent dinucleotide models (PD Dinuc) consistently outperformed other feature classes in predicting sgRNA efficiency and were close

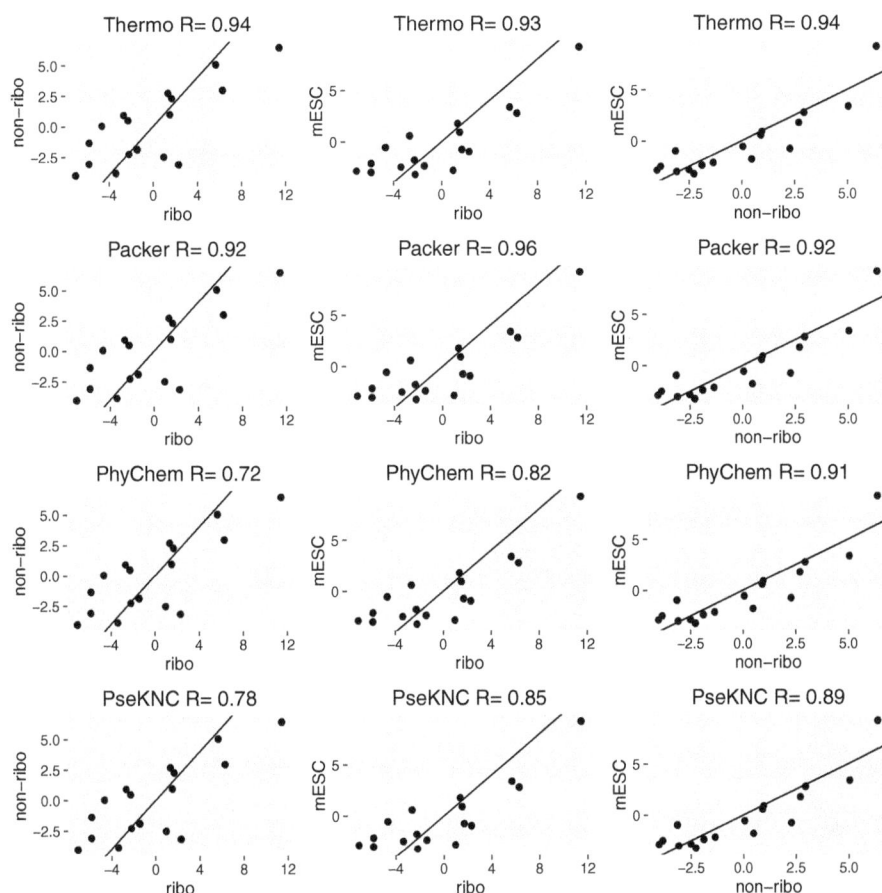

**Fig. 2** Pairwise correlation plot for each class of features. *Left* column is the pairwise correlation plot between ribosomal and non-ribosomal genes from [10]. *Middle* column is the pairwise correlation plots between ribosomal genes from [10] and mESC essential genes from [11]. *Right* column is the pairwise correlation plots between non-ribosomal genes from [10] and mESC essential genes from [11]. Each point is a feature

to results from the combination of feature classes models in all 3 scenarios. A similar pattern was also observed in [6, 7], in which they showed that position dependent dinucleotide features yielded the largest average Gini importance among the set of features considered in their dataset [4, 7].

We also compared the results using the random forest and boosted regression to construct the predictive model. Random forest [21] was implemented in the R package randomForest, whereas the boosted regression based on extensions to AdaBoost [24] and gradient boosted machine [25] was implemented in the R package gbm. The results were shown in Additional file 1: Tables S1, S2 and S3 (randomforest) and Additional file 1: Tables S4, S5 and S6 (gbm). These results were comparable to the results from elastic net.

Related work for predicting CRISPR/Cas9 guide efficiency based on nucleotide properties and melting temperatures includes azimuth [4, 6, 7], which constructed a predictive model based on gradient-boosted regression

trees as described earlier. This method was recommended by [26] for in-vivo (U6) transcribed guides. In contrast, the sgRNA scorer of [27] was a predictive model based on the support vector machine (SVM) algorithm using position dependent mono-nucleotide on 5' flanking (5 bp), spacer target and 3' flanking (NGG + 5 bp) region. We included these two methods for comparison in Table 3 and Fig. 4. In this comparison, each method was trained on different datasets, but the performance was evaluated on the same test dataset generated by an independent research group, i.e., [11] dataset. The statistical significance for pairwise AUC comparisons was based on DeLong's test [28]. Our proposed predictive algorithm achieved higher AUC compared to both azimuth and sgRNA scorer ($p < 0.001$ in both cases). On the other hand, azimuth had better performance than sgRNA scorer ($p < 0.001$). We have also implemented azimuth (based on continuous outcome gbm model) and sgRNA scorer (based on binary outcome SVM model) using the sequence features identified by [6, 7] and [27],

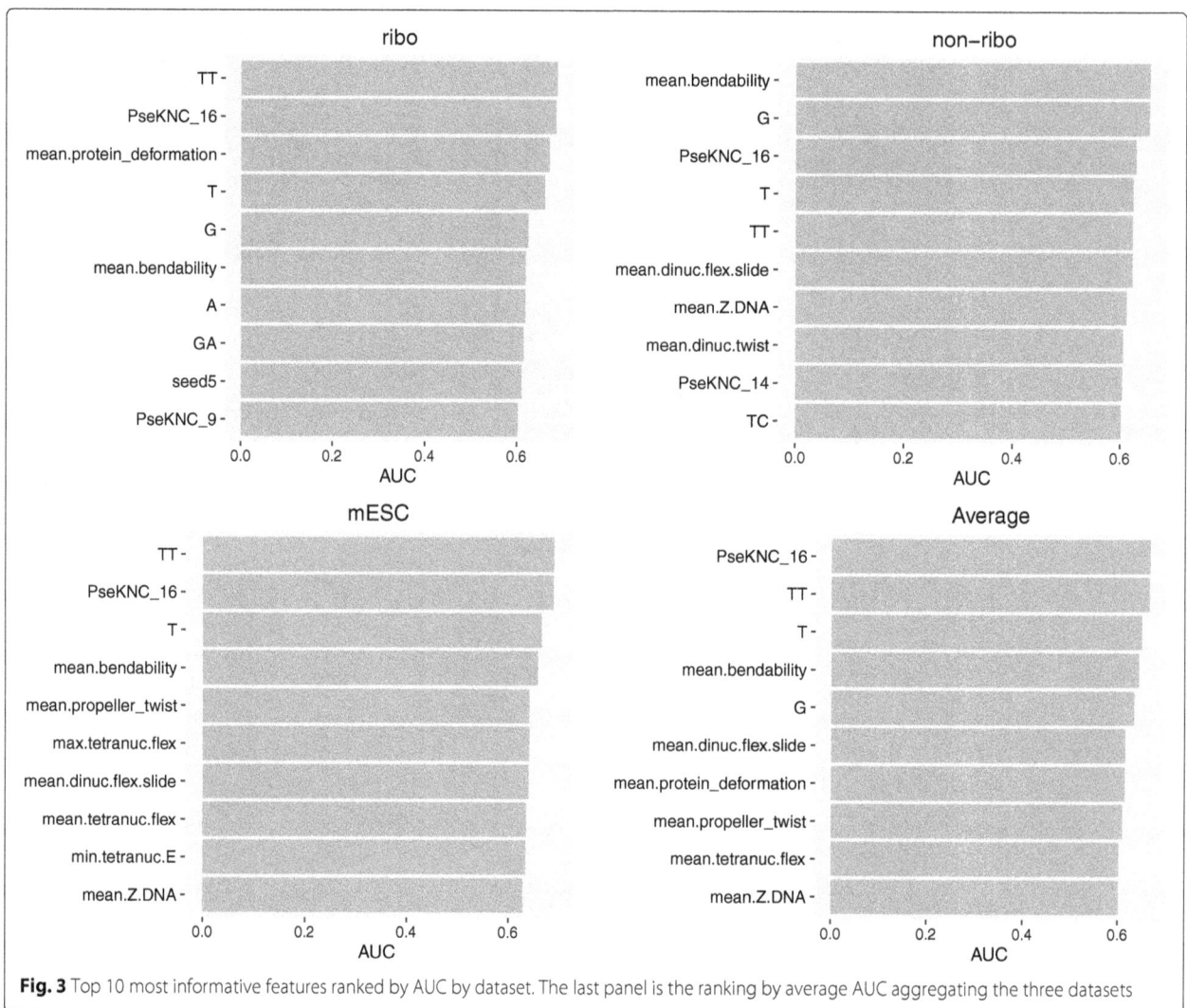

**Fig. 3** Top 10 most informative features ranked by AUC by dataset. The last panel is the ranking by average AUC aggregating the three datasets

respectively on the same training data (i.e., [10] ribosomal and non-ribosomal genes) (Table 3). As expected, the performance of sgRNA scorer was comparable to the model using position dependent mono-nucleotide (Table 3), whereas the performance of azimuth was comparable to the gbm results in Additional file 1: Table S15. Our proposed predictive algorithm achieved higher AUC compared to the refitted sgRNA scorer ($p = 0.048$) and comparable performance to the refitted azimuth ($p > 0.1$).

We also included comparison using a regression model based on (1) the average log2 fold change (12 cell doublings vs initial seeding states) of HL-60 and KBM-7 cell lines for [10] data and (2) the average log2 fold change (mESC vs plasmid control) of replicate 1 and replicate 2 of mouse ESC JM8 cell lines for [11] data. We compared the performance of the sequence properties in prediction in terms of AUC, Pearson correlation coefficient, Spearman rank correlation coefficient and mean squared

error on the test data. The results were presented in Additional file 1: Tables S7, S8 and S9. In addition, similar to the binary outcome model as described above; position-dependent dinucleotide models (PD Dinuc) consistently outperformed other feature classes in predicting sgRNA efficiency and were comparable to results from the combination of feature classes models in all 3 scenarios. Fusi et al. [6] and Doench et al. [7] showed that the regression model outperformed classification model using their dataset [4, 7]. However, we observed that the regression model and the classification model yielded comparable performance in both [10] and [11] datasets. The combination feature prediction model from the regression model (Comb Feature) exhibited larger AUC than both azimuth and sgRNA scorer ($p < 0.001$ for all pairwise AUC comparisons using DeLong's test [28]), but no difference using Spearman rank correlation coefficient for Comb Feature versus azimuth ($p = 0.88$ from Fisher's $Z$-transformation test [29, 30]) as shown in Additional file 1:

**Table 1** AUC, Youden index (*J*), Sensitivity (Se) and Specificity (Sp) from the 3-way cross validation within dataset 1 (ribosomal genes)

| Feature class | AUC | J | Se | Sp |
|---|---|---|---|---|
| PD Mono | 0.826 | 0.535 | 0.855 | 0.680 |
| PD Dinuc | 0.848 | 0.575 | 0.788 | 0.787 |
| Freq | 0.778 | 0.441 | 0.677 | 0.764 |
| Align | 0.613 | 0.188 | 0.746 | 0.442 |
| Thermo | 0.525 | 0.086 | 0.812 | 0.273 |
| Packer | 0.601 | 0.186 | 0.634 | 0.551 |
| PhyChem | 0.722 | 0.380 | 0.711 | 0.669 |
| PseKNC | 0.731 | 0.376 | 0.683 | 0.693 |
| Comb Feature | 0.867 | 0.618 | 0.826 | 0.792 |

Comb Feature: PD Mono+PD Dinuc+*Freq+Thermo+Packer+PhyChem*+PseKNC. We reported the average performance from the 3-way cross validation over 10 iterations of random sampling

**Table 3** AUC, Youden index (*J*), Sensitivity (Se) and Specificity (Sp) from inter-platform comparison (training set: ribosomal and non-ribosomal genes, test set: mESC essential genes)

| Feature class | AUC | J | Se | Sp |
|---|---|---|---|---|
| PD Mono | 0.797 | 0.486 | 0.751 | 0.735 |
| PD Dinuc | 0.832 | 0.544 | 0.792 | 0.752 |
| Freq | 0.751 | 0.382 | 0.716 | 0.667 |
| Align | 0.574 | 0.131 | 0.490 | 0.641 |
| Thermo | 0.641 | 0.261 | 0.817 | 0.444 |
| Packer | 0.667 | 0.241 | 0.514 | 0.726 |
| PhyChem | 0.726 | 0.351 | 0.718 | 0.632 |
| PseKNC | 0.733 | 0.370 | 0.660 | 0.709 |
| Comb Feature | 0.848 | 0.566 | 0.843 | 0.722 |
| azimuth | 0.795 | 0.463 | 0.857 | 0.607 |
| sgRNA Scorer | 0.669 | 0.288 | 0.548 | 0.739 |
| azimuth (retrained) | 0.833 | 0.543 | 0.787 | 0.756 |
| sgRNA Scorer (retrained) | 0.804 | 0.474 | 0.786 | 0.688 |

Comb Feature: PD Mono+PD Dinuc+*Freq+Align+Thermo+Packer+PhyChem*+PseKNC. azimuth and sgRNA Scorer were the results based on the softwares by [7] and [27], respectively developed using different training datasets. azimuth (retrained) and sgRNA Scorer (retrained) were the results obtained by refitting the algorithms on the current training set (ribosomal and non-ribosomal genes)

Table S9. The results from random forest and boosted regression were presented in Additional file 1: Tables S10, S11 and S12 (`randomforest`) and Additional file 1: Tables S13, S14 and S15 (`gbm`). These results were comparable to the results from elastic net.

Following [6, 7], we also included the results from leave-one-gene out prediction framework to obtain a generalization of our prediction model to new genes in Additional file 1 (Section 5 and Tables S19 and S20). The conclusion remained the same, i.e., Comb Feature yielded the largest AUC and PD Dinuc followed closely. Additional results including performance evaluation using 30 bp sequence [6, 7] instead of 40 bp sequence were presented in Additional file 1: Tables S16, S17 and S18. The results indicated that the performance of the prediction models were comparable regardless whether a 40 bp or 30 bp sequence was used.

**Table 2** AUC, Youden index (*J*), Sensitivity (Se) and Specificity (Sp) from intra-platform comparison (training set: ribosomal genes, test set: non-ribosomal genes)

| Feature class | AUC | J | Se | Sp |
|---|---|---|---|---|
| PD Mono | 0.785 | 0.443 | 0.717 | 0.726 |
| PD Dinuc | 0.792 | 0.478 | 0.765 | 0.713 |
| Freq | 0.700 | 0.332 | 0.779 | 0.553 |
| Align | 0.594 | 0.159 | 0.881 | 0.278 |
| Thermo | 0.616 | 0.222 | 0.639 | 0.580 |
| Packer | 0.637 | 0.207 | 0.431 | 0.776 |
| PhyChem | 0.659 | 0.241 | 0.633 | 0.608 |
| PseKNC | 0.647 | 0.243 | 0.694 | 0.549 |
| Comb Feature | 0.806 | 0.492 | 0.851 | 0.641 |

Comb Feature: PD Mono+PD Dinuc+*Thermo + Packer*+PhyChem

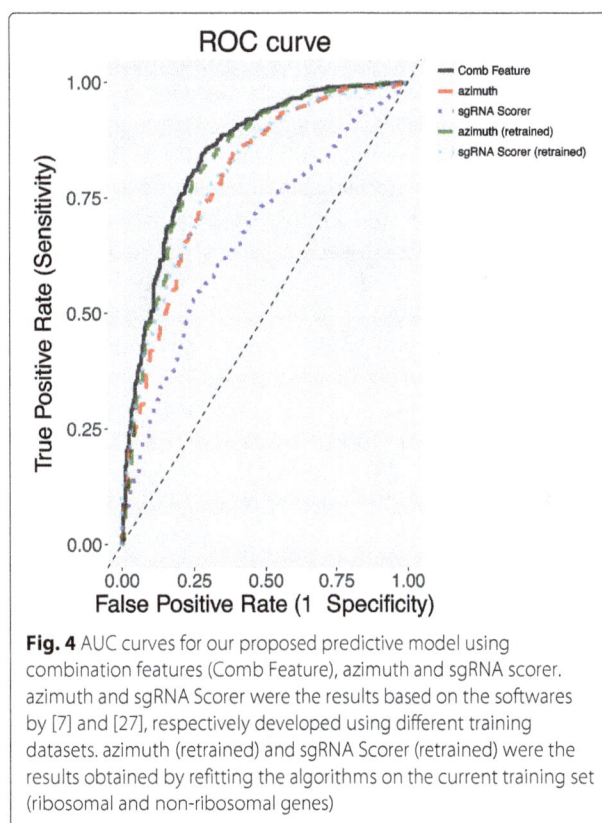

**Fig. 4** AUC curves for our proposed predictive model using combination features (Comb Feature), azimuth and sgRNA scorer. azimuth and sgRNA Scorer were the results based on the softwares by [7] and [27], respectively developed using different training datasets. azimuth (retrained) and sgRNA Scorer (retrained) were the results obtained by refitting the algorithms on the current training set (ribosomal and non-ribosomal genes)

We created an R package `predictSGRNA` implementing the proposed predictive algorithm based on position-dependent dinucleotide model, available at http://www.ams.sunysb.edu/~pfkuan/softwares.html#predictsgrna.

## Conclusions

In this paper, we explored various aspects of nucleotide compositions including position dependent models, secondary structure and thermodynamics to gain better understanding of the nucleotide properties on CRISPR sgRNA design efficiency in a similar way as [6, 7]. Candidate feature ranking in terms of association with sgRNA effiency identified features which characterize the flexibility of the underlying DNA structure. Specifically, we found that the frequency of T and TT dinucleotide exhibited the strongest negative association with sgRNA efficiency. Packer et al. [14] illustrated that TT dinucleotide has the most rigid step and least flexible in terms of the ability to slide and shift, which could explain the decreased efficiency of sgRNA with higher abundance of TT dinucleotides. The results from the different predictive algorithms showed that across datasets, the position dependent mono-nucleotide model [5] achieved good operating characteristics while the prediction algorithm trained on position dependent dinucleotide model offered additional improvement in terms on AUC. The advantage of position dependent dinucleotide model in predicting sgRNA efficiency was also observed in [6, 7].

One factor that may guide improvement of future predictive algorithms is chromatin structure. Chromatin accessibility (packed vs unpacked) has been shown to be the major determinant of genome-wide binding of dCas9-sgRNA in [16]. Examples of epigenetic marks which are implicated in chromatin remodeling and accessibility include DNase I hypersensitive sites, transcription factor binding, DNA methylation and histone modification. Future work will include integrating both the nucleotide composition features and chromatin structures as features in the predictive model to characterize the binding efficiency of sgRNA.

In this study, we used datasets of size 3141 and achieved AUC of > 0.8. Prior efforts to improve the efficiency of RNAi design utilized high-throughput functional testing of the efficacy of different RNAi sequences to generate large (2182) [31] and very large datasets (~250000) [32]. These large datasets in turn were used to develop improved prediction algorithms using machine-learning approaches similar to those used here [33, 34]. It is generally accepted that the first large test set (2182) was very useful for improving RNAi design, there is still uncertainty regarding the utility of examining very large datasets [34]. Part of the unresolved issues are the degree to which different prediction algorithms are dependent upon the vector used for shRNA expression [35] as well as the

sequence context in the genome outside of the immediate target [36]. Therefore, as more CRISPR/Cas9 screens datasets are becoming available, we anticipate that the specificity of sgRNA efficacy prediction can be further improved by considering the vector-dependent level of expression of the sgRNA.

## Additional file

**Additional file 1:** Supplementary Information. The pdf document that contains all supplementary notes, figures and tables. Figures S1-S2 plot the top 10 most informative features ranked by BIC and variable importance scores, respectively. Tables S1-S3 contain the results from `randomforest` in binary outcome model. Tables S4-S6 contain the results from `gbm` in binary outcome model. Tables S7-S9 contain the results from elastic net in continuous outcome model. Tables S10-S12 contain the results from `randomforest` in continuous outcome model. Tables S13-S15 contain the results from `gbm` in continuous outcome model. Tables S16-S18 contain the results comparing 30bp and 40bp sequences. Tables S19-S20 contain the results from leave-one-gene out prediction.

## Abbreviations
AUC: Area under the receiver operating characteristic curve; Align: Optimal pairwise alignment between spacer target and scaffold; BIC: Bayesian information criterion; CRISPR: Clustered regularly interspaced short palindromic repeats; Freq: Frequencies of mono and dinucleotides; LSL: Length of a potential stem-loop; MEF: Minimum energy folding; mESC: Mouse embryonic stem cell; PAM: Protospacer adjacent motif; PD Dinuc: Position-dependent dinucleotide; PD Mono: Position-dependent mono-nucleotide; PhyChem: Physiochemical properties of [17]; PseKNC: Pseudo k-tuple nucleotide composition of [18]; Packer: Secondary structures of [14, 15]; sgRNA: Single guide RNA; Thermo: Thermodynamics and secondary structures of [9]

## Acknowledgements
Not applicable.

## Funding
This work was supported in part by NIH grant U01CA168409 to S.P. The funding body had no role in the design, collection, analysis or interpretation of this study.

## Authors' contributions
PK conceived and designed the study. PK, SH and KL carried out analyses and wrote the software. PK, SP, SH, KL, XZ and BH wrote the paper. PK, SP, XZ and BH critically read the manuscript and contributed to the discussion of the whole work. All authors read and approved of the final version of the manuscript.

## Competing interests
The authors declare that they have no competing interests.

## Author details
[1]Department of Applied Mathematics and Statistics, Stony Brook University, 100 Nicolls Road, 11794 Stony Brook, USA. [2]Department of Pathology, Stony Brook University, 100 Nicolls Road, 11794 Stony Brook, USA. [3]Oncology Business Unit, Pfizer Inc., 558 Eastern Point Rd, 06340 Groton, USA.

## References

1. Barrangou R, Fremaux C, Deveau H, Richards M, Moineau P, Romero D, Horvath P. CRISPR provides acquired resistance against viruses in prokaryotesy. Science. 2007;315(5819):1709–12.
2. Jinek M, Chylinski K, Fonfara I, Hauer M, Doudna J, Charpentier E. A programmable dual-rna-guided dna endonuclease in adaptive bacterial immunity. Science. 2012;337:816–21.
3. Hsu P, Lander E, Zhang F. Development and applications of CRISPR-Cas9 for genome engineering. Cell. 2014;157:1262–78.
4. Doench J, Hartenian E, Graham D, Tothova Z, Hegde M, Smith I, Sullender M, Ebert B, Xavier R, Root D. Rational design of highly active sgrnas for CRISPR-Cas9–mediated gene inactivation. Nat Biotechnol. 2014;32(12):1262–67.
5. Xu H, Xiao T, Chen C, Li W, Meyer C, Wu Q, Wu D, Cong L, Zhang F, Liu J, Brown M, Liu S. Sequence determinants of improved CRISPR sgRNA design. Genome Res. 2015;25:1147–57.
6. Fusi N, Smith I, Doench J, Listgarten J. In silico predictive modeling of CRISPR/Cas9 guide efficiency. bioRxiv. 2015;1:021568.
7. Doench J, Fusi N, Sullender M, Hegde M, Vaimberg E, Donovan K, Smith I, Tothova Z, Wilen C, Orchard R, Virgin H, Listgarten J, Root D. Optimized sgrna design to maximize activity and minimize off-target effects of CRISPR-Cas9. Nat Biotechnol. 2016;34(2):184–91.
8. Wu Z, Irizarry R, Gentleman R, Martinez-Murillo F, Spencer F. A model-based background adjustment for oligonucleotide expression arrays. J Am Stat Assoc. 2004;99(468):909–17.
9. Wei H, Kuan P, Tian S, Yang C, Nie J, Sengupta S, Ruotti V, Jonsdottir G, Keles S, Thomson J, Stewart R. A study of the relationships between oligonucleotide properties and hybridization signal intensities from nimblegen microarray datasets. Nucleic Acids Res. 2008;36(9):2926–38.
10. Wang T, Wei J, Sabatini D, Lander E. Genetic screens in human cells using the CRISPR-Cas9 system. Nature. 2014;343:80–4.
11. Koike-Yusa H, Li Y, Tan E, Mdel CV-H, Yusa K. Genome-wide recessive genetic screening in mammalian cells with a lentiviral CRISPR-guide rna library. Nat Biotechnol. 2014;32(3):267–73.
12. Sambrook J, Fritsch EF, Maniatis T. Molecular Cloning: a laboratory manual. Cold Spring Harbor Laboratory Press; 1989. https://www.cabdirect.org/cabdirect/abstract/19901616061.
13. SantaLucia J. A unified view of polymer, dumbbell, and oligonucleotide DNA nearest-neighbor thermodynamics. Proc Natl Acad Sci. 1998;95(4):1460–5.
14. Packer M, Dauncey M, Hunter C. Sequence-dependent dna structure: Dinucleotide conformational maps. J Mol Biol. 2000;295:71–83.
15. Packer M, Dauncey M, Hunter C. Sequence-dependent dna structure: Tetranucleotide conformational maps. J Mol Biol. 2000;295:85–103.
16. Wu X, Scott D, Kriz A, Chiu A, Hsu P, Dadon D, Cheng A, Trevino A, Konermann S, Chen S, Jaenisch R, Zhang F, Sharp P. Genome-wide binding of the CRISPR endonuclease Cas9 in mammalian cells. Nat Biotechnol. 2014;32(7):670–5.
17. Chen W, Lin H, Feng P, Ding C, Zuo Y, Chou K. iNuc-PhysChem: a sequence-based predictor for identifying nucleosomes via physiochemical properties. PLoS One. 2012;7(10):47843.
18. Guo S, Deng E, Xu L, Ding H, Lin H, Chen W, Chou K. iNuc-PseKNC: a sequence-based predictor for predicting nucleosome positioning in genomes with pseudo k-tuple nucleotide composition. Bioinformatics. 2014;30(11):1522–9.
19. Zou H, Hastie T. Regularization and variable selection via the elastic net. J R Stat Soc Ser B. 2005;67:301–20.
20. Needleman S, Wunsch C. A general method applicable to the search for similarities in the amino acid sequence of two proteins. J Mol Biol. 1970;48(3):443–53.
21. Breiman L. Random forests. J Mach Learn. 2001;45(1):5–32.
22. Diaz-Uriarte R, de Andres SA. Gene selection and classification of microarray data using random forest. BMC Bioinforma. 2006;7(3):10–11861471210573.
23. Nicodemus K, Malley J, Strobl C, Ziegler A. The behaviour of random forest permutation-based variable importance measures under predictor correlation. BMC Bioinforma. 2010;11(110):10–11861471210511110.
24. Freund Y, Schapire R. A short introduction to boosting. J-Jpn Soc Artif Intell. 1999;14(771–780):1612.
25. Friedman JH. Greedy function approximation: a gradient boosting machine. Ann Stat. 2001;1:1189–232.
26. Haeussler M, Schonig K, Eckert H, Eschstruch A, Mianne J, Renaud J, Schneider-Maunoury S, Shkumatava A, Teboul L, Kent J, Joly J, Concordet J. Evaluation of off-target and on-target scoring algorithms and integration into the guide rna selection tool crispor. Genome Biol. 2016;17(1):148.
27. Chari R, Mali P, Moosburner M, Church G. Unraveling crispr-cas9 genome engineering parameters via a library-on-library approach. Nat Methods. 2015;12(9):823–6.
28. DeLong ER, DeLong DM, Clarke-Pearson DL. Comparing the areas under two or more correlated receiver operating characteristic curves: a nonparametric approach. Biometrics. 1988;1:837–45.
29. Fisher RA. On the probable error of a coefficient of correlation deduced from a small sample. Metron. 1921;1:3–2.
30. Myers L, Sirois MJ. Spearman correlation coefficients, differences between. Wiley StatsRef: Statistics Reference Online. 2006.
31. Huesken D, Lange J, Mickanin C, Weiler J, Asselbergs F, Warner J, Meloon B, Engel S, Rosenberg A, Cohen D, Labow M, Reinhardt M, Natt F, Hall J. Design of a genome-wide sirna library using an artificial neural network. Nat Biotechnol. 2005;23(8):995–1001.
32. Fellmann C, Zuber J, McJunkin K, Chang K, Malone C, Dickins R, Xu Q, Hengartner M, Elledge S, Hannon G, Lowe S. Functional identification of optimized rnai triggers using a massively parallel sensor assay. Mol Cel. 2005;41(6):733–46.
33. Vert J, Foveau N, Lajaunie C, Vandenbrouck Y. An accurate and interpretable model for sirna efficacy prediction. BMC Bioinforma. 2006;7(520):10–1186147121057520.
34. Knott S, Maceli A, Erard N, Chang K, Marran K, Zhou X, Gordon A, Demerdash OE, Wagenblast E, Kim S, Fellmann C, Hannon G. A computational algorithm to predict shrna potency. Mol Cel. 2014;56(6):796–807.
35. Watanabe C, Cuellar T, Haley B. Quantitative evaluation of first, second, and third generation hairpin systems reveals the limit of mammalian vector-based rnai. RNA Biol. 2016;13(1):25–33.
36. Liu L, Li Q, Lin H, Zuo Y. The effect of regions flanking target site on sirna potency. Genomics. 2013;102(4):215–22.

# 7

# Evaluation of high-throughput isomiR identification tools: illuminating the early isomiRome of *Tribolium castaneum*

Daniel Amsel[1]*⑩, Andreas Vilcinskas[1,2] and André Billion[1]

## Abstract

**Background:** MicroRNAs carry out post-transcriptional gene regulation in animals by binding to the 3' untranslated regions of mRNAs, causing their degradation or translational repression. MicroRNAs influence many biological functions, and dysregulation can therefore disrupt development or even cause death. High-throughput sequencing and the mining of animal small RNA data has shown that microRNA genes can yield differentially expressed isoforms, known as isomiRs. Such isoforms are particularly relevant during early development, and the extension or truncation of the 5' end can change the profile of mRNA targets compared to the original mature sequence. We used the publicly available small RNA dataset of the model beetle *Tribolium castaneum* to create the first comparative isomiRome of early developmental stages in this species. Standard microRNA analysis software does not specifically account for isomiRs. We therefore carried out the first comparative evaluation of the specialized tools isomiRID, isomiR-SEA and miraligner, which can be downloaded for local use and can handle next generation sequencing data.

**Results:** We compared the performance of isomiRID, isomiR-SEA and miraligner using simulated Illumina HiSeq2000 and MiSeq data to test the impact of technical errors. We also created artificial microRNA isoforms to determine the effect of biological variants on the performance of each algorithm. We found that isomiRID achieved the best true positive rate among the three algorithms, but only accounted for one mutation at a time. In contrast, miraligner reported all variations simultaneously but with 78% sensitivity, yielding isomiRs with 3' or 5' deletions. Finally, isomiR-SEA achieved a sensitivity of 25–33% when the seed region was mutated or partly deleted, but was the only tool that could accommodate more than one mismatch. Using the best tool, we performed a complete isomiRome analysis of the early developmental stages of *T. castaneum*.

**Conclusions:** Our findings will help researchers to select the most suitable isomiR analysis tools for their experiments. We confirmed the dynamic expression of 3' non-template isomiRs and expanded the isomiRome by all known isomiR modifications during the early development of *T. castaneum*.

**Keywords:** Insectomics, microRNA, Small RNA sequencing, isomiRID, isomiR-SEA, Miraligner

## Background

MicroRNAs (miRNAs) are post-transcriptional regulators of gene expression that influence a wide range of biological processes [1]. In insects, the dysregulation of miRNA expression during metamorphosis is often lethal [2–4]. Mature miRNAs are ~22 nucleotides in length and the 3′ end binds to a member of the Argonaute protein family to form an RNA-induced silencing complex (RISC)

[5]. The RISC binds target mRNAs within the 3′ untranslated region (UTR) or in the coding sequence via complementary base pairing with the miRNA seed region (nucleotides 1–8) and in some cases also the compensatory region (nucleotides 13–16) [6]. RISC binding inhibits further processing of the mRNA, thus blocking translation or promoting degradation [1].

The biogenesis of miRNAs can involve the production of isoforms known as isomiRs [7]. These are thought to be produced deliberately as separate products with defined roles in the cell, and do not represent errors of transcription or errors of sequencing [8]. The isomiRs may be extended or truncated

* Correspondence: Daniel.Amsel@ime.fraunhofer.de
[1]Fraunhofer Institute for Molecular Biology and Applied Ecology, Department of Bioresources, Winchester Str. 2, 35394 Giessen, Germany
Full list of author information is available at the end of the article

at either end compared to the mature miRNA, presumably due to imperfect cleavage by Drosha or Dicer [9]. Recent studies indicate that 5′ isomiRs undergo a seed region shift which changes the set of target mRNAs compared to the original miRNA [10]. The set of target mRNAs can also be changed by nucleotide editing [11, 12]. Mature miRNAs may also acquire non-templated polynucleotide 3′ tails generated by nucleotidyltransferases [13]. This phenomenon has been observed during early insect development as part of maternal transcriptome regulation [14, 15].

The results described above show that miRNAs and isomiRs play important roles during animal development, especially insect morphogenesis. To gain more insight into the prevalence of isomiRs in insects we screened the publicly available small RNA dataset of the model beetle *Tribolium castaneum* originally focusing exclusively on 3′ non-templated isomiRs in the early development stages [15]. The data had already undergone a conservative form of isomiR investigation by iteratively truncating the non-templated 3′ ends until a certain minimal length was reached or the sequence perfectly matched a known miRNA. We investigated the performance of tools for isomiR identification that account for more than non-templated 3′ tails. Several such tools have been developed but no comparative benchmarks are available. We selected a set of three candidate tools that are suitable for the analysis of high-throughput sequencing data and compared their performance to identify the best software. Using a simulated test set of Illumina reads and a set of artificial isomiRs, we investigated the influence of technical errors and biological variations on each type of software and determined the sensitivity and specificity for each case. From these values, we calculated a final weighted performance score for each tool. Taken individually, the two cases also provide detail information on the eventual need of post system error correction, considering the system error test case and possible detection leaks of isomiR types, uncovered by the biological variant test set.

## Methods
### IsomiR analysis software
Seven isomiR mining and alignment tools are currently available as non-proprietary software (Table 1). Three of them are command line tools that can be downloaded and integrated into high-throughput pipelines, and these are described in more detail below. We used these three methods for a comparative benchmark of their individual performance on simulated reads. If adjustable, we used the default settings in each tool without read abundance cutoffs. We wanted each tool to utilize its entire search space and therefore did not set the parameters to a common minimum in the case of mismatches, additions and deletions.

### isomiR-SEA
The C++ program isomiR-SEA focuses on the seed region of miRNAs. It is a standalone executable file without dependencies and can be run with parameters in the command line. It requires the mature miRNA file from miRBase and the sequence reads. The reads must be collapsed and reformatted with the unique read and its abundance in one line. The algorithm extracts the seed regions from the mature miRNAs and groups them together. At first, the reads are screened for seed regions. When found, the seed region is extended without gaps in both directions and the correct position of the seed block is checked. The algorithm continues the extension towards the 3′ end and allows a second mismatch if the distance between the two mismatches falls within a user-defined threshold. The alignment is then extended further until either the third mismatch or the end of the read is encountered. Then the scores for each aligned read are computed. The output files are grouped into unique mapping reads, ambiguous reads that map more than once, and ambiguous selected reads that also map to various miRNAs but can be assigned to a unique one due to an internal scoring function (Table 2). There are also "unique", "ambiguous" and "ambiguous selected" output files, referring to the miRNA instead of the read.

**Table 1** List of non-proprietary isomiR alignment programs

| Program | Usage | Alignment method | Publisher |
|---|---|---|---|
| isomiR-SEA 1.60 | Command line<br>isomiR-SEA_1_6 -s tca -l 10 -b 4 -i < in_path ><br>−p < out_path > −ss 6 -h 11 -m < mature_mir_file > −t < countfile> | User-defined seed size (default 6) | Urgese et al. [21] |
| isomiRID 0.53 | Command line<br>standard config file | bowtie1 | de Oliveira et al. [22] |
| miraligner<br>3. Feb 2016 | Command line<br>java -jar miraligner.jar -sub 1 -trim 3 -add 3 -s tca -freq | 8 nt seed | Pantano et al. [23] |
| IsomiRage | Desktop GUI | bowtie1 | Muller et al. [24] |
| DeAnnIso | Webapp | bowtie1 and BLAST | Zhang et al. [25] |
| isomiRex | Webapp | bowtie1 | Sablok et al. [26] |
| miR-isomiRExp | Webapp – offline | bowtie1 | Guo et al. [27] |

The three command line tools were used for our comparative evaluation. The others were discarded because they were incompatible with local high-throughput pipelines

**Table 2** Result files generated by isomiR-SEA

| Unique | Tag_unique |
|---|---|
| Unique_ambigue_selected | |
| Ambigue | Tag_ambigue |
| Ambigue_ambigue_selected | Tag_ambigue_selected |

The *tag* files focus on the read, whereas the others report the variants of the miRNA

### *isomiRID*

The Python 2.7 script isomiRID uses bowtie [16] to map small RNA sequencing reads against reference precursor miRNAs. The script uses a configuration file in which the user can specify the paths of the executables, the data and the parameters. In the first round, perfect matches against the precursors are identified. An optional filtering step of the unaligned reads against the corresponding transcriptome or genome can be performed to filter reads not from miRNAs. In the second step, reads with one mismatch are taken into account. Iterative trimming of the 5′ and 3′ ends is used to seek potential non-templated miRNA isoforms. The findings are filtered according to user-defined abundance cutoffs and the results are concatenated into output files, allowing for reads with more than one mapping location. The output is a tab separated file in which every mapped read is aligned under the assigned precursor sequence together with the identified type of isoform and the abundance of the read.

### *Miraligner*

The Java tool miraligner, originally from the SeqBuster package but now independent, is a single jar file without dependencies. It uses a collapsed read file and the miRNA hairpin FASTA file from miRBase [17] together with the hairpin secondary structure file. The reads are mapped to the hairpin sequences via seeds of eight nucleotides, allowing one mismatch within the sequence. It allows up to three non-templated nucleotide additions at the 3′ end, as well as up to three nucleotides that differ from the mature 3′ or 5′ ends. This allows a slight shift of the precursor compared to the annotated position in the hairpin secondary structure file from miRBase. We used the default settings with a maximum substitution of one and a trimming/adding of three. The output is a tab separated file. It shows a result for each mutation type, the read sequence together with the number of its assignments, as well as the names of the miRNA.

### Technical error simulation

We evaluated the effect of Illumina sequencing errors on the accuracy of isomiR identification by each tool. The small RNA sequencing data were simulated using ART [18] (version Mount Rainier 2016–06-05) with the Illumina HiSeq2000 and MiSeq-v1 sequencing system in single-strand mode: art_illumina -c 1000 -ss [HS20|MSv1] -i < pattern_file_with_miR_length_X > –l < miR_-length_X > –o < output>. We grouped all miRNAs with the same length into one file and ran the command for each file separately. Afterwards, the files were merged into one. These sequencing systems are widely used for small RNA sequencing and mirror the most recently analyzed biological data. To ensure traceability, the simulated sequences must be uniquely assignable to their source. In case of isomiRID and miraligner, this can be achieved by the sequence header. The results of isomiR-SEA lack this header and a traceability can only be provided by sequence identity. Therefore, we had to ensure a uniqueness of miRNAs and their reads. We used the 430 *T. castaneum* mature miRNAs from miRBase v21 and merged identical sequences. This new set of 422 sequences was then used as the pattern for the two simulations, with a coverage of 1000 reads per sequence. Due to the nature of the simulation program, about half of the 422,000 reads were sequenced as a reverse complement and were therefore omitted from further analysis. The remaining reads, 210,753 for HiSeq2000 and 210,961 for MiSeq-v1, were then filtered for redundancy. This resulted in 13,850 unique reads for HiSeq2000 and 5964 unique reads for MiSeq-v1. This ensured a coverage of 14–32 read variants per original miRNA and therefore a broad variety of technical errors. The correct assignment of erroneous reads to its source was treated as true positive, because the tools cannot distinguish between error and mutation. An additional analysis after the identification step might be of use, depending on the investigation.

### Biological variation simulation

In order to evaluate the isomiR programs comprehensively using biological data, we created custom sequences based on the mature *T. castaneum* miRNAs from miRBase v21. This mirrored seven different types of isoforms (Fig. 1). Both the 5′ and 3′ template isoforms were divided into truncated and extended variants. For the truncated variants, we created three different 5′ and three different 3′ isomiRs per mature microRNA, by iteratively trimming one nucleotide from the 5′ or from the 3′ end respectively. For the three 5′ and three 3′ extended variants, we added one nucleotide to the particular end of the mature miRNA, using the precursor miRNA as the template, until a maximum of three additions was reached. The 12 3′ non-templated isoforms per mature miRNA were created by adding one nucleotide of the same type to the mature miRNA, until a total of three nucleotides were added. We divided the single nucleotide polymorphism (SNP) isoforms into two distinct classes: the seed-SNPs and the tail-SNPs. We replaced each nucleotide from position 1 to 8 with the remaining three nucleotides for the seed-SNPs dataset and from position 9 to the end for the tail-SNPs dataset,

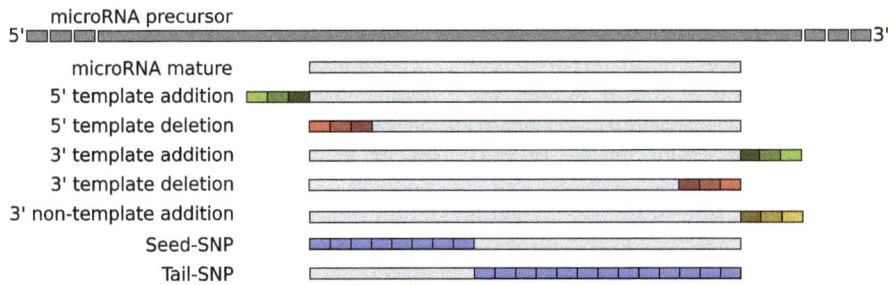

**Fig. 1** The seven types of isomiR custom mutations. The green boxes represent nucleotide additions. The red boxes represent nucleotide deletions. The yellow boxes represent non-template additions. The blue boxes show the positions of SNPs

resulting in three SNP isoforms per miRNA nucleotide position. This allowed us to distinguish the performance of seed-based search algorithms between seed and tail SNPs. We again kept the created reads non-redundant to ensure the traceability of the mapped reads by sequence identity. Our resulting test set finally mirrored each possible variation and therefore provided a general unbiased condition.

### Performance evaluation

We evaluated each algorithm using the simulated technical and biological *T. castaneum* reads. The results were classified as true positives (TP), false positives (FP) and false negatives (FN). True negatives (TN) were excluded because they were not needed for further calculations. Correctly assigned reads were treated as true positives. A wrongly assigned read was treated as false positive and a missing assignment to the correct miRNA was treated as false negative. We also calculated the sensitivity (TP/(TP + FN)) and the specificity (TP/(TP + FP)) of each isomiR software. Three possible approaches can be used to evaluate small RNA sequencing reads with more than one mapping location. One is to ignore multi-mapping reads completely and focus on distinct results. The second option is to group the miRNAs with the same read together. The third is to distribute the abundance of the read among the number of mapped miRNAs [19]. We decided to use the third approach because the other options would modify the isomiRome.

### *Tribolium castaneum* small RNA sequencing data

Recent studies have indicated the presence of abundant non-templated 3′ isomiRs during the early development stages of *T. castaneum* and *Drosophila melanogaster* [14, 15]. We used the publicly available *T. castaneum* small RNA sequencing data from the GSE63770 project (Table 3) for our analysis. Those datasets monitor the development of *T. castaneum* from the egg (including the switch from maternal to zygotic transcription after 5 h) until hatching (144 h) [15].

### Adapter trimming and quality filter

The *T. castaneum* small RNA sequencing data was trimmed with cutadapt [20] v1.8.3, using -m 17 as the minimum read length, –M 30 as the maximum read length and –trim-n, to trim potential N characters at the ends of the reads. We excluded reads with at least one N character in their sequence.

### Results

We selected three high-throughput isomiR analysis tools suitable for command line use and investigated the effects of biological variation and sequencing-derived errors on the results produced by each tool (Additional file 1: Figure S1). The technical test sets were created with ART, using a copy rate of 1000 reads per miRNA. We additionally created biological test sets geared to known miRNA isoforms and again reduced them to a non-redundant set, allowing us to measure the effects of biological variation on the results produced by each tool. We finally generated scores for each tool and selected the appropriate software for the analysis of the *T. castaneum* isomiRome.

**Table 3** List of publicly available *T. castaneum* small RNA datasets representing different developmental stages

| ID | Sample | Transcription |
|---|---|---|
| GSM1556886 | Oocyte small RNA replicate 1 | Maternal |
| GSM1556887 | Oocyte small RNA replicate 2 | Maternal |
| GSM1556888 | Embryo small RNA 0–5 h replicate 1 | Maternal |
| GSM1556889 | Embryo small RNA 0–5 h replicate 2 | Maternal |
| GSM1556890 | Embryo small RNA 8–16 h | Zygotic |
| GSM1556891 | Embryo small RNA 16–20 h | Zygotic |
| GSM1556892 | Embryo small RNA 20–24 h | Zygotic |
| GSM1556893 | Embryo small RNA 24–34 h | Zygotic |
| GSM1556894 | Embryo small RNA 34–48 h | Zygotic |
| GSM1556895 | Embryo small RNA 48–144 h | Zygotic |

After ~5 h, the maternal transcription phase ends and zygotic transcription commences [15]

## Effect of technical errors on isomiR analysis

We created simulated HiSeq2000 and MiSeq-v1 reads based on mature miRNA templates from miRBase v21 with ART [18]. The multiple isomiR-SEA result files were divided into two distinct evaluations. We distinguished between the total results reported by isomiR-SEA (unique - reads that mapped only once and ambigue - reads that mapped more than once) on one hand and the selected results, already filtered by isomiR-SEA (unique - reads that mapped only once and ambigue_selected - reads that mapped more than once, but were disambiguated through isomiR-SEA internal scorings) on the other. The number of isomiR-SEA false positives was lower in the selected set compared to the total results, falling by more than 15% for MiSeq-v1 and more than 18% for HiSeq2000 (Fig. 2a). However, the false negative rate increased by nearly 7% for both HiSeq2000 and MiSeq-v1 in the selected set. This is also reflected in the increased specificity (+23.15% for HiSeq2000 and +21.97% for MiSeq-v1) and weaker sensitivity (−1.95% for HiSeq2000 and −1.37% for MiSeq-v1) (Fig. 2b). The results produced by miraligner and IsomiRID were almost identical for this benchmark: miraligner achieved ~1.60% and ~0.78% more true positives than IsomiRID for the HiSeq2000 and MiSeq-v1 data, respectively, ~0.50% fewer false positives for both HiSeq2000 and MiSeq-v1, as well as 1.13% and 0.21% fewer false negatives for HiSeq2000 and MiSeq-v1, respectively.

## Effect of biological variation on isomiR analysis

We tested the three tools for their ability to process artificially mutated miRNAs representing isomiR variations. Although isomiRID achieved a true positive rate of at least 98.4%, the false positive rate was 0.7–1.6% for every variant, except 3′ additions with 0.08% false positives (Fig. 3a). In contrast, miraligner achieved a true positive rate of >99.5% and a false negative rate of ≤0.5% for all variants except 3′ and 5′ deletions, where the false negative rate was ~21% (Fig. 3b). We again distinguished between total and selected isomiR-SEA results, attempting to eliminate multi-mapping reads. For the total results (Fig. 3c) we observed for nearly every type of mutation a false positive rate of ~25%, with the exception of seed-SNPs and 5′ deletions where the false positive rates ranged from ~7% to ~10%. We also observed false negative rates of 60% and 70% in these two variants. For the selected results (Fig. 3d) the false positive rate ranged from 0% for 3′ non-templated additions to 1.5% for 5′ deletions. The false negative rates for 3′ and 5′ template additions, 3′-non-templated additions and variants covering mutations outside the seed region were all approximately 2%. However, the false negative rate increased to 7.8% for 3′ truncations, 66% for 5′ truncations and 77% for seed-SNPs.

The sensitivity of isomiRID was >99% for every variant and 100% for truncations and extensions at either end of the sequence (Fig. 4a). In contrast, the sensitivity of miraligner for deletion variants was 79% and ~99% for every

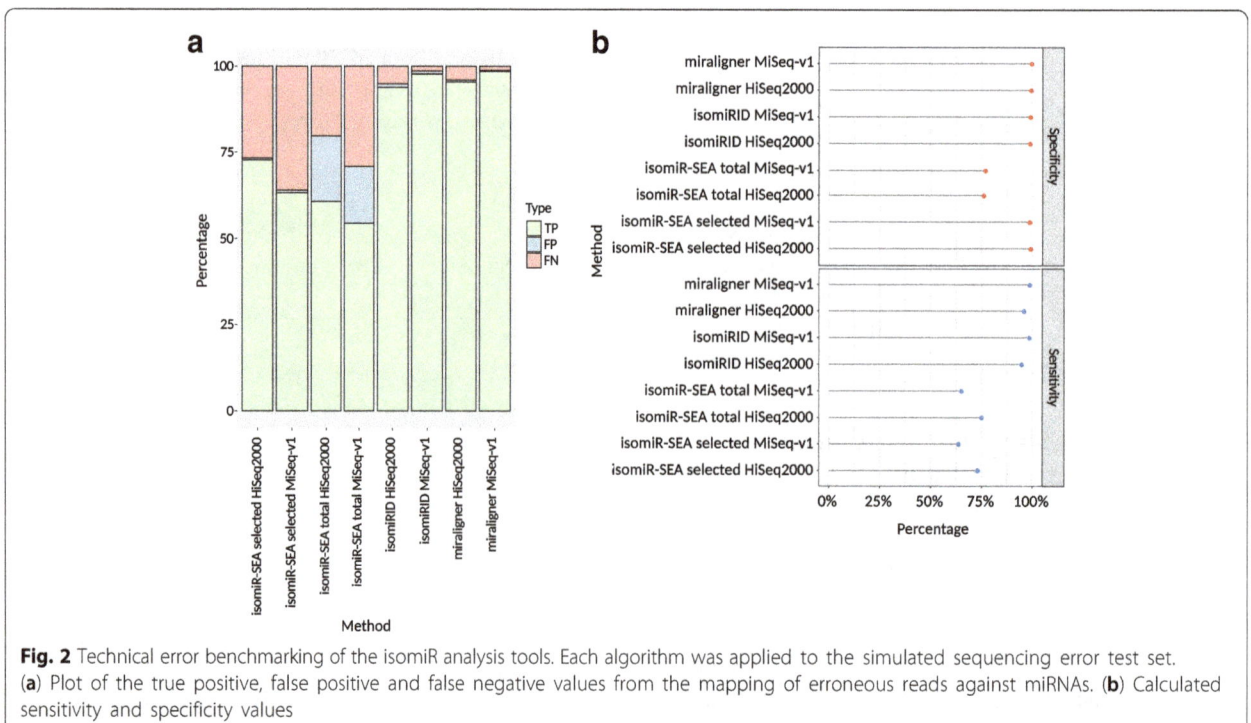

**Fig. 2** Technical error benchmarking of the isomiR analysis tools. Each algorithm was applied to the simulated sequencing error test set. (**a**) Plot of the true positive, false positive and false negative values from the mapping of erroneous reads against miRNAs. (**b**) Calculated sensitivity and specificity values

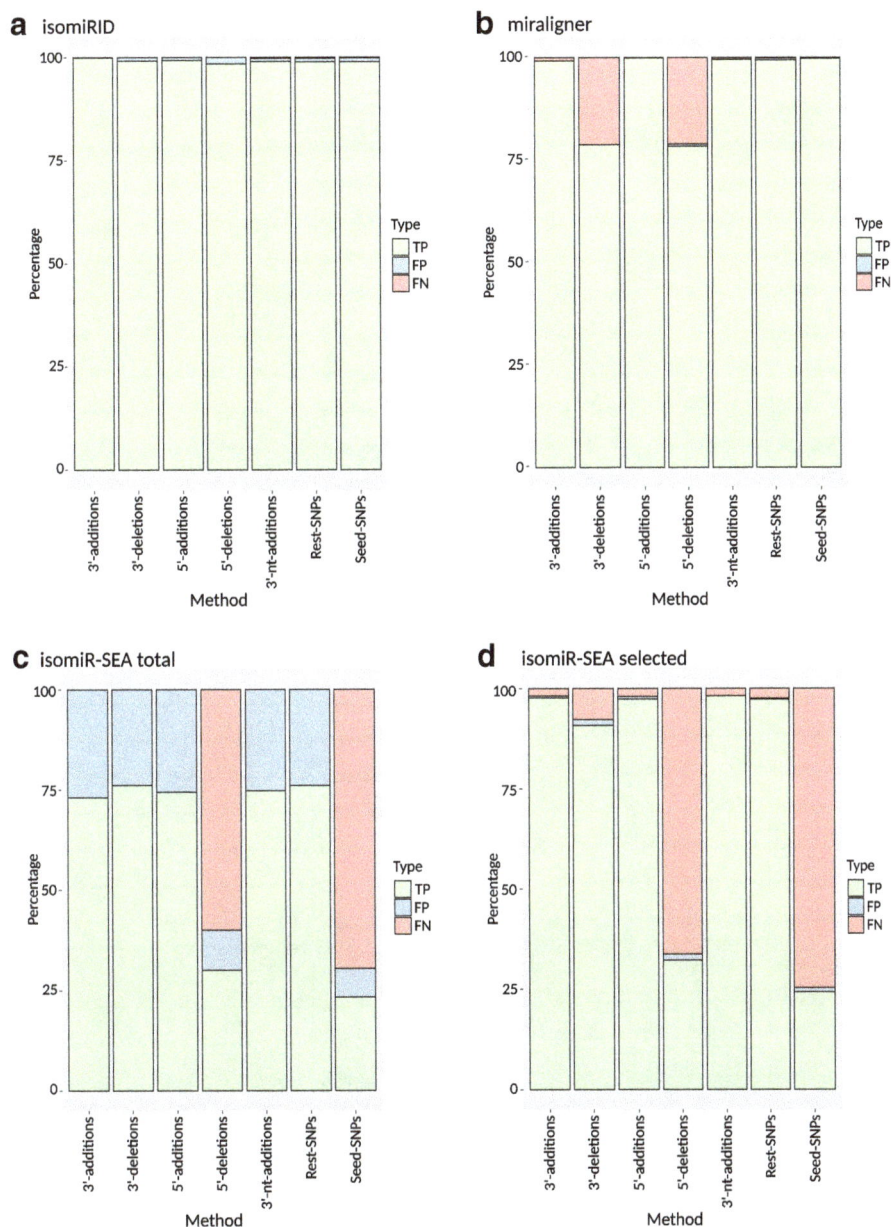

**Fig. 3** True positive, false positive and false negative results generated by isomiR analysis tools. The algorithms isomiRID (**a**), miraligner (**b**), isomiR-SEA total (**c**) and isomiR-SEA selected (**d**), were applied to the simulated biological variation test set

other variant (Fig. 4a). When considering the total results, the sensitivity of isomiR-SEA was 100% for every variant except seed-SNPs and 5′ deletions, where the sensitivity fell to 33% and 25%, respectively (Fig. 4c). When considering the filtered results, the sensitivity of isomiR-SEA ranged from 92% to 98% for most variants but again showed a lower sensitivity for seed-SNPs and 5′ deletions, with values almost identical to the total results (Fig. 4d). The specificity of isomiRID ranged from 98% for 5′ truncations to 99% for 3′ templated additions (Fig. 4a). The specificity of miraligner was 100% for templated 3′ and 5′ additions and 3′ truncations, and 99% for 5′ truncations

(Fig. 4b). The specificity of isomiR-SEA (total results) was 73–76% (Fig. 4c) whereas the selected results improved the specificity to 95–98% (Fig. 4d).

In order to exclude a possible influence of the read length to the result, we tested the effect of artificial read lengths on the method detection efficiency (Additional file 1: Figures S2 and S3). IsomiRID had a weak anti-correlation between read length and false positive rate of –0.36. Its highest false negative rate was at the length of 18 nt. Miraligner had a moderate anti-correlation between read length and false negative rate of –0.53. This was mainly caused by read lengths between 15 and 17 nt. The two

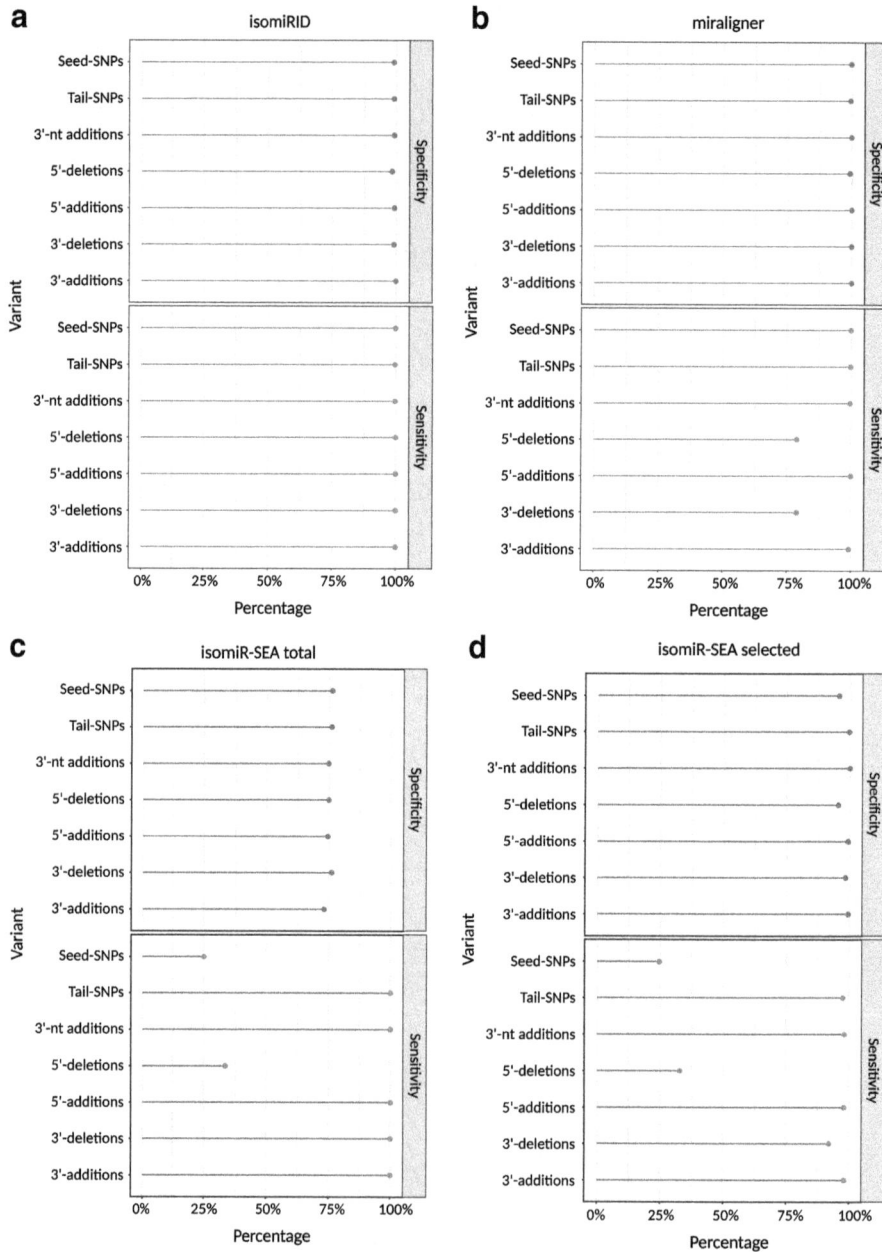

**Fig. 4** Sensitivity and specificity of the isomiR analysis tools isomiRID (**a**), miraligner (**b**), isomiR-SEA total (**c**) and isomiR-SEA selected (**d**). The values were calculated using the TP, FP and FN metrics from the analysis of the biological variation test set

variations of isomiR-SEA performed equally, concerning the correlations. They show an anti-correlating value of −0.24 and −0.22 for false negatives, caused by read lengths between 18 and 26 nt.

**Overall performance scores for isomiR analysis software**

Each of the analysis tools was scored according to its performance when handling technical errors and biological variations as described above, resulting in the overall ranking presented in Fig. 5. We calculated the f-scores for each tool and weighted them depending on their impact on real

data. The highest score of 12.90 points was achieved by isomiRID, followed by miraligner with 12.59 points and isomiR-SEA with 9.13 and 10.25 points for the total and selected data, respectively.

We calculated the f-scores for each testing variant. Then each f-score was weighted regarding to its impact on the targeting mechanism of the miRNA isoform. We assigned a weighting of 1 to the templated 3′ additions and truncations as well as the tail-SNPs because these do not affect the seed region and therefore the range of mRNA targets is unchanged. However, variants that affect the seed region

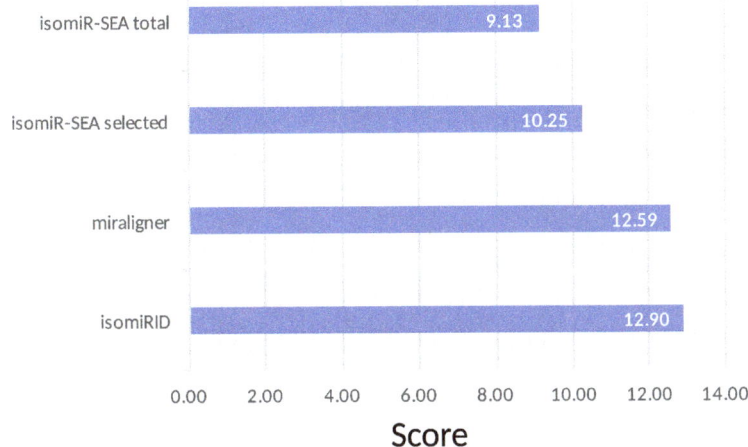

**Fig. 5** Overall ranking of the isomiR analysis tools. The points were calculated by weighting true positives, false positives and false negatives together with the impact on the seed region

such as seed-SNPs and 5′ additions and truncations were weighted with a multiplier of 2, because changes in this region can modify the mRNA target range and are more biologically significant. We also assigned a multiplier of 2 to the 3′ non-templated additions because of their impact during early development. Finally, every score was summed up for each tool and set as final score for the evaluation.

In selecting a method for analysis of the *T. castaneum* isomiRome, we also considered aspects of general usability. For example, isomiRID uses precursor sequences and calculates a dot alignment for every matching read, but the number of dots is sometimes incorrect. This results in a visually shifted mature sequence alignment. Furthermore, isomiRID also reports only one mutation at a time and does not mark 5p and 3p miRNAs. In contrast, miraligner can report all isoforms simultaneously but replaces reads with the same name. We also observed that the precursors tca-miR-3811c-1 and tca-miR-3851a-1 were not reported in the test output even though they were provided in the input file, whereas the precursors tca-miR-3811c-2 and tca-miR-3851a-2 were present. We compared each pair and found that those precursors share the same mature sequence.

We nevertheless selected miraligner for the further analysis of the *T. castaneum* isomiRome, using the same settings as in the test cases. It scored 0.31 fewer points than isomiRID but 2.34 more than isomiR-SEA using the filtered data. It reported all variations for each read and generated fewer false positives than isomiRID, which reports only one mutation at a time and therefore cannot be used for comprehensive isomiRome profiling. Precursor overwriting was ignored because we focused on the mature sequences.

## The isomiRome of *Tribolium castaneum*

We calculated the number of reads that matched each type of isomiR variant in counts per million (CPM). The multi-mapping reads were normalized by the number of assigned microRNAs to avoid overrepresentation (Fig. 6). We observed an increase in the number of 3′ non-templated additions (add) during the maternal transcription phase (oocyte replicates 1 and 2, embryo 0–5 h replicates 1 and 2) which agreed with previous studies in *T. castaneum* [15] and *D. melanogaster* [14]. We also observed an initial increase in the number of templated 3′ additions (t3) peaking during the embryonic phase 16–20 h and declining thereafter. The mature sequences showed an opposing expression profile, with the lowest point at 16–20 h and an increase thereafter. The final phase had a higher CPM than the templated 3′ additions. The 5′ templated additions (t5) were present at constantly low levels with the exception of the 34–48 h phase. The SNP isoforms (mism) ranked second highest in expression value in the oocytes, which is even higher than previously reported for non-templated 3′ additions [15]. The expression of SNP isoforms dropped to one of the lowest values of all variants in the post-oocyte phases although there was a second significant peak during the 20–24 h phase before falling to minimal levels thereafter.

We next scanned for all non-templated nucleotide additions at the 3′ end. We confirmed that isomiRs with poly-adenylate tails are strongly expressed in the oocyte and during the first embryonic stage; then expression weakens at the beginning of the first zygotic transcription phase (8 h). This reproduced the findings of the original study using the same dataset [15] (Additional file 1: Figure S4). Templated 3′ additions and deletions occurred very frequently in these datasets, although the expression level

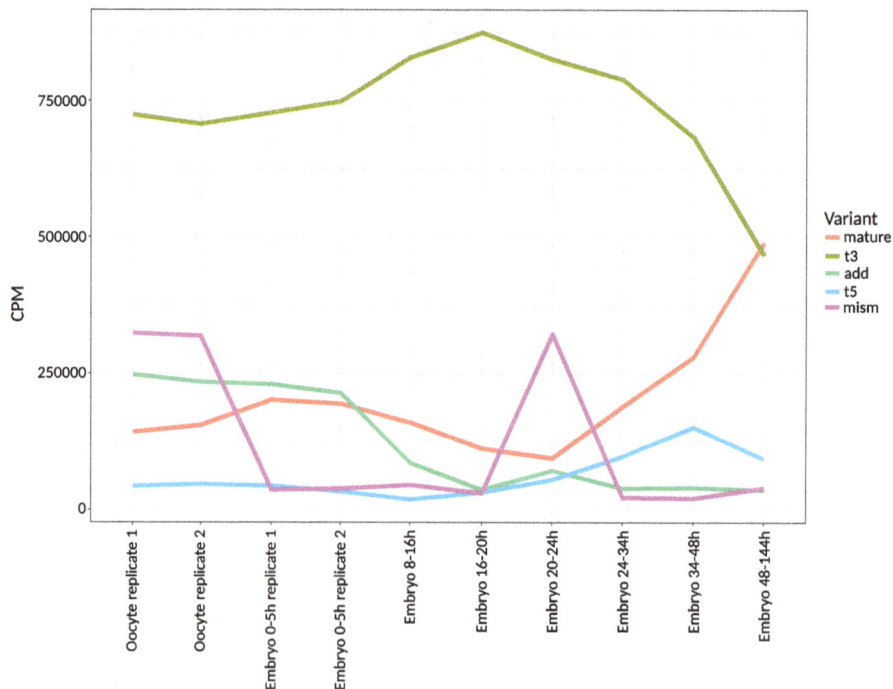

**Fig. 6** Counts per million reads per condition, normalized by the number of multi-mapping reads. This shows the 3' non-templated additions (add), the mature sequence (mature), the mismatches (mism), templated 3' additions and deletions (t3) and templated 5' additions and deletions (t5)

dropped below that of the unmodified mature microRNA in the final phase (48–144 h). In most cases, the 3' end was shortened by two or three nucleotides compared to the original miRNA, but we also observed isomiRs that were elongated by two or three nucleotides during the 8–16 h and 24–34 h phases (Fig. 7). We observed a steady low level of 5' isomiR expression with the exception of the penultimate and antepenultimate phases, where a single nucleotide 5' extension was prevalent.

During embryonic development, we observed a significant increase in the abundance of single-nucleotide mismatches during the 20–24 h stage, with a rapid decline immediately afterwards. We therefore characterized this phase in more detail, revealing frequent A-to-C mutations especially at position 5–7 in the microRNA seed region, and at positions 10 and 17–21 (Fig. 8). The latter segment lies directly behind the 3' compensatory region (nucleotides 13–16) of the microRNA [6]. In addition, we observed an increase in T-to-C, T-to-A and G-to-T transitions before the compensatory region, spanning positions 10–13.

We observed an increase in the expression of mature microRNAs during the last four phases, including tca-miR-10-5p (Additional file 1: Figure S5). Furthermore, we observed an abrupt increase in the expression of tca-miR-376-3p, tca-bantam-3p and tca-miR-281-5p (among others) between the 34-48 h and 48-144 h phases. We observed an increase in the number of different mature miRNAs accumulating during each successive phase.

## Discussion

We evaluated the performance of three algorithms for the identification of isomiRs in small RNA sequencing data (isomiR-SEA, isomiRID and miraligner) and used the most suitable of the three (miraligner) to generate an overview of the isomiRome of the red flour beetle *Tribolium castaneum*. All three tools found it difficult to process technical errors, probably because we clustered the identical reads. This step reduced the number of correct reads to single copies, shrinking the majority of reads. All the unique mutations and mutations with few copies were also reduced to a non-redundant set. Therefore, only one copy of each original miRNA remained in the data along with multiple variants with one or more sequencing errors. This may have increased the number of false negatives because the missed sequences presumably lay outside the scope of the algorithms due to the higher error rate as expected from isomiRs. False negatives were therefore weighted as neutral for the scoring process. Although a sequencing error can mislead the results of the study, we considered is a benefit, when the tools were able to assign it. Later analysis may then filter out possible erroneous reads to improve the investigation results.

The evaluation of biological variants characterized the partially strong effects of sequence variations on the accuracy of isomiR identification. Both isomiRID and miraligner performed well, although miraligner was unable to identify all isomiRs with 3' and 5' deletions probably reflecting the

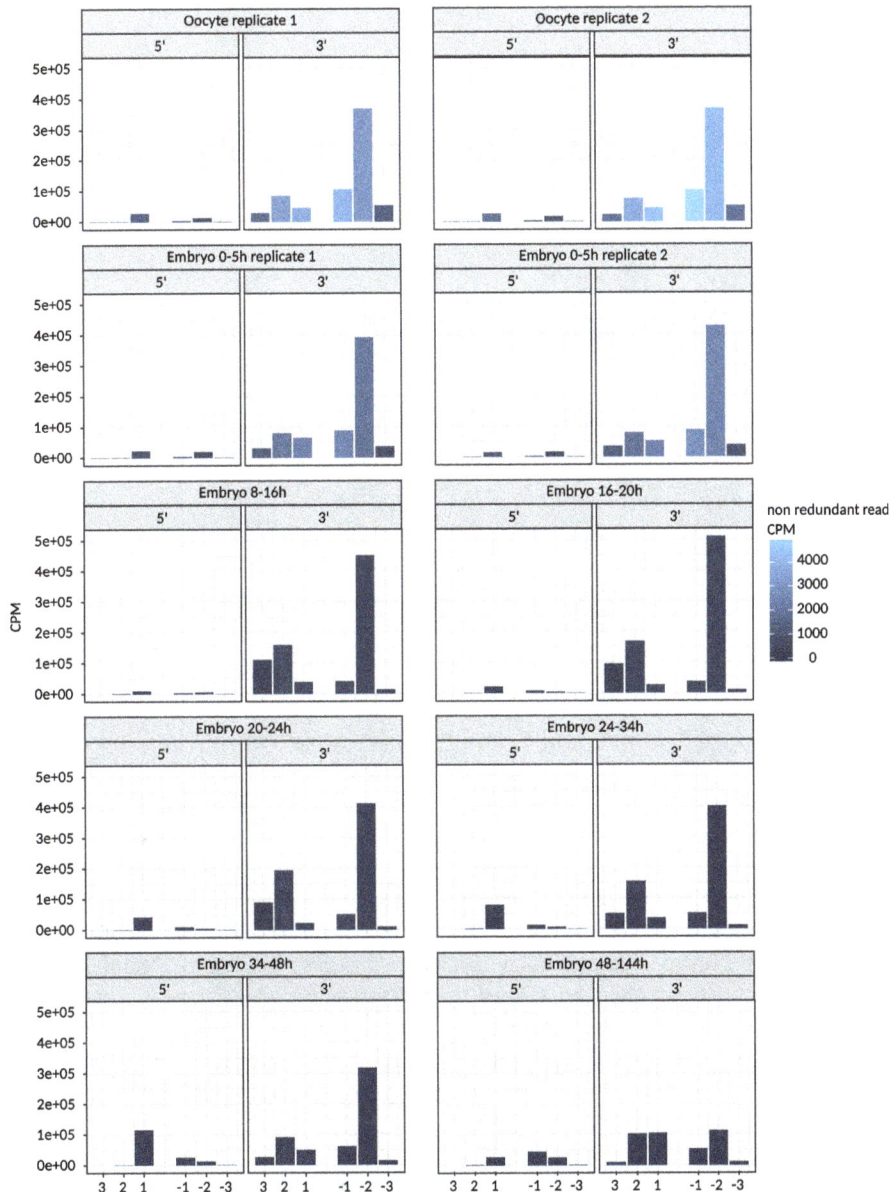

**Fig. 7** Templated 3' and 5' additions and deletions. The x-axis shows truncation in −1 steps and elongation in +1 steps and the y-axis shows the counts per million reads. The bar color displays the counts per million values of non-redundant reads supporting each miRNA variant

seed-based search method. In contrast, isomiR-SEA performed poorly when mapping 5′ deletions and seed-mutated isoforms, but this was expected because the algorithm uses seed-based clustering for every miRNA and builds its entire analysis on these sets.

Each of the algorithms demonstrated particular strengths for specific applications. Although isomiR-SEA achieved the weakest overall evaluation score, it is likely to be the most promising tool to screen for diverse and highly mutated isomiRs because it is the only software that supports more than one mismatch. It is also the only tool that uses just the read sequences and a single sequence file with all already known mature microRNAs.

This makes it ideal for non-model organisms, especially compared to isomiRID, which requires a genome file in addition to the files from miRBase. We assume that the visual output of isomiRID is designed for the manual evaluation of a small set of microRNAs. Because it is based on the bowtie1 aligner, it can only report one type of isoform per read and will not recognize combined mutations such as a mismatch combined with a templated 3′ addition. This can be checked visually but such combinations are not easily parsed by a pipeline. Finally, miraligner offered the best features of the other algorithms. It had a structured output comparable to isomiR-SEA, and scored nearly as much as isomiRID in terms of performance. It

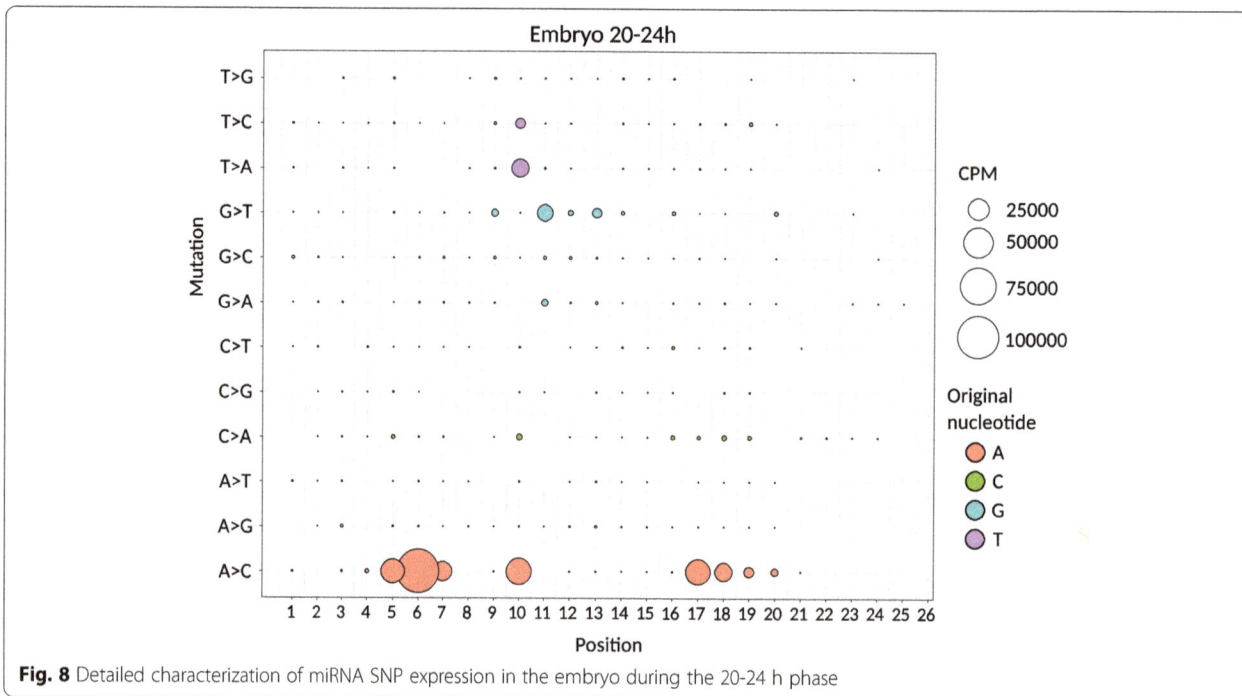

**Fig. 8** Detailed characterization of miRNA SNP expression in the embryo during the 20-24 h phase

also makes use of miRBase files, but does not need a genome reference like isomiRID.

Having evaluated and compared all three algorithms, we then used miraligner to characterize the *T. castaneum* isomiRome during embryonic development. Our analysis revealed that the isomiRome is more diverse and dynamic than previously reported. We were able to reproduce earlier reports that polyadenylated miRNAs are expressed in the oocyte and during the first embryonic phase. We found that the number of isomiRs with 5' extensions increases during the 24–34 h and 34–48 h phases, which may cause a seed shift in the miRNAs and therefore modify the range of mRNA targets. We also observed a high mutation rate within the seed region during the 20–24 h phase which would also have a strong effect on the range of mRNA targets. Many miRNAs showed a surge in expression during the last four phases, suggesting a greater need for those miRNAs before hatching. Those observations would now need to be investigated by target verification methods such as cross-linking immunoprecipitation.

## Conclusions

We evaluated the isomiR detection algorithms isomiR-SEA, isomiRID and miraligner, which are freely available and suitable for integration with local pipelines. We found that each program has advantages and disadvantages. Although isomiRID achieved the best performance against our evaluation criteria, the detailed visual output is more suitable for smaller datasets or the selected analysis of a few miRNAs. In contrast, isomiR-SEA gained a

low score overall, but it allows the analysis of diverse mutations in large datasets because it accounts for more than one mutation in each miRNA, and because it can be run with only one file of mature miRNAs it is ideal for non-model organisms. Finally, we selected miraligner because it achieved a high-performance score and its clear output is ideal for pipeline integration. We used miraligner to screen the publicly available small RNA dataset of early development stages from *T. castaneum*, revealing the dynamic expression of isomiRs at each phase. These isomiRs must now be investigated in more detail to determine their biological functions.

## Additional file

**Additional file 1:** Supplemental figures. **Figure S1.** Analysis scheme for artificial test set evaluation. **Figure S2.** Pearson correlation of the length against the true positive, false positive and false negative rate. IsomiRID has a weak anti-correlation of length and false positive rate. Miraligner has a moderate anti-correlation of length and false negative rate. IsomiR-SEA has in both variations a weak anti-correlation of length and false negative rate. **Figure S3.** Detail view on the various lengths and their individual TP, FP and FN rates. **Figure S4.** Non-templated 3' additions over all conditions. Strong expression of isomiRs with polyadenylate tails was observed in the oocyte and during the first embryonic phase. **Figure S5.** Expression of mature miRNAs during the last four embryonic phases. The number of mature miRNAs increases between the 20–24 h and 48–144 h phases.

**Abbreviations**
isomiR: MicroRNA isoform; miRNA: MicroRNA; NGS: Next-generation sequencing; RISC: RNA-induced silencing complex; SNP: Single-nucleotide polymorphism; UTR: Untranslated region

## Acknowledgements
We thank Dieter Quapil, Heiko Herrmann, Roman Szimanski and Niklas Pfeifer for the technical support of our bioinformatics environment and Richard M. Twyman for professional editing of the manuscript.

## Funding
The authors acknowledge the generous funding from the Hessen State Ministry of Higher Education, Research and the Arts (HMWK) via the LOEWE Center for Insect Biotechnology and Bioresources.

## Authors' contributions
DA designed the evaluation, chose the programs to be evaluated, designed the experiments, analyzed the results and created the draft manuscript. AB and AV supervised the work and critically revised the paper. All authors read and approved the final manuscript.

## Competing interests
The authors declare that they have no competing interests.

## Author details
[1]Fraunhofer Institute for Molecular Biology and Applied Ecology, Department of Bioresources, Winchester Str. 2, 35394 Giessen, Germany. [2]Institute for Insect Biotechnology, Heinrich-Buff-Ring 26-32, 35392 Giessen, Germany.

## References
1. Bartel DP. MicroRNAs: genomics, biogenesis, mechanism, and function. Cell. 2004;116:281–97.
2. Agrawal N, Sachdev B, Rodrigues J, Sree KS, Bhatnagar RK. Development associated profiling of chitinase and microRNA of Helicoverpa Armigera identified chitinase repressive microRNA. Sci Rep. 2013;3:2292. Nature Publishing Group; [cited 2017 Feb 9]. Available from: http://www.nature.com/articles/srep02292
3. Zhang YL, Huang QX, Yin GH, Lee S, Jia RZ, Liu ZX, et al. Identification of microRNAs by small RNA deep sequencing for synthetic microRNA mimics to control Spodoptera Exigua. Gene. 2015;557:215–21.
4. Jayachandran B, Hussain M, Asgari S. An insect trypsin-like serine protease as a target of microRNA: utilization of microRNA mimics and inhibitors by oral feeding. Insect Biochem Mol Biol. 2013;43:398–406.
5. Bernstein E, Caudy AA, Hammond SM, Hannon GJ. Role for a bidentate ribonuclease in the initiation step of RNA interference. Nature. 2001;409: 363–6. Nature Publishing Group; [cited 2017 Jan 5]. Available from: http://www.nature.com/doifinder/10.1038/35053110
6. Bartel DP. MicroRNAs: target recognition and regulatory functions. Cell. 2009;136:215–33.
7. Morin RD, O'Connor MD, Griffith M, Kuchenbauer F, Delaney A, Prabhu A-L, et al. Application of massively parallel sequencing to microRNA profiling and discovery in human embryonic stem cells. Genome Res. 2008;18:610–21. [cited 2015 Sep 16]; Available from: http://genome.cshlp.org/content/18/4/610.abstract
8. Lee LW, Zhang S, Etheridge A, Ma L, Martin D, Galas D, et al. Complexity of the microRNA repertoire revealed by next-generation sequencing. RNA. 2010;16:2170–80. Cold Spring Harbor Laboratory Press; [cited 2017 Jan 19]. Available from: http://www.ncbi.nlm.nih.gov/pubmed/20876832
9. Kuchenbauer F, Morin RD, Argiropoulos B, Petriv OI, Griffith M, Heuser M, et al. In-depth characterization of the microRNA transcriptome in a leukemia progression model. Genome Res. 2008;18:1787–97. Cold Spring Harbor Laboratory Press; [cited 2017 Jan 19]. Available from: http://www.ncbi.nlm.nih.gov/pubmed/18849523
10. Tan GC, Chan E, Molnar A, Sarkar R, Alexieva D, Isa IM, et al. 5' isomiR variation is of functional and evolutionary importance. Nucleic Acids Res. 2014;42:9424–35. Oxford University Press; [cited 2016 Nov 7]. Available from: http://www.ncbi.nlm.nih.gov/pubmed/25056318
11. Luciano DJ, Mirsky H, Vendetti NJ, Maas S. RNA editing of a miRNA precursor. RNA. 2004;10:1174–7. Cold Spring Harbor Laboratory Press; [cited 2017 Jan 19]. Available from: http://www.ncbi.nlm.nih.gov/pubmed/15272117
12. Sun G, Yan J, Noltner K, Feng J, Li H, Sarkis DA, et al. SNPs in human miRNA genes affect biogenesis and function. RNA. 2009;15:1640–51. Cold Spring Harbor Laboratory Press; [cited 2017 Jan 27]. Available from: http://www.ncbi.nlm.nih.gov/pubmed/19617315
13. Wyman SK, Knouf EC, Parkin RK, Fritz BR, Lin DW, Dennis LM, et al. Post-transcriptional generation of miRNA variants by multiple nucleotidyl transferases contributes to miRNA transcriptome complexity. Genome Res. 2011;21:1450–61. Cold Spring Harbor Laboratory Press; [cited 2016 Oct 31]. Available from: http://www.ncbi.nlm.nih.gov/pubmed/21813625
14. Fernandez-Valverde SL, Taft RJ, Mattick JS. Dynamic isomiR regulation in Drosophila development. RNA. 2010;16:1881–8. [cited 2015 Aug 20]. Available from: http://rnajournal.cshlp.org/content/16/10/1881.abstract
15. Ninova M, Ronshaugen M, Griffiths-Jones S. MicroRNA evolution, expression, and function during short germband development in Tribolium castaneum. Genome Res. 2016;26:85–96. [cited 2016 Aug 3]. Available from: http://www.ncbi.nlm.nih.gov/pubmed/26518483
16. Langmead B, Trapnell C, Pop M, Salzberg SL. Ultrafast and memory-efficient alignment of short DNA sequences to the human genome. Genome Biol. 2009;10(3):R25.
17. Griffiths-Jones S, Grocock RJ, van Dongen S, Bateman A, Enright AJ. miRBase: microRNA sequences, targets and gene nomenclature. Nucleic Acids Res. 2006;34:D140–4. Oxford University Press; [cited 2016 Aug 24]. Available from: http://nar.oxfordjournals.org/lookup/doi/10.1093/nar/gkj112
18. Huang W, Li L, Myers JR, Marth GT. ART: a next-generation sequencing read simulator. Bioinformatics. 2012;28:593–4. Oxford University Press; [cited 2016 Aug 31]. Available from: http://www.ncbi.nlm.nih.gov/pubmed/22199392
19. Landgraf P, Rusu M, Sheridan R, Sewer A, Iovino N, Aravin A, et al. A mammalian microRNA expression atlas based on small RNA library sequencing. Cell. 2007;129:1401–14.
20. Martin M. Cutadapt removes adapter sequences from high-throughput sequencing reads. EMBnet J. 2011;17:10. [cited 2016 Aug 24]. Available from: http://journal.embnet.org/index.php/embnetjournal/article/view/200
21. Urgese G, Paciello G, Acquaviva A, Ficarra E, Bartel D, Bartel D, et al. isomiR-SEA: an RNA-Seq analysis tool for miRNAs/isomiRs expression level profiling and miRNA-mRNA interaction sites evaluation. BMC Bioinformatics. 2016;17:148. BioMed Central; [cited 2016 Aug 3]. Available from: http://bmcbioinformatics.biomedcentral.com/articles/10.1186/s12859-016-0958-0
22. de Oliveira LFV, Christoff AP, Margis R. isomiRID: a framework to identify microRNA isoforms. Bioinformatics. 2013;29:2521–3. Oxford University Press; [cited 2016 Aug 3]. Available from: http://www.ncbi.nlm.nih.gov/pubmed/23946501
23. Pantano L, Estivill X, Martí E. SeqBuster, a bioinformatic tool for the processing and analysis of small RNAs datasets, reveals ubiquitous miRNA modifications in human embryonic cells. Nucleic Acids Res. 2010;38:e34. Oxford University Press; [cited 2016 Aug 3]. Available from: http://www.ncbi.nlm.nih.gov/pubmed/20008100
24. Muller H, Marzi MJ, Nicassio F. IsomiRage: from functional classification to differential expression of miRNA Isoforms. Front Bioeng Biotechnol. 2014;2: 38. Frontiers; [cited 2016 Aug 3]. Available from: http://journal.frontiersin.org/article/10.3389/fbioe.2014.00038/abstract
25. Zhang Y, Zang Q, Zhang H, Ban R, Yang Y, Iqbal F, et al. DeAnnIso: a tool for online detection and annotation of isomiRs from small RNA sequencing data. Nucleic Acids Res. 2016;44:W166–75. Oxford University Press; [cited 2016 Aug 3]. Available from: http://nar.oxfordjournals.org/lookup/doi/10.1093/nar/gkw427
26. Sablok G, Milev I, Minkov G, Minkov I, Varotto C, Yahubyan G, et al. isomiRex: Web-based identification of microRNAs, isomiR variations and differential expression using next-generation sequencing datasets. FEBS Lett. 2013;587: 2629–34. [cited 2016 Aug 3]. Available from: http://doi.wiley.com/10.1016/j.febslet.2013.06.047
27. Guo L, Yu J, Liang T, Zou Q. miR-isomiRExp: a web-server for the analysis of expression of miRNA at the miRNA/isomiR levels. Sci Rep. 2016;6:23700. Nature Publishing Group; [cited 2016 Aug 3]. Available from: http://www.ncbi.nlm.nih.gov/pubmed/27009551

# 8

# TnseqDiff: identification of conditionally essential genes in transposon sequencing studies

Lili Zhao[1]* ⓘ, Mark T. Anderson[2], Weisheng Wu[3], Harry L. T. Mobley[2] and Michael A. Bachman[4]

## Abstract

**Background:** Tn-Seq is a high throughput technique for analysis of transposon mutant libraries to determine conditional essentiality of a gene under an experimental condition. A special feature of the Tn-seq data is that multiple mutants in a gene provides independent evidence to prioritize that gene as being essential. The existing methods do not account for this feature or rely on a high-density transposon library. Moreover, these methods are unable to accommodate complex designs.

**Results:** The method proposed here is specifically designed for the analysis of Tn-Seq data. It utilizes two steps to estimate the conditional essentiality for each gene in the genome. First, it collects evidence of conditional essentiality for each insertion by comparing read counts of that insertion between conditions. Second, it combines insertion-level evidence for the corresponding gene. It deals with data from both low- and high-density transposon libraries and accommodates complex designs. Moreover, it is very fast to implement. The performance of the proposed method was tested on simulated data and experimental Tn-Seq data from *Serratia marcescens* transposon mutant library used to identify genes that contribute to fitness in a murine model of infection.

**Conclusion:** We describe a new, efficient method for identifying conditionally essential genes in Tn-Seq experiments with high detection sensitivity and specificity. It is implemented as TnseqDiff function in R package Tnseq and can be installed from the Comprehensive R Archive Network, CRAN.

**Keywords:** Transposon sequencing, Essential gene, Differential test, Tn-Seq, InSeq, CD function

## Background

Large scale transposon mutagenesis coupled with high throughput sequencing (Tn-Seq, also known as INseq, HITS and TraDIS) [1–4] has become a powerful tool to simultaneously assess the essentiality of all genes under experimental conditions. There are mainly two types of data analysis in such experiments: 1) To identify genes required under any growth condition (absolutely essential genes) and 2) to identify conditionally essential genes between conditions (i.e., a differential test). In this paper, we focus on the second analysis. With Tn-Seq, a library of tens of thousands of bacterial mutants is constructed. The location of each insertion mutation and the number

of bacteria with that mutation is determined by massively parallel sequencing. By comparing the mutant counts before and after an experimental condition, the fitness contribution (i.e., conditional essentiality) of each gene can be assessed.

To date, analysis of Tn-Seq data has relied on over-simplified $t$-tests or their nonparametric alternatives [5–11]. Recently, several papers considered statistical methods developed for RNA-Seq data [12–14]. These studies applied edgeR [12, 15] to the overdispersed count data to either identify differentially represented (DE) mutants (i.e., the insertion-level inference) [12, 14] or DE genes based on the sum of insertion counts in each gene [13]. For the gene-level inference, however, they ignored special features of the Tn-Seq data. One distinct feature is that each gene is disrupted at multiple locations, where each insertion site represents a unique mutant. When the

*Correspondence: zhaolili@umich.edu
[1]Department of Biostatistics, University of Michigan, 1415 Washington Heights, Ann Arbor, USA
Full list of author information is available at the end of the article

library is subjected to a selective condition, such as an animal model of infection, each mutant with an insertion in the gene is expected to have decreased abundance in the output samples if that gene is important for fitness. Hence, each insertion site into a particular gene provides independent evidence to prioritize that gene as being conditionally essential in that condition.

Recently the hidden Markov modeling (HMM) has been adapted to identify conditionally essential genes using the insertion-level data [16]. The HMM is a probabilistic statistical model that decodes whether genomic regions belong to a particular biological category given the fold changes in read counts at every insertion site in the genome. A major drawback of the HMM is that it relies on a high-density transposon library to determine whether a gene or region is truly essential (the density is required to be greater than 50%).

Another method that considers the insertion-level data to assess the gene essentiality is the permutation test implemented in software TRANSIT [17]. The permutation test does not require a high-density library, and it identifies essential genes between conditions using a resampling approach. Although the resampling is done on the insertion-level by randomly reshuffling the observed counts at sites in the gene among all the samples, the statistics are based on the total read counts at all the sites for each gene. Additionally, the permutation test has some disadvantages compared to a parametric approach, including 1) a low power with a small number of replicates, 2) misleading results when the samples are correlated or of unequal precision, and 3) inability to accommodate complex design and quality weights [18].

To address all the above limitations, we propose an efficient, parametric method to identify conditionally essential genes based on insertion-level data. The proposed method deals with data from both low- and high-density libraries and is able to accommodate complex designs with multiple inoculum pools and even with multiple conditions. The proposed method was implemented as R package Tnseq (https://CRAN.R-project.org/package= Tnseq).

## Methods
### Data preprocessing
Before applying TnseqDiff, the raw sequence reads need to be processed (e.g., align transposon-flanking sequence reads to genome, filter reads mapped to multiple loci, remove reads from transposons inserted in the 3' end of a gene that cause loss of function, filtering out spurious insertions by removing insertions with low read counts). The final dataset for analysis contains the read counts of all the insertions in each gene for each sample in the Tn-Seq study. The data processing step can be done using pipelines [13, 17, 19]. The resulting data for analysis is a count matrix, where each column represents a sample from a particular inoculum pool under a specific condition, and each row represents an insertion site in a particular gene in the bacterial genome (see the hypothetical data in Table 1). The default normalization method in TnseqDiff is TMM (trimmed mean of M values) [20]. TnseqDiff also takes the read count data that was already normalized by other methods (see a discussion of normalization methods in [17]).

TnseqDiff allows the user to visually evaluate the bias caused by replication process for each sample. Because of asynchronous initiation of DNA replication and cell division, insertions near the origin of replication (ORI) typically are represented as a higher proportion of DNA than insertions farther from the ORI. This is a primary problem when identifying essential genes in a single library and is less of a concern when identifying conditionally essential genes since replication processes are likely to be similar between samples. TnseqDiff provides a method similar to [13] to correct the replication bias when replication processes are different between samples.

TnseqDiff utilizes two steps to estimate the conditional essentiality for each gene in the genome. First, it collects evidence of conditional essentiality for each insertion by

**Table 1** Each column represents a sample (S) from the input or output condition. Each row represents an insertion site in a particular gene in the bacterial genome. Each entry is the read counts mapped to a particular insertion site in a particular gene for a particular sample

| Gene | Location | Pool I | | | | | Pool II | | | | |
| | | Input | | Output | | | Input | | Output | | |
| | | S1 | S2 | S1 | S2 | S3 | S1 | S2 | S1 | S2 | S3 |
|---|---|---|---|---|---|---|---|---|---|---|---|
| 1 | 110 | 478 | 500 | 90 | 100 | 121 | 0 | 0 | 0 | 0 | 0 |
| 1 | 150 | 810 | 910 | 120 | 10 | 5 | 810 | 910 | 120 | 10 | 5 |
| 1 | 350 | 910 | 700 | 50 | 80 | 37 | 0 | 0 | 0 | 0 | 0 |
| 1 | 400 | 1522 | 1544 | 142 | 150 | 124 | 1522 | 1544 | 142 | 150 | 124 |
| 1 | 520 | 320 | 240 | 50 | 1170 | 132 | 320 | 240 | 50 | 1170 | 132 |
| 1000 | 3110 | 100 | 120 | 20 | 10 | 30 | 210 | 190 | 20 | 0 | 70 |

comparing read counts of that insertion between conditions. Second, it combines insertion-level evidence to infer the essentiality for the corresponding gene.

### Step 1: collect evidence of conditional essentiality for each insertion

A normal linear modeling is used in TnseqDiff to obtain the insertion-level information. Specifically, log2-counts per million (logcpm) at each insertion site are modelled as a linear function of the condition (i.e., $y_{ij} = \alpha_i + \beta_i x_j$, where $y_{ij}$ is the logcpm for insertion $i$ in sample $j$, and $x_j$ takes 0 if sample $j$ is in output and 1 if it is in input). The slope coefficient, $\beta_i$, in the model represents the log fold-change (logFC), which is the key parameter for the estimation of conditional essentiality. For example, a large logFC (input over output) might indicate stronger evidence for that insertion being conditionally essential. To consider the over-dispersion of the count data, a precision weight is estimated for each observation from the mean-variance relationship of the logcpm and is then entered into the linear modeling [21]. TnseqDiff relies on the Limma package [18, 22] for the above estimation.

To collect evidence of conditional essentiality for each insertion, we construct a confidence distribution (CD) [23–25] for the logFC at each insertion site using estimates from the above linear model. The CD has attracted a surge of attention in recent years. A CD function contains a wealth of information for inferences; much more than a point estimator or a confidence interval. It is a "frequentist" analogue of a Bayesian posterior. Furthermore, it provides a framework to combine evidence through combining CD functions (in our case, combining insertion-level CD functions to make inference for the gene).

The CD function for the $i^{th}$ insertion, $H(\beta_i)$, is defined as

$$H(\beta_i) = F_{t_{d_i}}\left(\frac{\beta_i - \hat{\beta}_i}{s_i}\right),$$

where $\hat{\beta}_i$ is the mean estimate of the logFC, $s_i$ is the standard error and $d_i$ is the degrees of freedom. $F_{t_{d_i}}$ is the cumulative distribution function of the $t_{d_i}$ distribution. When $\beta_i$ varies, $H(\beta_i)$ forms a function on the parameter space of $\beta_i$, which contains a wealth of information about the $\beta_i$, including point estimates (such as mean, median and mode), confidence intervals of various levels and significance testing (see details in [23, 25] and Figure 1 in [25] graphically illustrates the above estimates).

Alternatively we can replace $s_i$ and $d_i$ by the corresponding moderated estimates based on the empirical Bayes method [18, 22]. The CD function constructed based on the moderated estimates is a moderated CD function, which efficiently borrows information from similar insertions to aid inference for any single insertion.

### Step 2: combine insertion-level evidence

TnseqDiff combines the insertion-level CD functions to obtain a single CD function for the corresponding gene. This is accomplished by the use of a simple formula

$$H_g(\beta) = \Phi\left(\frac{1}{\sqrt{\sum_{i=1}^{N} w_i^2}}\left[w_1\Phi^{-1}(u_1) + \cdots + w_N\Phi^{-1}(u_N)\right]\right), (1)$$

where $\Phi$ is the cumulative distribution function of the standard normal distribution, $u_i$ is the CD function for insertion $i$ and $w_i$ ($w_i \geq 0$) is its weight. If $w_i = 0$, insertion $i$ is not included in the combined CD function. The combined CD function, $H_g(\beta)$, contains essentiality information from all $N$ insertions. Here, subscript "$g$" is used to indicate that the combined CD function is on the gene level.

It is important to note that the combined CD function, $H_g(\beta)$, automatically puts more weight on the insertion-level CD function containing more information even when $w_i$'s are all equal. The idea of combining CD functions is illustrated with a simple example in the Fig. 1. In this figure, three insertion-level CD density functions (black curves) have different means and variances (variances increase from the left to the right curve). The blue curve is the combined CD density function using formula (1) with equal weights. As shown in this figure, the combined function is located near the insertion-level function with less spread (i.e., a smaller variance).

Furthermore, TnseqDiff allows unequal weights for the combination. Insertions with low read counts ($\approx 0$) in the input condition might suggest that they are essential for growth in any given condition, therefore, the analysis

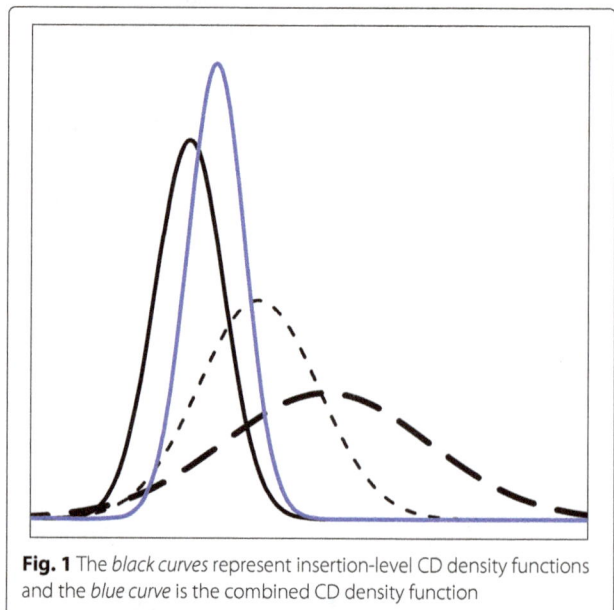

**Fig. 1** The *black curves* represent insertion-level CD density functions and the *blue curve* is the combined CD density function

should exclude or consider a small weight for these insertions. TnseqDiff identifies insertions with "low" counts using a fast dynamic programming algorithm for optimal univariate 2-means clustering [26]. The insertions that are clustered into the group with a smaller mean are assigned weights less than one, specifically, these weights are estimated from an exponential function (the smallest count gets a weight close to zero, while the largest count gets a weight close to one). We call this weight function hc. TnseqDiff also takes weights specified by the user. For example, the probability of an insertion being absolutely essential can be used as the weight for that insertion and obtained from a separate method (such as the method in [16] or [27]).

### Identify conditionally essential genes based on the combined CD function

TnseqDiff estimates the conditional essentiality for a particular gene using the combined CD function $H_g(\beta)$. As shown in [25], the median logFC is estimated based on $H_g^{-1}(\frac{1}{2})$. Specifically, TnseqDiff uses a numeric algorithm to solve for $\beta_g$ in the equation

$$\sum_{i=1}^{N} w_i \Phi^{-1} \left( F_{t_{d_i}} \left( \frac{\beta_g - \hat{\beta}_i}{s_i} \right) \right) = 0$$

In a simple case where $w_1 = \cdots = w_N \equiv 1$ and the $t$ distribution can be approximated by a normal distribution, the median logFC is simplified as

$$\text{Median logFC} = \frac{\sum_i^N \hat{\beta}_i / s_i}{\sum_i^N 1/s_i}$$

In this case, the median logFC is a weighted average of the insertion-level logFC estimates, with the weight inversely proportional to the standard error.

Similarly, the lower and upper bound of a level $100(1 - a)\%$ confidence interval can be calculated by solving equation

$$\sum_{i=1}^{N} w_i \Phi^{-1}(F_{t_{d_i}}((\beta_g - \hat{\beta}_i)/s_i)) - \left( \sum_i^N w_i \right)^{\frac{1}{2}} \Phi^{-1}(a/2) = 0$$

and

$$\sum_{i=1}^{N} w_i \Phi^{-1} \left( F_{t_{d_i}} \left( \left( \beta_g - \hat{\beta}_i \right) / s_i \right) \right) - \left( \sum_i^N w_i \right)^{\frac{1}{2}} \Phi^{-1}(1-a/2) = 0,$$

respectively.

For testing if a gene is conditionally non-essential versus essential, the hypotheses are $H_0 : \beta_g \leq 0$ vs. $H_1 : \beta_g > 0$. As defined in [25], the one-sided $p$-value is simply $H_g(0)$, where

$$H_g(0) = \Phi \left( \frac{1}{\sqrt{\sum_i^N w_i}} \sum_{i=1}^{N} w_i \Phi^{-1} \left( F_{t_{d_i}} \left( -\frac{\hat{\beta}_i}{s_i} \right) \right) \right).$$

The two-sided $p$-value is $2 \times \min\{H_g(0), 1 - H_g(0)\}$ (TnseqDiff provides a two-sided $p$-value). These $p$-values are then adjusted for multiple testing using the Benjamini-Hochberg Procedure [28].

In real applications, differentially represented genes are generally selected based on both the adjusted $p$-value and the fold-change (FC). Tnseqdiff uses the median logFC as defined above, that is, FC= $2^{\text{median logFC}}$. It is important to note that TnseqDiff calculates the $p$-value and median logFC from the combined CD function. If only interested in identifying conditionally essential genes (i.e., identifying genes with decreased counts in output), we can set the rule as the FC (input over output) $\geq 2$ and the adjusted $p$-value $< 0.025$ in a two-sided test (or $p$-value $< 0.05$ in a one-sided test).

In addition to the above estimates, TnseqDiff also provides descriptive statistics for each gene, including the number of (unique) insertions in input samples and averaged counts in input and output samples (after accounting for the differences in library sizes).

Our proposed method is much simpler to implement than a model-based approach and it can be easily extended to analyze more complex designs.

### Analyze designs with multiple inoculum pools

Mutant pools are often too large ($\sim$ 50,000 random mutants for a 5 Mbp gemome) to be tested in one mouse, or an experimental "bottleneck" would cause random loss of mutants from a large inoculum. In these cases, the mutant library is split and smaller pools are used to inoculate separate sets of mice. Hence, different mutants within a particular gene are tested in different mice. It would be inaccurate to sum over the insertion counts that are observed in different mice due to the loss of biological variability. However, our method is directly applicable to such designs since samples at each insertion site in different pools are independent (the only requirement for combining CD functions). TnseqDiff first combines insertion-level CD functions to obtain a CD function for each gene in a given pool, and then it combines CD functions from multiple pools for each gene to obtain a single CD function for identifying conditionally essential genes.

### Results and discussion
#### Simulation studies

We ran simulation studies to investigate our proposed methods and compared them to 1) the permutation test in the TRANSIT software [17] and 2) the negative binomial test in the ESSENTIALS software [13]. In the permutation test, the read counts at all the sites and all samples in each condition are summed for each gene. The difference in the sum between conditions was calculated. The significance of this difference was evaluated by comparing to a resampling distribution generated from randomly reshuffling

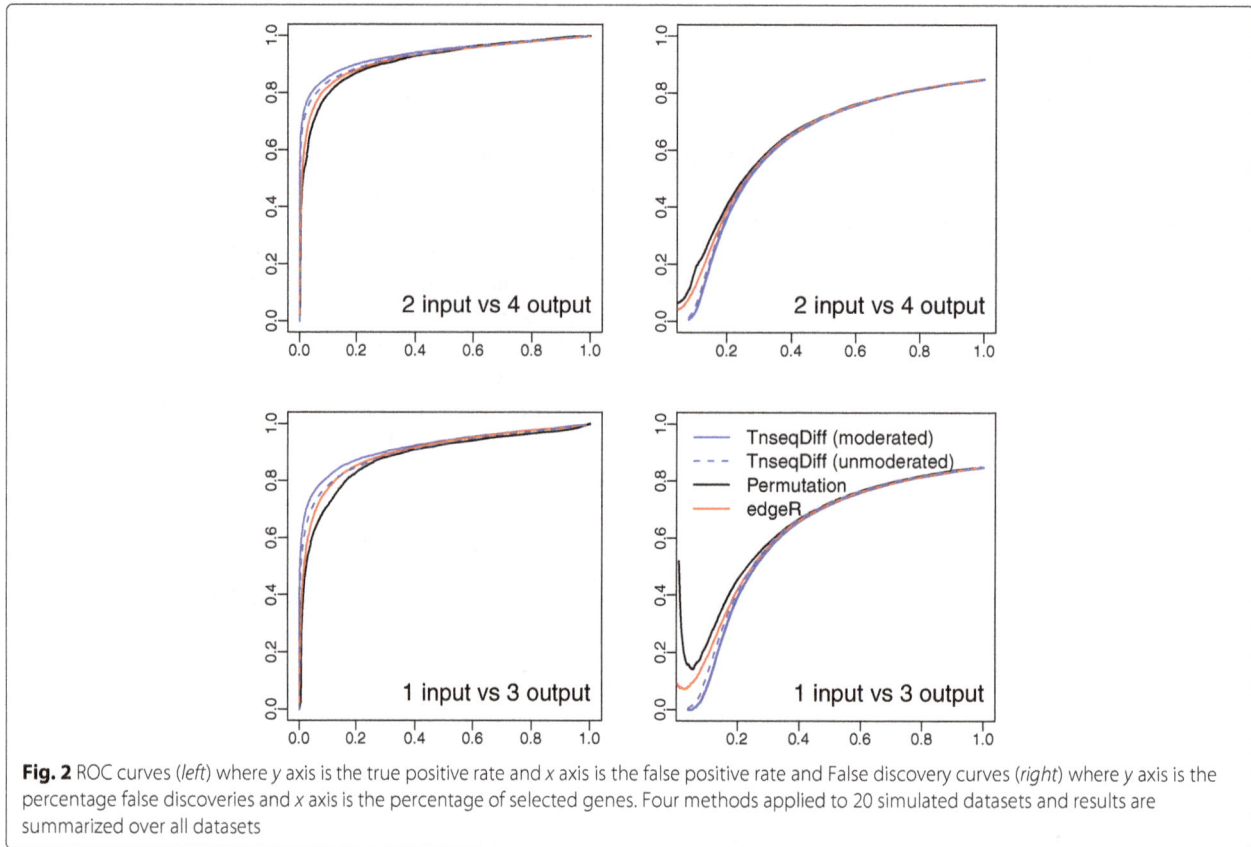

**Fig. 2** ROC curves (*left*) where *y* axis is the true positive rate and *x* axis is the false positive rate and False discovery curves (*right*) where *y* axis is the percentage false discoveries and *x* axis is the percentage of selected genes. Four methods applied to 20 simulated datasets and results are summarized over all datasets

the observed counts at sites in the gene among all the samples. A *p*-value was then derived from the proportion of 10,000 reshuffled samples that have a difference more extreme than that observed in the actual experimental data. ESSENTIALS used the method in edgeR to identify DE genes based on the total gene counts, therefore, we directly applied edgeR to the datasets after obtaining the total gene counts by summing over the insertion counts for each gene.

To make simulation studies more realistic, the data and insertion distributions in simulated datsets were similar to a real dataset. The real dataset was generated from a *Serratia marcescens* transposon mutant library with the objective of identifying bacterial genes that contribute to fitness in a murine model of bloodstream infection [29] (details are shown in the next section). It consists of five inoculum pools with 2 input and 4 output samples per pool. We merged data from five pools and assumed that insertions

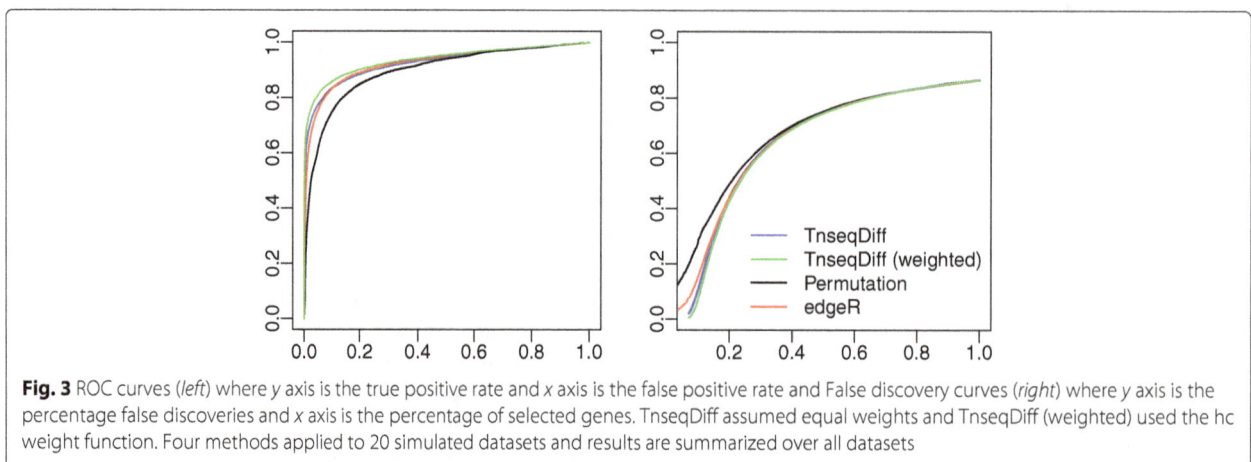

**Fig. 3** ROC curves (*left*) where *y* axis is the true positive rate and *x* axis is the false positive rate and False discovery curves (*right*) where *y* axis is the percentage false discoveries and *x* axis is the percentage of selected genes. TnseqDiff assumed equal weights and TnseqDiff (weighted) used the hc weight function. Four methods applied to 20 simulated datasets and results are summarized over all datasets

at the same genomic location in different pools were different insertions. After data normalization, we averaged the two input samples and excluded insertions with an averaged count < 5 (remaining insertions were considered as true insertions). The final dataset consists of 4,075 genes with 42,639 insertions. The number of insertions per gene ranged from 1 to 202 (median is 8, the first and third quartile is 4 and 14, respectively). This insertion distribution was assumed in the first two simulation studies. Input data were generated from Poisson distributions because input samples (in vitro) are technical replicates, while output data were generated from negative binomial (NB) distributions because the output samples (in vivo) are biological replicates.

*The first simulation study: all insertions are genuine insertions*
In this study, we focused on identifying conditionally essential genes based on true insertion data and assumed

that absolutely essential genes and spurious insertions (in vitro) have been removed. Given the insertion distribution in the real dataset, we first generated the input data for each insertion from a Poisson distribution with the mean parameter equal to the averaged count. Then we randomly selected 10% of the genes to be under-represented (i.e., conditionally essential) and 5% to be over-represented in the output samples. For insertions in under-represented genes, logFCs were generated from a left truncated standard normal distribution, while insertions in over-represented genes were generated from a right truncated standard normal distribution. For non-DE genes, logFCs were fixed to be zero. Finally, we generated the output data from a NB distribution with the mean equal to the product of the input mean and the FC. Rather than fixing the dispersion parameter to be the same for all insertions, we generated dispersion parameters from a gamma distribution with a shape = 1, scale

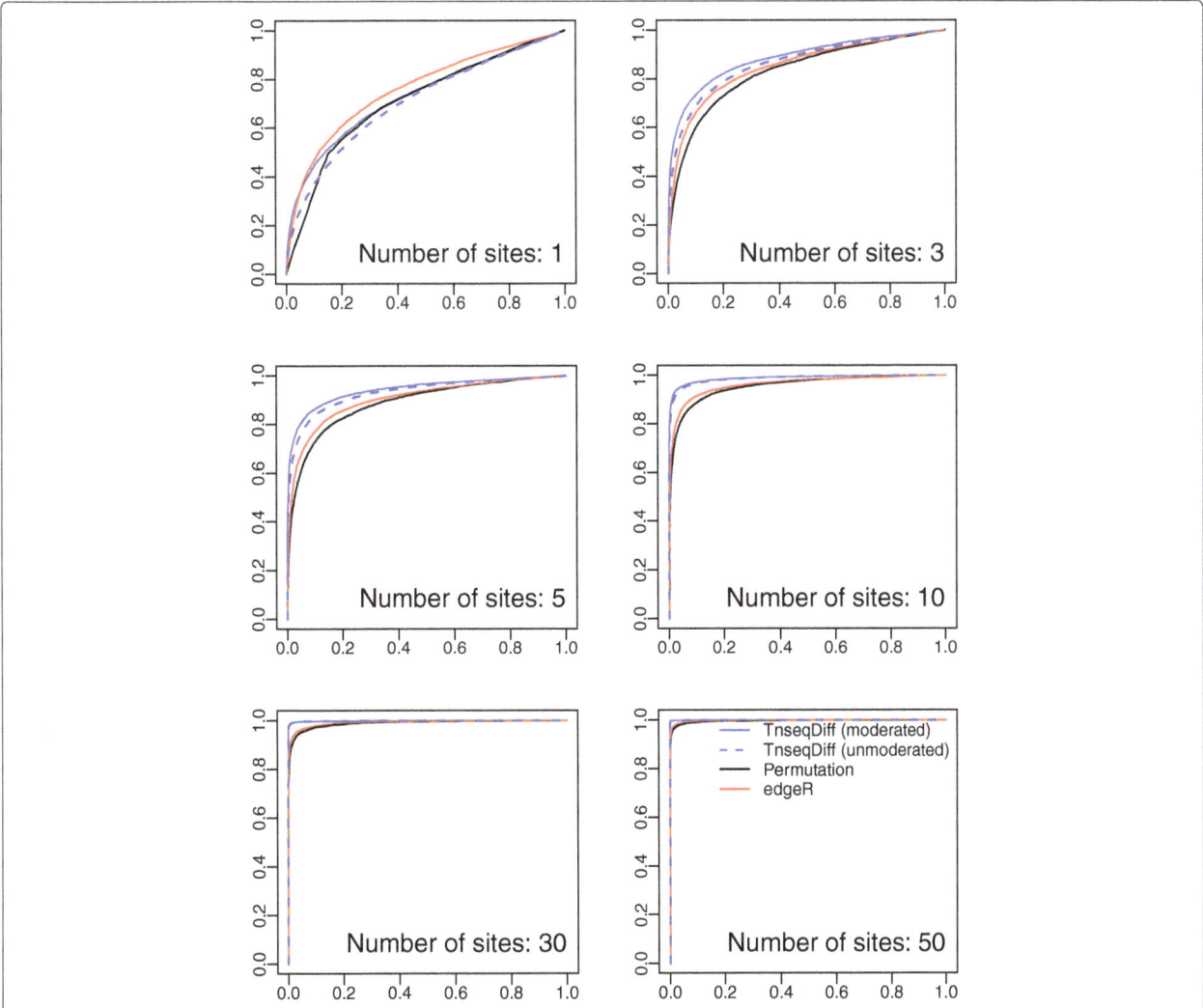

**Fig. 4** ROC curves where y axis is the true positive rate and x axis is the false positive rate. Four methods applied to datasets consisting of 2 input and 4 output samples. Each plot presents a scenario for a fixed number of sites per gene

= 0.5 (these two parameters were determined based on the real dataset). In this study, we tried two sample sizes: 1) 2 input vs 4 output samples, and 2) 1 input sample vs 3 output samples.

We applied TnseqDiff to 20 simulated datasets as described above and assumed equal weights for combining the insertion-level CD functions. We considered both moderated and unmoderated CD functions in TnseqDiff and call them moderated and unmoderated TnseqDiff.

*Simulation results:* As shown in Fig. 2, TnseqDiff performed significantly better than edgeR and the permutation test under the two studied sample sizes, as evidenced by improved accuracy in separating the truly DE and non-DE genes and a much smaller false discovery rate given the same number of selected genes. Moreover, moderated TnseqDiff performed slightly better than the unmoderated TnseqDiff. Similar conclusions can be reached for the

conditionally essential gene detection (i.e., the one-sided test) except that the unmoderated TnseqDiff is similar to the moderated TnseqDiff (ROC and False discovery curves were shown in Additional file 1).

### The second simulation study: some insertions are spurious insertions

In this study, we included 500 (about 10% of bacterial genome) absolutely essential genes in each simulated dataset. Since an absolutely essential gene should not contain any real insertion, we generated low read count data for these spurious insertions from a Poisson distribution with rate = 3 (1-14 "insertions" were assumed within each absolutely essential gene). Additionally, 2,132 spurious insertions (5% of the total 42,639 insertions) were randomly added to the bacterial genome such that a DE gene may contain false insertions. The rest of the simulations

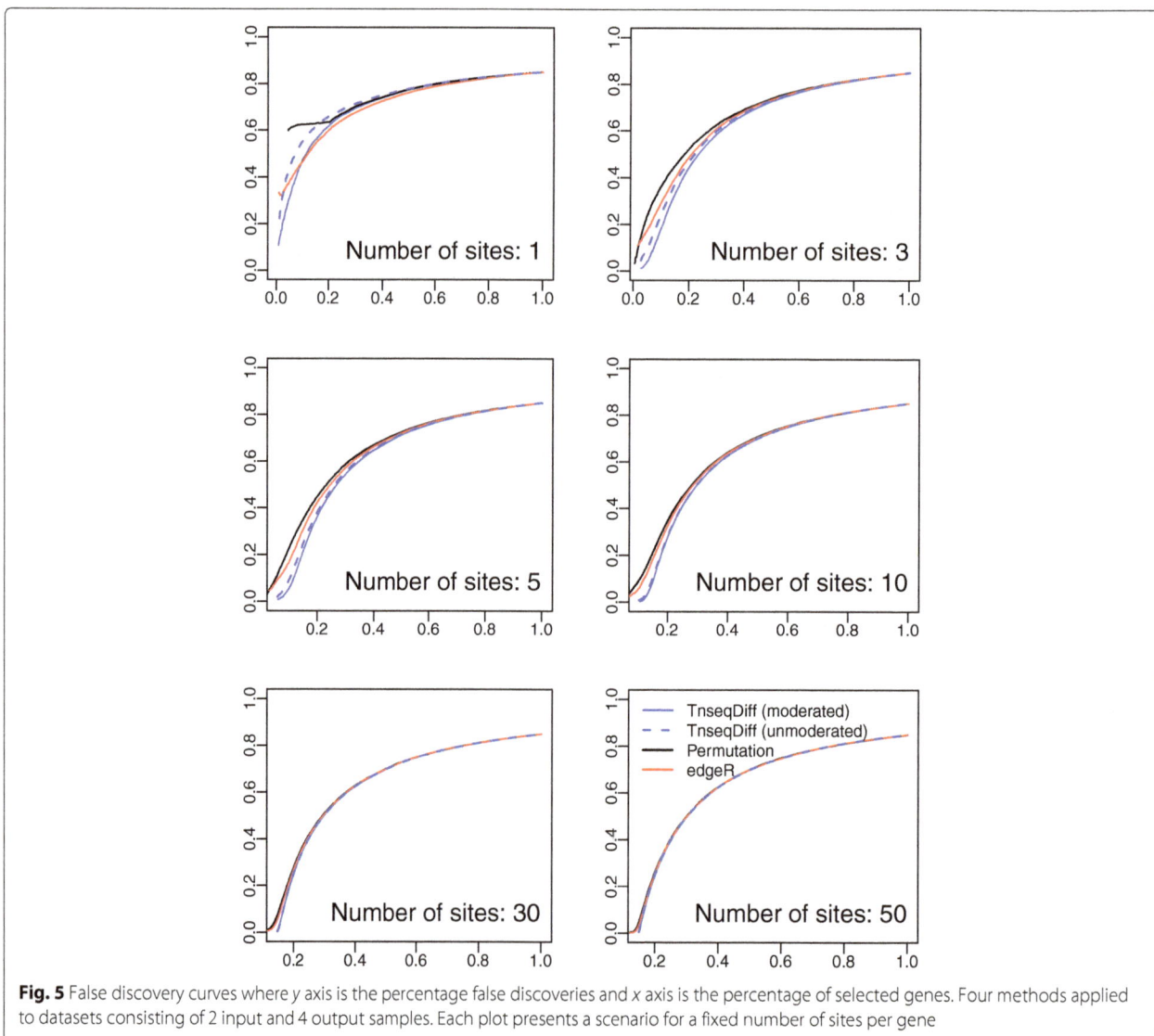

**Fig. 5** False discovery curves where *y* axis is the percentage false discoveries and *x* axis is the percentage of selected genes. Four methods applied to datasets consisting of 2 input and 4 output samples. Each plot presents a scenario for a fixed number of sites per gene

were the same as in the first simulation study. This study has 2 input vs 4 output samples.

We applied moderated TnseqDiff to 20 simulated datasets as described above and considered the equal and the hc weight function. The hc weight function down-weighs spurious insertions in the analysis (see details in step 2 of the Method section).

*Simulation results:* As shown in Fig. 3, TnseqDiff with equal weights performed similarly, or slightly better in terms of the false discovery rate, than edgeR and the permutation test. The TnseqDiff with the hc weight function performed better than the TnseqDiff with the equal weight function. Furthermore, we found that all absolutely essential genes were correctly identified as non-DE genes in the weighted TnseqDiff and edgeR, while 68 (13.6%) absolutely essential genes were wrongly identified as DE genes in the permutation test.

### The third simulation study: each gene has a fixed number of insertions

To investigate the effect of number of insertions per gene on the model performance, we assumed that each gene has a fixed number of insertions (denoted by $n$). Each simulated dataset consists of 5000 genes with $n = 1, 3, 5, 10, 20, 30$, or 50. We first sampled 5000 genes containing at least $n$ sites from the above 4075 genes with replacement (the sampling weight for each gene is proportional of the number of sites in that gene). Then we sampled $n$ mean parameters from each gene with replacement and these parameters were used in the Poisson distribution to generate the input data. The rest of the simulations are the same as in the first simulation study. This study has 2 input vs 4 output samples.

We applied both the moderated and unmoderated TnseqDiff to 10 simulated datasets as described above. Since all insertions are true insertions, we assumed equal weight in TnseqDiff.

*Simulation results:* As shown in Figs. 4 and 5, TnseqDiff performed significantly better than edgeR and the permutation test when the number of insertions is > 1. When there is just one insertion per gene, TnseqDiff is equivalent to Limma for detecting DE genes (no CD function combining in this case), and the moderated TnseqDiff performed better than the unmoderated TnseqDiff since the moderated estimates borrowed information from similar insertions across all genes. Furthermore, all methods had increased accuracy when the number of insertions per gene was increased. In other words, a gene with a larger number of insertions contains more information and is more likely to be identified as a DE or non-DE gene correctly.

To our surprise, the permutation test performed the worst in all studied scenarios. This could be due to the fact that the permutation test requires that the two distributions are identical [30], however, Tn-Seq studies generally have very different distributions for the input and output data.

Furthermore, TnseqDiff is much faster to implement than the permutation test especially when the number of insertion sites per gene is small (see Fig. 6).

### Application to a real transposon dataset

We applied TnseqDiff to a published Tn-Seq dataset [29]. The Tn-Seq dataset was generated from a *Serratia marcescens* transposon mutant library with the objective of identifying bacterial genes that contribute to fitness in a murine model of bloodstream infection. A mariner-based transposon encoded in suicide plasmid pSAM-Cm [1] was used to generate a random library of transposon insertion mutants in strain UMH9. An initial mutant library of > 32,000 unique transposon insertion mutants was equally split into five inoculum pools. Each pool was used to infect 4 mice and spleens from infected mice were collected after 24 hrs. The insertion sites from input and output pools were PCR-amplified and then sequenced via the Illumina HiSeq platform using 50 cycle single-end reads [31]. Sequence reads were mapped to the UMH9 annotated genome using the ESSENTIALS pipeline with default parameter settings. One output sample from each of pools 3-5 was eliminated from the analysis due to mice that succumbed to infection or insufficient PCR product for sequencing. The final dataset consisted of 4106 genes with at least one transposon insertion, and the number of insertions for a given gene ranged from 1 to 322, with

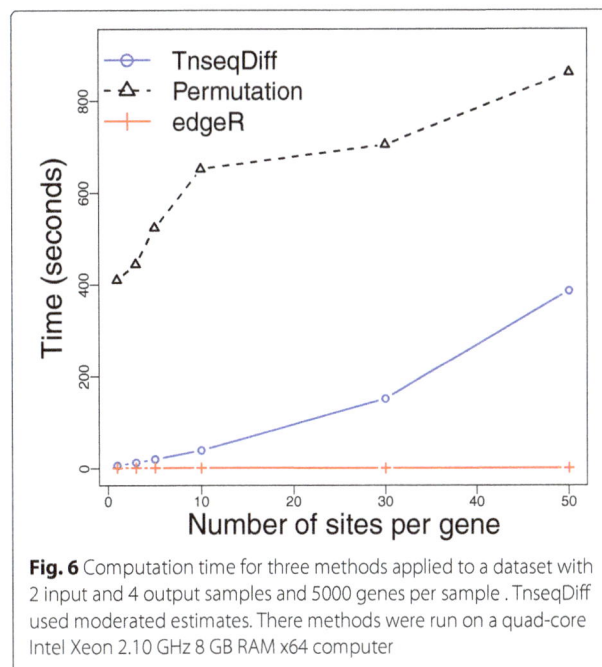

**Fig. 6** Computation time for three methods applied to a dataset with 2 input and 4 output samples and 5000 genes per sample . TnseqDiff used moderated estimates. There methods were run on a quad-core Intel Xeon 2.10 GHz 8 GB RAM x64 computer

over 50% of the genes having 12 or less insertions. HMM approach is not appropriate for analyzing this dataset since the density of the transposon library is not high.

In TnseqDiff (moderated or unmoderated), equal weights were assumed because the data has been pre-processed using the ESSENTIALS to exclude absolutely essential gene detection. Conditionally essential genes were determined based on the fold-change (input over output) $\geq 2$ and the adjusted $p$-value $< 0.025$. We also applied ESSENTIALS to the same dataset. As shown in Fig. 7, majority of fitness genes were identified by both TnseqDiff and ESSENTIALS and moderated TnseqDiff identified 21 more genes than the unmoderated TnseqDiff. Seven of these genes, encoding a wide range of biological functions and identified by TnseqDiff (moderated and unmoderated) and ESSENTIALS, were chosen for validation of the Tn-Seq screen. Deletion-insertion mutations were constructed for each of the genes and the resulting strains were tested for in vivo fitness defects in competition with the wild-type strain using the murine bacteremia model. The results from these experiments confirmed that six of the seven tested genes contribute to *S. marcescens* fitness in the mammalian host. Importantly, none of the seven mutants exhibited a general growth defect when cultured in vitro. Figure 8 shows four genes that were identified as conditionally essential by TnseqDiff but not by ESSENTIALS. Genes SmUMH9_0913 (*galF*) and SmUMH9_0917 (*neuA*) are both located in the 18-gene *S. marcescens* capsule biosynthesis locus, within which other genes are important for fitness [29]. Genes SmUMH9_1422 and SmUMH9_2227 are predicted to be co-transcribed with a functionally-related adjacent gene that was identified by both TnseqDiff and ESSENTIALS. Complete analysis results from ESSENTIALS and TnseqDiff were presented in Additional file 2.

## Conclusions

We developed methods that are specifically designed for analyzing Tn-Seq data and implemented these methods in the TnseqDiff function in R package Tnseq. TnseqDiff takes into account the unique features of Tn-Seq data and identifies conditionally essential genes using insertion-level data. TnseqDiff handles data from both low- and high-density transposon libraries. We have demonstrated its advantages over the existing methods, including 1) better performance in separating true DE and non-DE genes and a smaller false discovery rate, 2) a much faster computation time, and 3) the ability to accommodate complex designs (for example, designs with multiple pools). TnseqDiff can be easily extended to analyze data with multiple experimental conditions. In this case, data from all conditions will be included in the linear model, and coefficient estimates or estimates of interested contrasts can be used to construct the CD function for testing interested hypotheses.

It is worth noting that, unlike the HMM method, TnseqDiff does not rely on a high-density transposon library for inference. It focuses on identifying conditionally essential genes and is most efficient when absolutely essential genes and spurious insertions have been removed first. TnseqDiff with the hc weight function downweighed spurious insertions and it worked well in simulation studies where absolutely essential genes and spurious insertions were present in the bacterial genome. These weights can also be obtained using other existing softwares for the absolutely essential gene detection (such as ARTIST or TRANSIT). In these softwares, an estimated probability for an insertion to be absolutely essential can be considered as the weight for that insertion and incorporated into TnseqDiff for the differential test.

Unlike the HMM approach in ARTIST, TnseqDiff is annotation-dependent. It evaluates conditional essentiality for previously-annotated genomic features (e.g., ORFs, ncRNAs). However, TnseqDiff allows inference for intergenic regions and subdomains of ORFs if these regions are pre-defined in the dataset by combining the insertions within that region for inference.

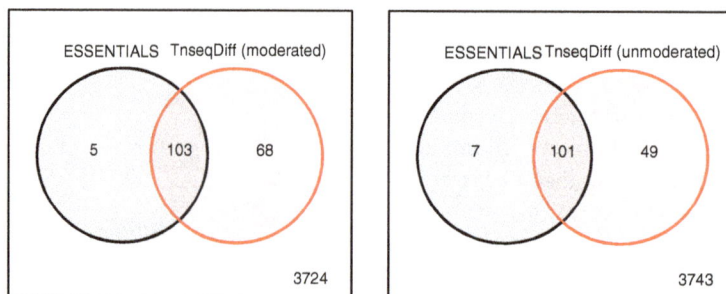

**Fig. 7** Overlap of conditionally essential genes from ESSENTIALS and TnseqDiff. A gene is essential if the fold-change (input over output) $\geq 2$ and the adjusted $p$-value $< 0.025$

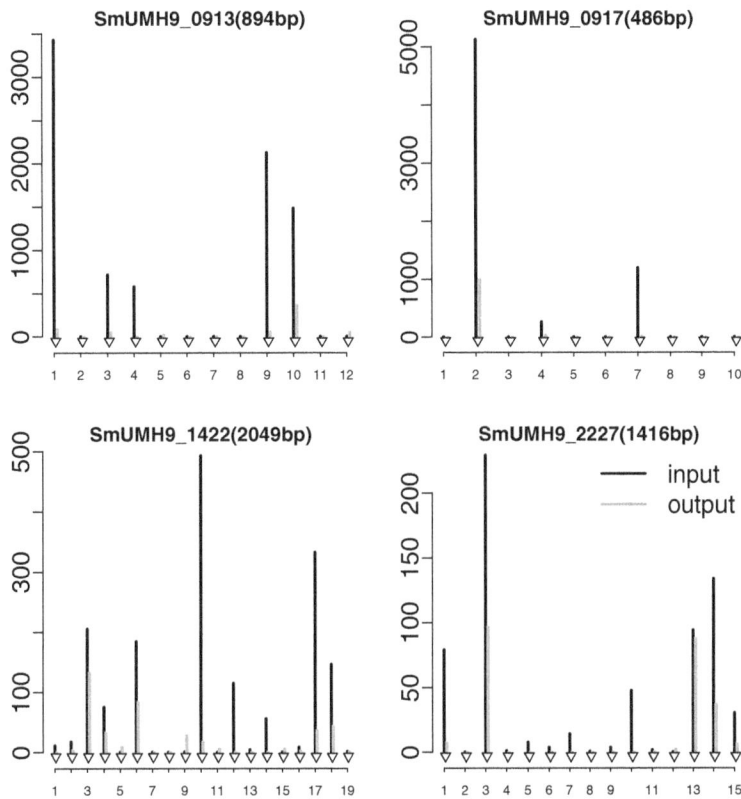

**Fig. 8** Distribution of insertion counts in four genes. The x-axis is the location and each insertion site is indicated by a black arrowhead. The y-axis is the averaged normalized read counts for input (*black*) and output (*orange*) samples

## Abbreviations
CD: Confidence distribution; DE: Differentially represented; FC: Fold change; HMM: Hidden Markov modeling; NB: Negative binomial; ORI: Origin of replication; TMM: Trimmed mean of M values Tn-Seq: Experiments with large scale transposon mutagenesis coupled with high throughput sequencing

## Acknowledgements
The authors gratefully acknowledge the constructive comments of referees and thank Dr. Yanming Li for helpful discussions in building the R package.

## Funding
This work was supported by the National Institutes of Health (grant P30 CA 046592-28). The funding body was not involved in the design of the study and collection, analysis, and interpretation of data or in writing the manuscript.

## Authors' contributions
LZ developed the algorithm and wrote the R package. MA performed the *S. marcescens* experiment. WW pre-processed the sequencing data and used ESSENTIALS for the real data analysis. LZ, MA, WW, HM and MB participated in writing the manuscript. All authors read and approved the final manuscript.

## Competing interests
The authors declare that they have no competing interests.

## Author details
[1]Department of Biostatistics, University of Michigan, 1415 Washington Heights, Ann Arbor, USA. [2]Department of Microbiology and Immunology, School of medicine, University of Michigan, Ann Arbor, USA. [3]BRCF Bioinformatics Core, University of Michigan, Ann Arbor, USA. [4]Department of Pathology, School of medicine, University of Michigan, Ann Arbor, USA.

## References
1. Bachman MA, Breen P, Deornellas V, Mu Q, Zhao L, Wu W, Cavalcoli JD, Mobley HLT. Genome-wide identification of klebsiella pneumoniae fitness genes during lung infection. mBio. 2015;6:00775–15.
2. Langridge GC, Phan MD, Turner DJ, Perkins TT, Parts L, Haase J, Charles I, Maskell DJ, Peters SE, Dougan G, Wain J, Parkhill J, Turner AK. Simultaneous assay of every salmonella typhi gene using one million transposon mutants. Genome Res. 2009;19:2308–16.
3. van Opijnen T, Bodi KL, Camilli A. Tn-seq: high-throughput parallel sequencing for fitness and genetic interaction studies in microorganisms. Nat Methods. 2009;6:767–72.
4. Gawronski JD, Wong SM, Giannoukos G, Ward DV, Akerley BJ. Tracking insertion mutants within libraries by deep sequencing and a genome-wide screen for haemophilus genes required in the lung. Proc Natl Acad Sci U S A. 2009;106:16422–7.

5.  Fu Y, Waldor MK, Mekalanos JJ. Tn-seq analysis of vibrio cholerae intestinal colonization reveals a role for t6ss-mediated antibacterial activity in the host. Cell Host Microbe. 2013;14:652–63.

6.  Kamp HD, Patimalla-Dipali B, Lazinski DW, Wallace-Gadsden F, Camilli A. Gene fitness landscapes of vibrio cholerae at important stages of its life cycle. PLoS Pathog. 2013;9:1003800.

7.  McDonough E, Lazinski DW, Camilli A. Identification of in vivo regulators of the vibrio cholerae xds gene using a high-throughput genetic selection. Mol Microbiol. 2014;92:302–15.

8.  Troy EB, Lin T, Gao L, Lazinski DW, Camilli A, Norrisand SJ, Hu LT. Understanding barriers to borrelia burgdorferi dissemination during infection using massively parallel sequencing. Infect Immun. 2013;81: 2347–57.

9.  Burghout P, Zomer A, CEvdG-d J, Janssen-Megens EM, K-J F, Stunnenberg HG, Hermans PWM. Streptococcus pneumoniae folate biosynthesis responds to environmental co2 levels. J Bacteriol. 2013;195: 1573–82.

10. de Vries SP, Eleveld MJ, Hermans PW, Bootsma HJ. Characterization of the molecular interplay between moraxella catarrhalis and human respiratory tract epithelial cells. PLOS ONE. 2013;8:72193.

11. Maria JPS, Sadaka A, Moussa SH, Brown S, Zhang YJ, Rubin EJ, Gilmore MS, Walker S. Compound-gene interaction mapping reveals distinct roles for staphylococcus aureus teichoic acids. Proc Natl Acad Sci. 2014;111: 12510–5.

12. Robinson DG, Chen W, Storey JD, Gresham D. Design and analysis of bar-seq experiments. G3 (Bethesda). 2014;4:11–18.

13. Zomer A, Burghout P, Bootsma HJ, Hermans PW, van Hijum SA. Essentials: software for rapid analysis of high throughput transposon insertion sequencing data. PLoS ONE. 2012;7:43012.

14. Dembek M, Barquist L, Boinett CJ, Cain AK, Mayho M, Lawley TD, Fairweather NF, Fagan RP. High-throughput analysis of gene essentiality and sporulation in clostridium difficile. mBio. 2015;6:02383–14.

15. McCarthy JD, Chen Y, Smyth KG. Differential expression analysis of multifactor RNA-seq experiments with respect to biological variation. Nucleic Acids Res. 2012;40:4288–97.

16. Pritchard JR, Chao MC, Abel S, Davis BM, Baranowski C, Zhang YJ, Rubin EJ, Waldor MK. ARTIST: High-resolution genomewide assessment of fitness using transposon-insertion sequencing. PLoS Genet. 2014;10: 1004782.

17. DeJesus MA, Ambadipudi C, Baker R, Sassetti C, Ioerger TR. TRANSIT - a software tool for Himar1 Tnseq analysis. PLoS Comput Biol. 2015;11: 1004401.

18. Ritchie ME, Phipson B, Wu D, Hu Y, Law CW, Shi W, Smyth GK. limma powers differential expression analyses for rna-sequencing and microarray studies. Nucleic Acids Res. 2015;43:47.

19. Goodman AL, McNulty NP, Zhao Y, Leip D, Mitra RD, Lozupone CA, Knight R, Gordon JI. Identifying genetic determinants needed to establish a human gut symbiont in its habitat. Cell Host Microbe. 2009;6:279–89.

20. Robinson MD, McCarthy DJ, Smyth GK. Edger: a bioconductor package for differential expression analysis of digital gene expression data. Bioinformatics. 2010;26:139–40.

21. Law CW, Chen Y, Shi W, Smyth GK. voom: precision weights unlock linear model analysis tools for RNA-seq read counts. Genome Biol. 2014;15:29.

22. Smyth GK. Linear models and empirical bayes methods for assessing differential expression in microarray experiments. Stat Appl Genet Mol Biol. 2004;3:3.

23. Singh K, Xie M, Strawderman WE. Combining information from independent sources through confidence distributions. Ann Statist. 2005;33:159–83.

24. Singh K, Xie M, Strawderman WE. Confidence distributions and a unifying framework for meta-analysis. J Am Statist Assoc. 2011;106:320–33.

25. Xie M, Singh K. Confidence distribution, the frequentist distribution estimator of a parameter: A Review. Int Stat Rev. 2013;81:3–39.

26. Wang H, Song M. Ckmeans.1d.dp: optimal k-means clustering in one dimension by dynamic programming. R Journal. 2011;3:29–33.

27. Liu F, Wang C, Wu Z, Zhang Q, Liu P. A zero-inflated poisson model for insertion tolerance analysis of genes based on Tn-seq data. Bioinformatics. 2016;32:1701–8.

28. Benjamini Y, Hochberg Y. Controlling the false discovery rate: a practical and powerful approach to multiple testing. J R Stat Soc Series B. 1995;57: 289–300.

29. Anderson MT, Mitchell LA, Zhao L, Mobley HLT. Capsule production and glucose metabolism dictate fitness during serratia marcescens bacteremia. mBio. 2017;8:00740–17.

30. Huang Y, Xu H, Calianand V, Hsu JC. To permute or not to permute. Bioinformatics. 2006;22:2244–8.

31. Goodman AL, Wu M, Gordon JI. Identifying microbial fitness determinants by insertion sequencing using genome-wide transposon mutant libraries. Nat Protoc. 2011;6:1969–80.

# 9

# Local sequence and sequencing depth dependent accuracy of RNA-seq reads

Guoshuai Cai[1,2*], Shoudan Liang[3], Xiaofeng Zheng[3] and Feifei Xiao[4*]

## Abstract

**Background:** Many biases and spurious effects are inherent in RNA-seq technology, resulting in a non-uniform distribution of sequencing read counts for each base position in a gene. Therefore, a base-level strategy is required to model the non-uniformity. Also, the properties of sequencing read counts can be leveraged to achieve a more precise estimation of the mean and variance of measurement.

**Results:** In this study, we aimed to unveil the effects on RNA-seq accuracy from multiple factors and develop accurate modeling of RNA-seq reads in comparison. We found that the overdispersion rate decreased when sequencing depth increased on the base level. Moreover, the influence of local sequence(s) on the overdispersion rate was notable but no longer significant after adjusting the effect from sequencing depth. Based on these findings, we propose a desirable beta-binomial model with a dynamic overdispersion rate on the base-level proportion of sequencing read counts from two samples.

**Conclusions:** The current study provides thorough insights into the impact of overdispersion at the position level and especially into its relationship with sequencing depth, local sequence, and preparation protocol. These properties of RNA-seq will aid in improvement of the quality control procedure and development of statistical methods for RNA-seq downstream analyses.

**Keywords:** RNA-seq, Non-uniformity, Bias, Base-level modeling, Overdispersion, Beta-binomial, Differential expression analysis

## Background

Today, RNA-seq is a common technique for surveying RNA expression. Because sequencing read counts from individuals often show dispersion of measurements significantly larger than that given by Poisson distribution, fine modeling on this so-called *overdispersion* is required for RNA-seq data analysis [1, 2]. Negative binomial based distributions have been used by edgeR, DESeq/ DESeq2, baySeq, and other methods to model overdispersed RNA-seq data for differential expression (DE) analysis [1–5]. Alternatively, beta-binomial distribution based methods have been proposed [6, 7]. However, these methods are still under development for more accurate model fitting, due to the elusive properties of

RNA-seq read counts, especially from the aspect of dispersion. Dispersion of RNA-seq was strongly related to the sequencing depth [1], which was found to be critical to the power of detection of all expressed genes and differentially expressed genes between groups [8–10]. Previously, we investigated the variance of RNA-seq reads between samples with no biological difference, such as runs of different library preparations from the same sample, and found strong dependency between overdispersion and sequencing depth [7]. In the current study, we continued to study this scenario that samples have the identical genetic background, such as identifying differentially expressed genes in the same cell line with stimulation by a ligand.

RNA-seq data has many biases and effects which make developing accurate methods challenging [11–17]. Li et al. demonstrated the non-uniformity of RNA-seq reads by showing that the number of reads per nucleotide might vary by 100-fold across the same gene, which was caused by random hexamer priming bias in the nucleotide

* Correspondence: Guoshuai.Cai@dartmouth.edu; xiaof@mailbox.sc.edu
[1]Department of Molecular and Systems Biology, Geisel School of Medicine at Dartmouth, Hanover, NH, USA
[4]Department of Epidemiology and Biostatistics, Arnold School of Public Health, University of South Carolina, Columbia, SC, USA
Full list of author information is available at the end of the article

composition at the beginning of transcriptome sequencing reads [12, 13]. Therefore, a naive Poisson model, which assumes counts from all base positions are independently sampled from a Poisson distribution with a single rate proportional to the expression, is not appropriate. Several methods have been proposed to model local sequence related RNA-seq biases for transcript abundance estimation. Li et al. [13] proposed a method to predict variable rates based on local sequence and correct the non-uniformity, alpine [18] used a Poisson generalized linear model to model RNA-seq fragment sequence bias related to fragment GC content and GC stretches, and Salmon [19] provided a fast method with sample-specific bias models to capture fragment GC content bias and other effects. However, capturing the fluctuation at each base position among replicates, which is critical for precise RNA-seq data modeling and accurate differential expression analysis, is out of the research scopes of those tools. In this study, we aim to achieve an accurate modeling of RNA-seq reads with fluctuation estimation at each base position for comparison by taking random hexamer primer effect into consideration.

Given the same influence from the same local sequence of one particular gene, it is reasonable to assume that the mean number of sequencing reads on each base in one experimental condition is consistently proportional to that in another experimental condition. This assumption is supported by the observation in the study of Li et al. that the patterns of sequencing reads mapped to the same local sequences were highly consistent, even across different tissue types [13]. Therefore, we modeled the proportions of base-level coverage comparing two samples based on beta-binomial distribution, assuming the proportions have different dispersion but the same mean. Thus, high variable Poisson rates only enter the process indirectly through the dispersion which is advantageous in modeling. We previously observed decreasing gene-level overdispersion corresponding to increasing sequencing depth [7], which is expected to be true on base pair level as well. Therefore, local sequence composition and sequencing depth might be confounders in estimating overdispersion rate, and this remains unstudied. To investigate this confounding effect, we evaluated and compared three beta-binomial models: a full model with effects of both local sequence and sequencing depth and two reduced models with one of effect each.

Here, we focused on studying the dependency of overdispersion with sequencing depth and local primer sequence at base level. Large-scale consortium-based RNA-seq studies, such as ENCODE [20], MAQC [21], SEQC [22] and others, provide opportunities to investigate the properties of RNA-seq data and evaluate proposed methodologies. We estimated the base-level

overdispersion rate of RNA-seq read count from EN-CODE spike-in dataset which has a large sample size [23]. Also, we investigated the potential biases introduced by library preparation protocols including fragmentation and strand synthesis. We evaluated the fitting performance of the proposed beta-binomial models with a dynamic overdispersion rate and compared them to binomial model and beta-binomial model with a consistent overdispersion rate. In application to DE analysis, we compared our models with widely used methods including binomial test, $t$ test, DESeq [1], edgeR [2] and limma-voom [24]. RNA-seq datasets related to the MAQC project with real-time PCR measurements were used in this comparison [25].

## Methods

### Datasets

Two datasets were used, the ENCODE spike-in dataset [23] and the MAQC dataset with real-time PCR data [25] (Table 1).

### ENCODE dataset

Long NonPolyA RNAs from whole cells were measured in the ENCODE dataset. Two replicates from each of 14 human cell lines (Gm12878, Ag04450, Bj, Huvec, A549, H1hesc, Hepg2, K562, Hsmm, Mcf7, Nhlf, Sknshra, Nhek, and Helas3) were used in this study. Synthetic spike-in standards from the External RNA Control Consortium (ERCC) were sequenced along with human samples following the dUTP strand-specific sequencing protocol [23]. Two primers, mate1 and mate2, were used to distinguish specific strands. The sequencing reads from the ERCC libraries were mapped to the ERCC reference using Bowtie version 0.11.3 with parameters −v2 −m1 [26]. Gene-level abundances were estimated by counting uniquely mapped reads. We used samples (underlined in Table 1) with approximately the same total counts to estimate accurate dispersion between replicates by avoiding bias from sequencing depth. We truncated 76 nucleotides from the end of each gene as no count of 76 base-pair-long read was available in this region.

### MAQC dataset

Bullard et al. measured two distinct MAQC reference samples, brain and UHR, using RNA-seq [25]. Four UHR libraries (A, B, C and D) and one brain library were prepared. RNAs were first fragmented and then converted into cDNAs using random hexamer priming approach. We used STAR [27] to align reads to the UCSC human genome hg19 assembly. Gene-level abundances were estimated by counting uniquely mapped reads in all exons. Additionally, 997 genes had previously been assayed by real-time PCR with high detection

**Table 1** Summary of the datasets used

| ENCODE | ERCC | GSM758567 GSM758572 GSM758573 GSM758577 GSM765389 GSM765391 GSM765396 GSM765398 GSM767845 GSM767847 | | | |
| | | GSM767851 GSM767854 GSM767855 GSM767856 | | | |
| MAQC | Brain | UHR library A | UHR library B | UHR library C | UHR Library D |
| | SRR037455 | SRR037466 | SRR037470 | SRR037473 | SRR037479 |
| | SRR037456 | SRR037467 | SRR037471 | SRR037474 | |
| | SRR037457 | SRR037468 | SRR037472 | SRR037475 | |
| | SRR037458 | SRR037469 | | SRR037476 | |

Training datasets were underlined

specificity and detection sensitivity, which can be used for validation of differential expression detection. We truncated 35 nucleotides from the end of each gene as no count of 35 base-pair-long read was available in this region.

**Estimation of Overdispersion rate $\theta_{ij}$ per base pair**

Let $n_{ij}$ and $m_{ij}$ be the number of mapped reads starting at the $j$-th nucleotide of the $i$-th gene for the two samples in comparison, respectively. The probability mass function for the beta-binomial distribution is

$$f\left(n_{ij}|\alpha_{ij},\beta_{ij},m_{ij}\right) = \binom{n_{ij}+m_{ij}}{n_{ij}} \frac{B\left(n_{ij}+\alpha_{ij},m_{ij}+\beta_{ij}\right)}{B\left(\alpha_{ij}+\beta_{ij}\right)} \tag{1}$$

where $\alpha_{ij}$ and $\beta_{ij}$ are two parameters of the beta-binomial distribution. The beta-binomial distribution can be represented using the following parameters: $p_{ij} = \frac{\alpha_{ij}}{\alpha_{ij}+\beta_{ij}}$ and $\theta_{ij} = \frac{1}{\alpha_{ij}+\beta_{ij}}$ for each $i$ and $j$. Based on our assumption that the proportion of counts per base pair across a gene comparing two samples is a constant, $p_{ij}$ is consistent for all positions on the $i$-th gene, as $p_i$. Analytically, for the $i$-th gene with $J_i$ base pairs, the true and unknown proportion $p_i$ can be estimated as $\frac{\sum_{j=1}^{J_i} n_{ij}}{\sum_{j=1}^{J_i} n_{ij}+\sum_{j=1}^{J_i} m_{ij}}$. Assuming most genes do not change, the neutral proportion of two samples $p_n$ can be estimated from all $(J_1, J_2, \ldots, J_i, \ldots, J_G)$ base pairs of all $G$ genes as $\frac{\sum_{i=1}^{G}\sum_{j=1}^{J_i} n_{ij}}{\sum_{i=1}^{G}\sum_{j=1}^{J_i} m_{ij}+\sum_{i=1}^{G}\sum_{j=1}^{J_i} m_{ij}}$. For any two replicates, the proportion of each gene should be equal to the neutral proportion, that $p_i = p_n$. Based on the beta-binomial distribution, $\theta_{ij}$ can be estimated from the variance calculated from replicates as

$$\hat{\theta}_{ij} = \frac{\frac{1}{R}\sum_{r=1}^{R}\left(\frac{\sigma_{p_{ijr}}}{p_{nr}(1-p_{nr})} - \frac{1}{n_{ijr}+m_{ijr}}\right)}{1-\frac{1}{R}\sum_{r}^{R}\frac{\sigma_{p_{ijr}}}{p_{nr}(1-p_{nr})}} \tag{2}$$

where $r$ denotes the $r$-th pair among $R$ total combination

pairs of replicates and $p_{nr}$ indicates the neutral proportion comparing the $r$-th pair. For the $j$-th nucleotide of the $i$-th gene from the $r$-th pair of replicates, $\sigma_{p_{ijr}}$ indicates the variance of proportion, $n_{ijr}$ and $m_{ijr}$ indicate read counts mapped in the current pair of replicates. We estimated $\sigma_{p_{ijr}}$ from base-level read counts per replicate pair separately and estimated $\theta_{ij}$ according to formula (2).

**Base-level model**

After reparametrizing by $p_i$ and $\theta_{ij}$, the log-likelihood of the beta-binomial (Eq. 1) for the $i$-th gene with $J_i$ base pairs was derived as

$$\log(\mathscr{L}_i) = \sum_{j=1}^{J_i}[\sum_{k=0}^{n_{ij}-1} \log(p_i + k\theta_{ij})$$
$$+ \sum_{k=0}^{m_{ij}-1} \log(1-p_i + k\theta_{ij})$$
$$- \sum_{k=0}^{n_{ij}+m_{ij}-1} \log(1 + k\theta_{ij})] \tag{3}$$

Previously, we proposed an efficient gene-level beta-binomial model for DE analysis with

$$\theta_i = \frac{D_i}{(n_i + m_i)^\gamma},$$

in which $\gamma$ represents the degree of dependency to sequencing depth [7]. $D_i$ is a gene specific factor. In the current study, we assumed $D_i$ to be consistent for all genes as $D$ based on our observation. To achieve a better data fit, we propose a full model here, taking the local sequence around the first nucleotide of a read into consideration:

$$\theta_{ij} = \frac{De^{\left\{\sum_{k=1}^{K}\sum_{h\in\{A,\ T,\ C\}}\beta_{kh}I\left(b_{ijk}=h\right)\right\}}}{(n_{ij}+m_{ij})^\gamma} \tag{4}$$

In this model, $K$ is the length of the surrounding sequence around the $j$-th nucleotide of the $i$-th gene. We set $K = 80$ as suggested in the study of Li et al. [13]

such that the surrounding sequence of 40 nucleotides before and 40 nucleotides after the $j$-th nucleotide was considered. Also, the indictor function $I(b_{ijk} = h)$ is 1 when the
$k$-th base pair is letter $h$, which is A, T, or C exclusively,
and 0 otherwise. $D$, $\beta_{kh}$, and $\gamma$ are unknown parameters
which require estimation. It is natural to assume $D$ varies
among sample pairs and thus pair-specific $D$ will be estimated based on the determined $\beta_{kh}$ and $\gamma$.

We took the logarithm of Eq. 4 and obtained the
following formula that facilitates model fitting:

$$\log(\theta_{ij}) = \log(D) + \sum_{k=1}^{K} \sum_{h \in \{A,T,C\}} \beta_{kh} I(b_{ijk} = h)$$
$$+ \gamma \log(n_{ij} + m_{ij}) \tag{5}$$

Based on the observation of Wu et al. that the distribution of the logarithm of sample dispersion is approximately Gaussian distributed [28], we assumed $\log(\theta_{ij})$
follows a Gaussian distribution and efficiently estimated
these parameters using the linear least-squares approach
in this study. In comparison to the sum of all the
positions in all the genes, the parameter size in Eq 5,
240, is very small.

In order to investigate the confounding effect of the
read depth and local primer sequence on the overdispersion rate, we further developed two reduced beta-
binomial models: primer-free model ($\beta_{kh} = 0$) and depth-
free model ($\gamma = 0$) in which the overdispersion rate was formulated as shown in the following Eqs. 6 and
7 respectively:

$$\log(\theta_{ij}) = \log(D) + \gamma \log(n_{ij} + m_{ij}) \tag{6}$$

$$\log(\theta_{ij}) = \log(D) + \sum_{k=1}^{K} \sum_{h \in \{A,T,C\}} \beta_{kh} I(b_{ijk} = h) \tag{7}$$

We refer to models shown in Eqs. 4, 5, 6, 7 as models
with a dynamic dispersion rate. Alternatively, a beta-
binomial model with a constant overdispersion rate was
obtained when $\gamma = 0$ and $\beta_{kh} = 0$.

## Model fitting

To validate the dependency between local sequence,
sequencing depth, and overdispersion, we set training
datasets and test datasets. Training datasets shown in
Table 1 were used to investigate the dependency of overdispersion, sequencing depth, and local sequence and
determine the parameters of $\gamma$ and $\beta_{kh}$. Then, the captured dependency was borrowed to achieve better data
fit and higher power of differential expression analysis
on the test datasets.

(a) Estimation of $\gamma$ and $\beta_{kh}$

1. Estimate $\hat{p}_n = \dfrac{\sum_{i=1}^{G} \sum_{j=1}^{J_i} n_{ij}}{\sum_{i=1}^{G} \sum_{j=1}^{J_i} n_{ij} + \sum_{i=1}^{G} \sum_{j=1}^{J_i} m_{ij}}$ on the
   training set.
2. Set $p_n$ as a known parameter and obtain $\hat{\theta}_{ij}$
   according to Eq. 2. The least-squares estimation
   method is then applied to the full model (Eq. 5),
   the primer-free model (Eq. 6) and the depth-free
   model (Eq. 7) to estimate $\gamma$ and $\beta_{kh}$.

(b) Modeling test samples

1. Initialize $\hat{p}_i = \hat{p}_n$ in the beta-binomial model
   (Eq. 3) on the test set.
2. Borrow the estimation of $\gamma$ and $\beta_{kh}$ from the
   training set for the full model and the primer-free
   model separately.
3. Set $p_i$ as a known parameter and maximize the
   beta-binomial log likelihood (Eq. 3) to estimate
   pair-specific $D$.
4. Set $\theta_{ij}$ according to Eq. 4 as a known parameter
   and maximize the beta-binomial log likelihood to
   update $\hat{p}_i$. This step is skipped when comparing
   replicates.
5. Proceed to step 3 unless the deviance decreases
   less than 1%. This step is skipped when
   comparing replicates.

## Likelihood ratio test

According to the likelihood ratio test, $-2 \ln \mathcal{L}(p_n) + 2 \ln \mathcal{L}(p_i)$ follows the $\chi^2$ distribution with 1 degree of freedom, where $p_i$ is the proportion for gene $i$ and $p_n$ is the
neutral proportion. Equation 3 models the proportion of
a pair of samples, which can be used to test samples
without replicates by borrowing information from
previously measured replicates. When replicates were
available, we calculated the sum of their pairwise $\chi^2$
scores comparing samples from two groups and obtained $p$-values with a summation of degrees of freedom.

## Model comparison

In this study, we evaluated the overall fitting of models.
First, we evaluated the fitting of linear models shown in
Eqs. 5 and 7 to study the confounding effect on overdispersion from sequencing depth and local sequence.
Second, we compared models on data fitting in comparing the sequencing read counts from two replicates.
Third, we assessed the performance of models in DE
analysis. The strategies of comparison were shown in
Fig. 1, including dataset usage, model fitting, test
statistic, and evaluation purpose. Detailed methods for
evaluating the models are as follows.

(a) Goodness of fit of the depth-free model (Eq. 7) and
   the full model (Eq. 5) on $\log(\theta_{ij})$.
   We calculated the coefficient of determination $R^2$.
   We utilized the 5-fold cross validation strategy.
   Each of the training sets (shown in Table 1) were

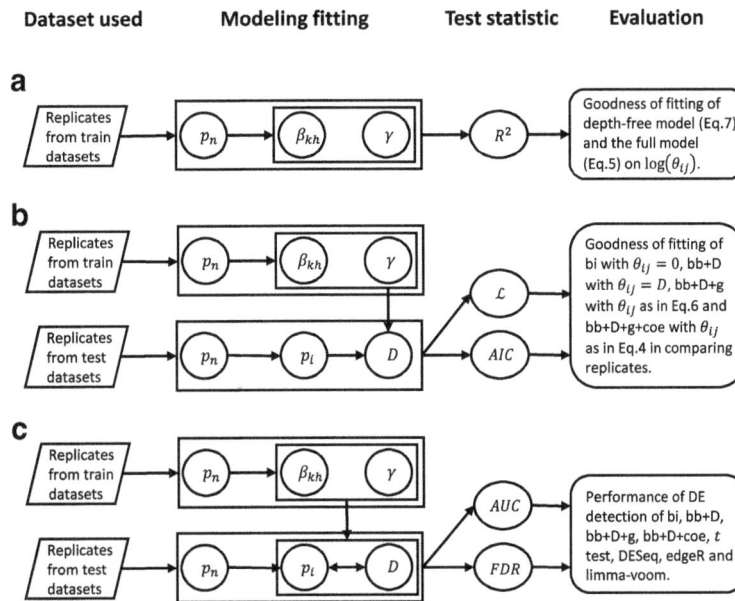

**Fig. 1** The strategy of model fitting and comparison

randomly split into five groups of equal size. In each round, we fit our model using four of these five groups, and then calculated $R^2$ on the remaining subset by the regression sum of squares divided by the total sum of squares. The process was repeated for 10 times and the overall cross-validation $R^2$ was determined by the mean.

(b) Goodness of fit of four models in comparing replicates, including the binomial model (bi) with $\theta_{ij} = 0$, the beta-binomial model (bb + D) with $\theta_{ij} = D$, the reduced primer-free model (bb + D + g) with $\theta_{ij}$ as in Equation 6, and the full model (bb + D + g + coe) with $\theta_{ij}$ as in Eq. 5.

*Likelihood value* We calculated the maximum likelihood values of pairwise comparisons of replicates to evaluate the goodness of fit. Proportion $p_i$ was estimated as $\hat{p}_n$ and fixed for all four models. Sequentially, other parameters were determined by our model fitting strategy (iterative fitting was skipped as $p_i$ was fixed), and likelihood values were calculated based on estimated parameters. The $\chi^2$ test was performed on $D = -2\ln(\mathscr{L}_{nested}) + 2\ln(\mathscr{L})$, where $\mathscr{L}$ and $\mathscr{L}_{nested}$ are likelihoods for a model and its nested model, respectively.

*AIC* Akaike information criterion (AIC) is a measure of the relative goodness of fit of a statistical model. AIC was calculated by definition as $2k - 2\ln(\mathscr{L})$, where $k$ was the number of parameters and $\mathscr{L}$ is the maximum-likelihood value. The overall AICs were determined by the mean of all AICs from pairwise replicates.

(c) Performance of DE detection of four models (bi, bb + D, bb + D + g, bb + D + coe) and widely used methods including $t$ test, DESeq, edgeR and limma-voom. Evaluation was performed on MAQC dataset which has standard data for validation.

*AUC* The area under the receiver operating characteristic curve (AUC) was determined by the method described in our previous study [7].

*False housekeeping gene detections* To test the false discovery control ability, we assumed that housekeeping genes detected as differentially expressed genes at a given $p$-value were false discoveries. We compares the numbers of falsely discovered housekeeping genes given specific numbers of significantly differentially expressed genes. A list of 3804 housekeeping genes identified by Eisenberg and Levanon were used in this study [29].

### DE analysis methods in comparison

We compared our models with $t$ test, binomial test, DESeq, edgeR and limma-voom on DE analysis. A two-tailed $t$ test was performed on total counts normalized and logarithm transformed RNA-seq read counts. Four brain samples (SRR037455, SRR037456, SRR037457 and SRR037458) were compared to four UHR samples (SRR037469, SRR037472, SRR037476 and SRR037479) in the test datasets. The DE analyses in this study were performed using R version 3.2.5 and we applied packages "DESeq 1.22.1", "edgeR 3.12.1" and "limma 3.26.9" to test the difference of sequencing read counts. "GLM" approach was used in DESeq and edgeR DE analysis.

Normalization and model fitting were performed using the default parameters. When estimating the dispersions by DESeq, "local" fitType, "maximum" sharingMode and "pooled" estimation methods were used. All other parameters were set to the default in all DESeq, edgeR and limma-voom analyses. Functions of our proposed methods are available in the github repository (https:// github.com/GuoshuaiCai/BBDG.git).

## Result

### Base-pair Overdispersion rate decreases with sequencing depth

We empirically investigated the effect of sequencing depth on the overdispersion rate of the measurement per base. Analyzing the ENCODE spike-in dataset, we calculated the variance of the proportion of the reads mapped to the $j$-th base pair of the $i$-th gene from replicates and then determined the overdispersion rate $\theta_{ij}$ (described in Methods). Figure 2 shows that the overdispersion rate was strongly inversely correlated with sequencing depth. That is, the overdispersion rate continually decreased as the sequencing depth increased without a sign of saturation. The correlation was sufficiently strong, causing the majority of the points to be concentrated along a line. This supported our assumption that all genes have consistent $D$ and the proposed linear model shown by Equation 6. Moreover, local sequences starting with GGGG were found to have more sequencing reads and larger overdispersion than those starting with AAAA, indicating that hexamer priming

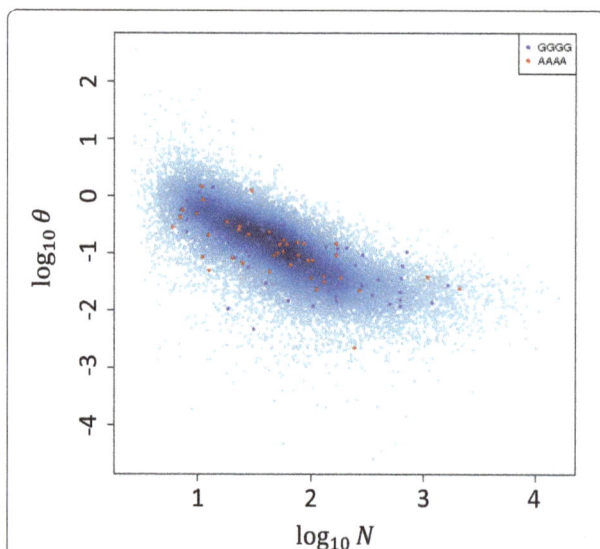

**Fig. 2** The relationship of overdispersion and sequencing depth. The base-level overdispersion rate of proportion $\theta_{ij}$ versus the mean tag counts in base 10 log scale. The $\theta_{ij}$ values were computed from replicates from the ENCODE spike-in training dataset. The *blue* and *red* points are for the positions with local sequences starting with GGGG and AAAA, respectively

might influence the overdispersion rate through affecting sequencing read counts. Therefore, local sequence and sequencing depth are not independent from each other and might be confounders.

### Sequencing procedure introduces extra noise

Elements of the sequencing procedure (e.g., fragmentation methods, random hexamer priming, etc.) can introduce types of bias to RNA-seq measurements [12]. We compared the overdispersion rates estimated from two datasets with different RNA-seq protocols (described in Methods) in Fig. 3. Interestingly, in the ENCODE dataset, the overdispersion rates were significantly larger at the tail (less than ~200 base pairs) of the genes. The same result was obtained in the calculation of the variance (Additional file 1: Figure S1). This may suggest a bias in ENCODE dataset. Therefore, we removed the reads mapped to the last 200 base pairs of each gene in our analyses to avoid this extra bias. However, no such difference was observed in MAQC UHR datasets.

This discrepancy might be explained by the different processes in sequencing library preparation of these two studies. In the ENCODE study, fragment selection after cDNA PCR amplification might lead to a loss of many fragments located at the transcript tails, thereby introducing an additional error. By contrast, according to the protocol used in the MAQC study, fragmentation was carried out prior to cDNA PCR amplification, leading to the same process of selection across the entirety of the gene.

### Models of the Overdispersion rate

To reveal the confounding effects of the local primer sequence and the sequencing depth on the overdispersion rate, we studied two models: the full model with parameters for both the local primer sequencing and the sequencing depth and the depth-free model without parameters for the sequencing depth (described in Methods). After the linear formula transformation (Eq. 5), 240 coefficients of 80 positions around the primers were estimated efficiently. Coefficients estimated from MAQC UHR data were plotted against their corresponding positions in Fig. 4. From the depth-free model, we observed a similar pattern to those reported by Hansen et al. and Li et al. [12, 13] (Fig. 4a, c). However, no such pattern was observed from the full model (Fig. 4b, d). We observed similar results from the ENCODE spike-in data as well (Additional file 1: Figure S2). Both Hansen et al. and Li et al. demonstrated an association between hexamer primer and measurement count number. Plus, we observed in this study that the overdispersion rate on base pair decreased with increasing sequencing depth (Fig. 2). These findings lead to an inference that a hexamer primer might influence the overdispersion rate by affecting the count number; consequently, upon adjustment by count

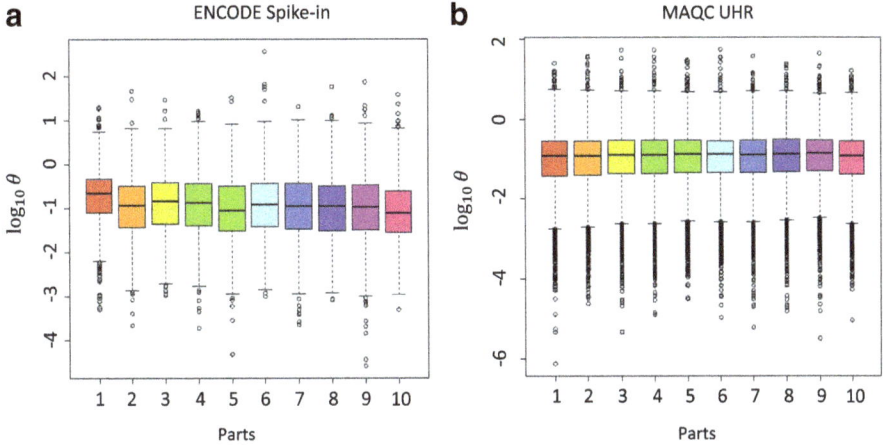

**Fig. 3** The pattern of overdispersion on parts of genes. The overdispersion rate was estimated on any position in 10 categories with equal data points according to the distance to the end of the genes. Part 1 is located on the gene tail and Part 10 is located on gene start. **a** ENCODE spike-in dataset. **b** MAQC UHR dataset. For strand-specific sequencing, only reads generated with mate2 primers on antisense strand were investigated. x-axis shows categories from the end of the genes

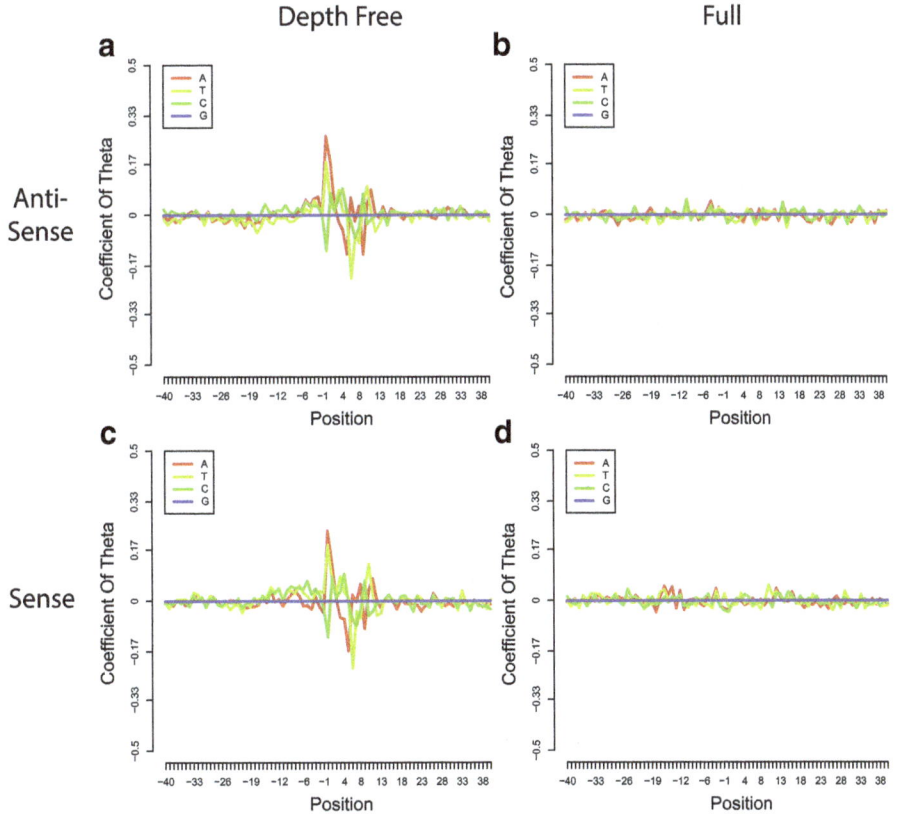

**Fig. 4** Coefficients of local sequence from the MAQC UHR dataset. x-axis shows the positions around the 5′ end of mapped reads, which was labelled as 0. Coefficients were calculated by two models on different strands: **a** Depth-free model on antisense strand, **b** Full model on antisense strand, **c** Depth-free model on sense strand and **d** Full model on sense strand

number, the relationship between the use of a hexamer primer and the overdispersion rate was no longer significant as observed in the full model (Fig. 4a, c). In addition, we calculated the coefficient of determination $R^2$ using a 5-fold cross-validation strategy (described in Methods). $R^2$ values of 0.481 and 0.488 were obtained for the depth-free model and the full model, respectively, from the MAQC UHR data; while values of 0.270 and 0.273, respectively, were obtained from the ENCODE spike-in data. Therefore, about half of the variance was explained by our models for the MAQC UHR dataset. Also, as expected, the depth-free model achieved a similar $R^2$ with the full model.

We investigated the influence of primers corresponding to the reads from the antisense and sense strands, respectively. We observed from the MAQC UHR dataset that reads mapped to antisense and sense strands showed quite similar patterns (Fig. 4a, c), which was consistent with the finding of Hansen et al. [12]. However, the reads on the sense strand should not be primer-related because they were synthesized by the RNase H niche method without hexamer priming. Hansen et al. [12] explained that the hexamer primer might not be completely digested. In contrast, this dependency was not observed on sense strands in the ENCODE spike-in dataset (Additional file 1: Figure S2). Its strand-specific protocol might be responsible for the different patterns on two strands, but further validation studies are required. In the present study, we estimated coefficients of local sequence separately for each strand in the present study.

## Comparison of four models
### Goodness of fit
Comparing likelihood values is a straightforward way to select statistical models. We calculated likelihood values from four models: bi, bb + D, bb + D + g and bb + D + g + coe (described in Methods). As expected, the models with additional parameters had higher maximum likelihood values. Figure 5a shows the increase of likelihood value of the ENCODE spike-in dataset. The bb + D model made a huge jump from the bi model (improved by 30% - 90%, Chi-square test $p$-value <0.001). And the parameter $\gamma$ in dynamic $\theta_{ij}$ in bb + D + g model also improved the fit by roughly 15% (Chi-square test $p$-value <0.001). However, the full model had no significant improvement from the primer-free model (Chi-square test $p$-value = 1), and the latter had the lowest AIC (Fig. 5c). We observed similar results in both training and test datasets and from the MAQC dataset as well (Fig. 5b, d; results for training dataset not shown). However, due to the small experimental library effect in the MAQC UHR dataset [7], increase of likelihood was not as significant as that shown in the ENCODE dataset. As expected, no

difference of data fit was observed on MAQC brain samples which were from the same library (Fig. 5b).

### DE detection
Further, we compared the AUC of DE analysis performance based on four models (bi, bb + D, bb + D + g, bb + D + coe) and widely used methods including $t$ test on logarithm transformed RNA-seq read counts, DESeq, edgeR and limma-voom (Fig. 6a). As a result of the small library effect, no significant difference was observed between these four binomial based models when comparing MAQC brain and UHR samples, which agreed with our previous gene level study [7]. However, our beta-binomial based models (bb + D, bb + D + g, bb + D + coe) had good performances close to DESeq, edgeR, and limma-voom, which are slightly better than binomial-test and significantly superior to Student's $t$ test. Similar results were observed on the false discovery control, but DESeq, edgeR and bb + D falsely identified the least number of housekeeping genes given a certain number of discoveries (Fig. 6b). Testing the different library preparations from a same sample, bb + D produced non-uniformly distributed $p$-values with insufficient small ones (Fig. 7b), whereas bi had an overabundance of small $p$-values (Fig. 7a). In contrast, the histogram of the $p$-values was more flat for the beta-binomial models with a dynamic overdispersion rate, bb + D + g and bb + D + g + coe (Fig. 7c, d), indicating that the errors between samples from different libraries were captured more accurately by these two models.

## Discussion
In this study, we accurately modeled of the non-uniformity of RNA-seq read counts at the base level. We investigated the relationship of overdispersion rate with sequencing depth, local sequence, and library preparation protocols to study the properties of overdispersion. Based on these properties, base-level models are proposed to estimate the overdispersion rate accurately.

To the best of our knowledge, this is the first study of the confounding effects from sequencing depth and local sequence on overdispersion rate. We found they are strongly associated with each other. First, the overdispersion rate decreases as the sequencing depth increases on the base level. Second, random hexamer priming can notably influence the overdispersion rate. However, with the count number as a covariate in the modeling, the local sequence showed little influence on the overdispersion rate. Consequently, it is preferable to use the primer-free model with less parameters for superior computing efficiency and power.

Together with various systematic errors that have been identified in differential RNA-seq protocols and platforms [30, 31], our new findings provide important

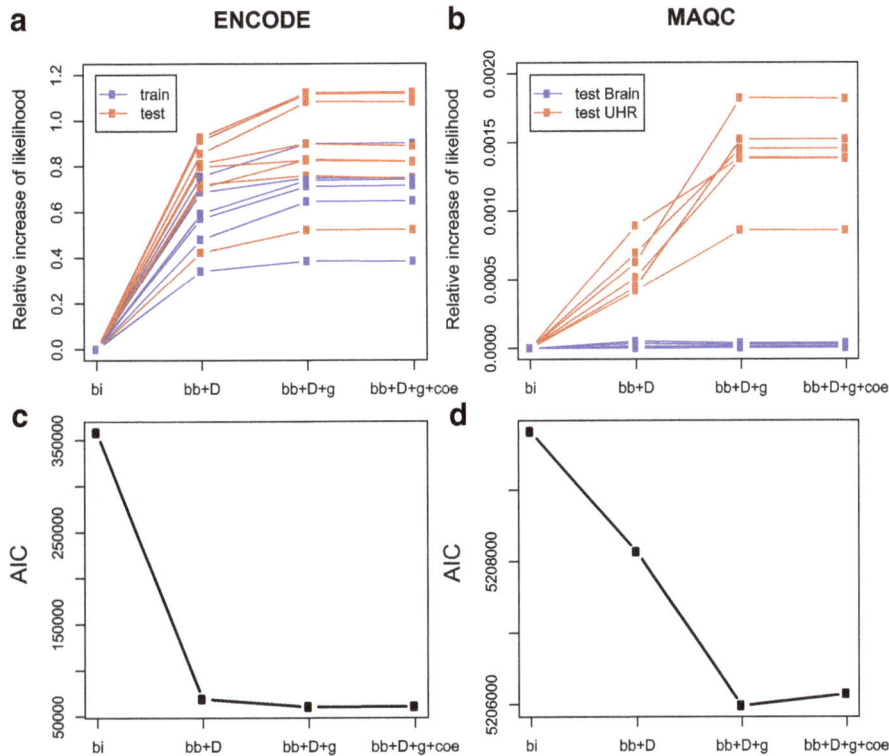

**Fig. 5** Goodness-of-fit on pairwise comparison of replicates. **a**, **c** ENCODE dataset. **b**, **d** MAQC dataset. **a**, **b** The mean percentages of change in the likelihood value compared to the nested model. **c**, **d** The mean AICs measured for four models

insights into the development of bias correction strategies in RNA-seq analyses. Based on the observation of extra noise on the tails of transcripts when fragmentation was performed before PCR, we concluded that experimental protocols before sequencing may influence the overdispersion rate of the RNA-seq reads and that the order of steps in the protocol matters. Therefore, we suggest removing the last 200 base pairs if fragmentation is performed before PCR in RNA-seq library preparation. Moreover, we suggest further studies of RNA-seq non-uniformity on sense and antisense strands separately.

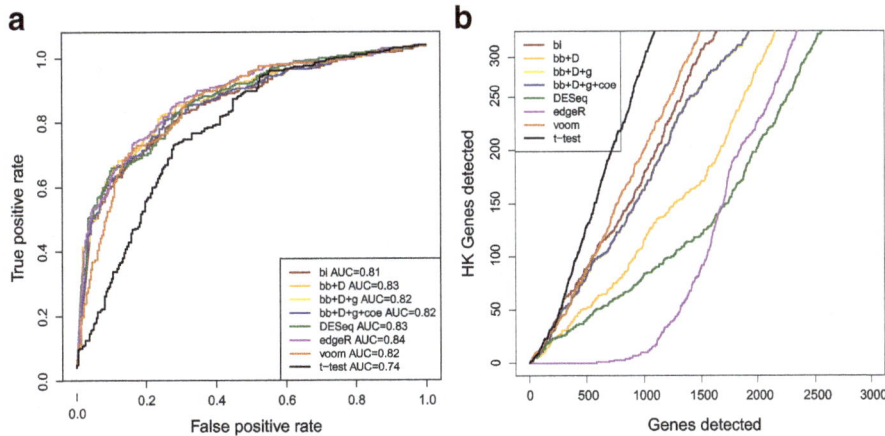

**Fig. 6** Performance of DE detection. DE genes were detected by comparing MAQC Brain samples to UHR samples. 7 methods (bb + D, bb + D + g, bb + D + coe, $t$ test on logarithm transformed RNA-seq read counts, DESeq, edgeR and limma-voom) were applied. **a** ROC and **b** false housekeeping gene detections were used to evaluate their performances

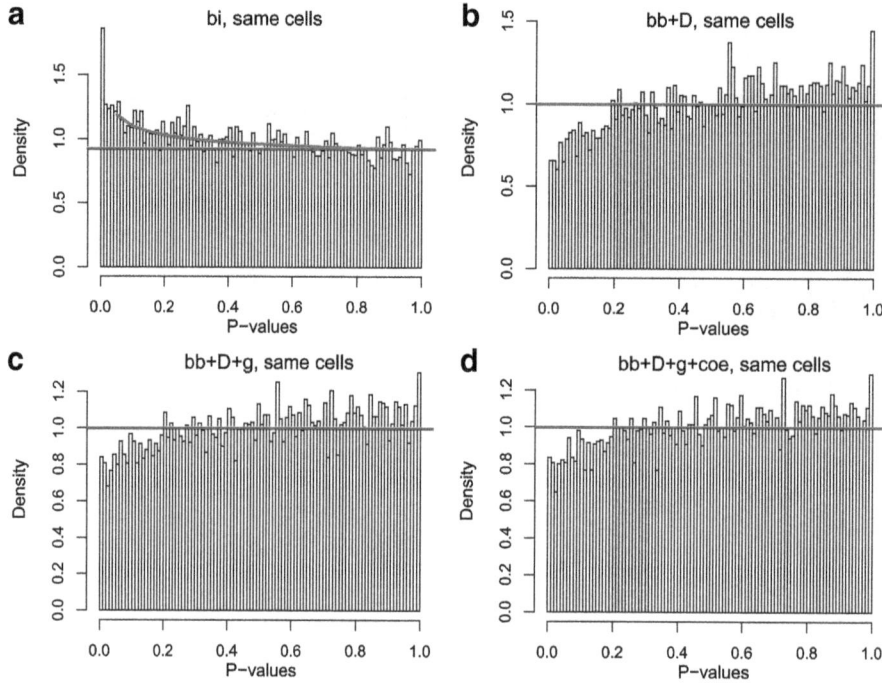

**Fig. 7** Histograms of *p*-values from comparison of replicates. *p*-values were calculated by **a** binomial model, **b** beta-binomial model with constant $\theta_{ij}$, **c** the primer-free beta-binomial model and **d** the full beta-binomial model. *Blue* line indicates an estimated uniform distributions; *green* line indicates a mixture distribution of beta distribution and uniform distribution

Compared with models which ignore the overdispersion rate or use a constant overdispersion rate, bb + D + g accounting for a dynamic overdispersion rate fits the RNA-seq counts best with the highest likelihood value and the lowest AIC. It produced a similar AUC to popular DE analysis methods including DESeq, edgeR and limma-voom. bb + D showed the best false discovery control among proposed models, which may result from its insufficient power to detect small alterations (Fig. 7b). Theoretically, our model has two main advantages compared to these widely used DE analysis tools: (1) the catastrophe-resistant ability. The gene-level read counts might be susceptible to positions with high counts but with high fluctuations. Our model addresses this issue by down-weighting those unreliable read counts with highly variable dispersion rate and (2) borrowing information from spike-in measurement. Usually few experimental replicates are performed due to the cost. Spike-in transcripts, measured along with the samples, can be used a cost-effective alternative to estimate overdispersion rate.

The current study investigated the dependency between the overdispersion rate and the sequencing depth using replicates with no biological variance. However, the relationship between replicates with biological variance and systematic effect remains elusive. SEQC dataset, which was specifically designed to test the intra- and inter-site reproducibility [22, 32], warrants the future studies of that relation in the context of systematic effects. Also, the current model can be used to detect any base-level changes including gene expression alteration and differential exon usage. The exon level or isoform level differential analysis is thus required to take different usage of exons between samples into consideration.

## Conclusions

In conclusion, the current study provides thorough insights into the property of the overdispersion rate on the position level, especially into its relationship with sequencing depth, local sequence, and preparation protocol. These properties of RNA-seq will aid in improvement of quality control procedures and the development of statistical methods for downstream RNA-seq data analyses. Based on these properties, we propose a method to model the non-uniformity measurement in comparison study. Still, new sequencing strategies and protocols are emerging rapidly, such as the PCR-free sequencing technique [33]. The properties of sequencing reads as well as the biases and effects vary among different platforms. Future studies on investigating these properties are necessary to improve the methods for modeling RNA-seq data.

## Abbreviations
AIC: Akaike information criterion; AUC: Area under the receiver operating characteristic curve; DE: Differential expression; ERCC: External RNA Control Consortium; PCR: Polymerase chain reaction

## Acknowledgements
We would like to thank Jennifer M. Franks and Stephanie C. Her for editing the draft.

## Funding
Not applicable.

## Authors' contributions
GC, SL and FX conceived the study. GC and FX developed the methods, performed the analysis, and drafted the manuscript. XZ assisted in data analysis and modeling. SL and XZ reviewed and revised the manuscript. All authors have read and approved the final manuscript.

## Competing interests
The authors declare that they have no competing interests.

## Author details
[1]Department of Molecular and Systems Biology, Geisel School of Medicine at Dartmouth, Hanover, NH, USA. [2]Department of Environmental Health Sciences, Arnold School of Public Health, University of South Carolina, Columbia, SC, USA. [3]Department of Bioinformatics and Computational Biology, The University of Texas MD Anderson Cancer Center, Houston, TX, USA. [4]Department of Epidemiology and Biostatistics, Arnold School of Public Health, University of South Carolina, Columbia, SC, USA.

## References
1. Anders S, Huber W. Differential expression analysis for sequence count data. Genome Biol. 2010;11(10):R106.
2. Robinson MD, McCarthy DJ, Smyth GK. edgeR: a Bioconductor package for differential expression analysis of digital gene expression data. Bioinformatics. 2010;26(1):139–40.
3. Love MI, Huber W, Anders S. Moderated estimation of fold change and dispersion for RNA-seq data with DESeq2. Genome Biol. 2014;15(12):550.
4. Hardcastle TJ, Kelly KA. baySeq: empirical Bayesian methods for identifying differential expression in sequence count data. BMC bioinformatics. 2010;11:422.
5. Trapnell C, Hendrickson DG, Sauvageau M, Goff L, Rinn JL, Pachter L. Differential analysis of gene regulation at transcript resolution with RNA-seq. Nat Biotechnol. 2013;31(1):46–53.
6. Zhou YH, Xia K, Wright FA. A powerful and flexible approach to the analysis of RNA sequence count data. Bioinformatics. 2011;27(19):2672–8.
7. Cai G, Li H, Lu Y, Huang X, Lee J, Muller P, Ji Y, Liang S. Accuracy of RNA-Seq and its dependence on sequencing depth. BMC Bioinformatics. 2012; 13(Suppl 13):S5.
8. Blencowe BJ, Ahmad S, Lee LJ. Current-generation high-throughput sequencing: deepening insights into mammalian transcriptomes. Genes Dev. 2009;23(12):1379–86.
9. Tauber S, von Haeseler A. Exploring the sampling universe of RNA-seq. Stat Appl Genet Mol Biol. 2013;12(2):175–88.
10. Tarazona S, Garcia-Alcalde F, Dopazo J, Ferrer A, Conesa A. Differential expression in RNA-seq: a matter of depth. Genome Res. 2011;21(12):2213–23.
11. Gao L, Fang Z, Zhang K, Zhi D, Cui X. Length bias correction for RNA-seq data in gene set analyses. Bioinformatics. 2011;27(5):662–9.
12. Hansen KD, Brenner SE, Dudoit S. Biases in Illumina transcriptome sequencing caused by random hexamer priming. Nucleic Acids Res. 2010;38(12):e131.
13. Li J, Jiang H, Wong WH. Modeling non-uniformity in short-read rates in RNA-Seq data. Genome Biol. 2010;11(5):R50.
14. Roberts A, Trapnell C, Donaghey J, Rinn JL, Pachter L. Improving RNA-Seq expression estimates by correcting for fragment bias. Genome Biol. 2011;12(3):R22.
15. Schwartz S, Oren R, Ast G. Detection and removal of biases in the analysis of next-generation sequencing reads. PLoS One. 2011;6(1):e16685.
16. Taub MA, Corrada Bravo H, Irizarry RA. Overcoming bias and systematic errors in next generation sequencing data. Genome Medicine. 2010;2(12):87.
17. Zheng W, Chung LM, Zhao H. Bias detection and correction in RNA-sequencing data. BMC Bioinformatics. 2011;12:290.
18. Love MI, Hogenesch JB, Irizarry RA. Modeling of RNA-seq fragment sequence bias reduces systematic errors in transcript abundance estimation. Nat Biotechnol. 2016;34(12):1287–91.
19. Patro R, Duggal G, Love MI, Irizarry RA, Kingsford C. Salmon provides fast and bias-aware quantification of transcript expression. Nat Methods. 2017;14(4):417–9.
20. Consortium EP. An integrated encyclopedia of DNA elements in the human genome. Nature. 2012;489(7414):57–74.
21. Consortium M, Shi L, Reid LH, Jones WD, Shippy R, Warrington JA, Baker SC, Collins PJ, de Longueville F, Kawasaki ES, et al. The MicroArray quality control (MAQC) project shows inter- and intraplatform reproducibility of gene expression measurements. Nat Biotechnol. 2006;24(9):1151–61.
22. Consortium SM-I. A comprehensive assessment of RNA-seq accuracy, reproducibility and information content by the sequencing quality control consortium. Nat Biotechnol. 2014;32(9):903–14.
23. Jiang L, Schlesinger F, Davis CA, Zhang Y, Li R, Salit M, Gingeras TR, Oliver B. Synthetic spike-in standards for RNA-seq experiments. Genome Res. 2011;21(9):1543–51.
24. Law CW, Chen Y, Shi W, Smyth GK. voom: Precision weights unlock linear model analysis tools for RNA-seq read counts. Genome Biol. 2014;15(2):R29.
25. Bullard JH, Purdom E, Hansen KD, Dudoit S. Evaluation of statistical methods for normalization and differential expression in mRNA-Seq experiments. BMC Bioinformatics. 2010;11:94.
26. Langmead B, Trapnell C, Pop M, Salzberg SL. Ultrafast and memory-efficient alignment of short DNA sequences to the human genome. Genome Biol. 2009;10(3):R25.
27. Dobin A, Davis CA, Schlesinger F, Drenkow J, Zaleski C, Jha S, Batut P, Chaisson M, Gingeras TR. STAR: ultrafast universal RNA-seq aligner. Bioinformatics. 2013;29(1):15–21.
28. Wu H, Wang C, Wu Z. A new shrinkage estimator for dispersion improves differential expression detection in RNA-seq data. Biostatistics. 2013;14(2):232–43.
29. Eisenberg E, Levanon EY. Human housekeeping genes, revisited. Trends in genetics : TIG. 2013;29(10):569–74.
30. Harismendy O, Ng PC, Strausberg RL, Wang X, Stockwell TB, Beeson KY, Schork NJ, Murray SS, Topol EJ, Levy S, et al. Evaluation of next generation sequencing platforms for population targeted sequencing studies. Genome Biol. 2009;10(3):R32.
31. Raz T, Kapranov P, Lipson D, Letovsky S, Milos PM, Thompson JF. Protocol dependence of sequencing-based gene expression measurements. PLoS One. 2011;6(5):e19287.
32. Li S, Labaj PP, Zumbo P, Sykacek P, Shi W, Shi L, Phan J, Wu PY, Wang M, Wang C, et al. Detecting and correcting systematic variation in large-scale RNA sequencing data. Nat Biotechnol. 2014;32(9):888–95.
33. Mamanova L, Andrews RM, James KD, Sheridan EM, Ellis PD, Langford CF, Ost TW, Collins JE, Turner DJ. FRT-seq: amplification-free, strand-specific transcriptome sequencing. Nat Methods. 2010;7(2):130–2.

# 10

# Optimal alpha reduces error rates in gene expression studies

J. F. Mudge[1], C. J. Martyniuk[2] and J. E. Houlahan[1*]

## Abstract

**Background:** Transcriptomic approaches (microarray and RNA-seq) have been a tremendous advance for molecular science in all disciplines, but they have made interpretation of hypothesis testing more difficult because of the large number of comparisons that are done within an experiment. The result has been a proliferation of techniques aimed at solving the multiple comparisons problem, techniques that have focused primarily on minimizing Type I error with little or no concern about concomitant increases in Type II errors. We have previously proposed a novel approach for setting statistical thresholds with applications for high throughput omics-data, optimal α, which minimizes the probability of making either error (i.e. Type I or II) and eliminates the need for post-hoc adjustments.

**Results:** A meta-analysis of 242 microarray studies extracted from the peer-reviewed literature found that current practices for setting statistical thresholds led to very high Type II error rates. Further, we demonstrate that applying the optimal α approach results in error rates as low or lower than error rates obtained when using (i) no post-hoc adjustment, (ii) a Bonferroni adjustment and (iii) a false discovery rate (FDR) adjustment which is widely used in transcriptome studies.

**Conclusions:** We conclude that optimal α can reduce error rates associated with transcripts in both microarray and RNA-seq experiments, but point out that improved statistical techniques alone cannot solve the problems associated with high throughput datasets – these approaches need to be coupled with improved experimental design that considers larger sample sizes and/or greater study replication.

**Keywords:** Microarrays, RNA-seq, Type I and II error rates, High throughput analysis, Multiple comparisons, Post-hoc corrections, Optimal α

## Background

Microarrays and next generation sequencing (NGS) have been described as technological advances that provide global insight into cellular function and tissue responses at the level of the transcriptome. Microarray and NGS are used in experiments in which researchers are testing thousands of single-gene hypotheses simultaneously. In particular, microarrays and NGS are often used to test for differences in gene expression across two or more biological treatments. These high-throughput methods commonly use *p*-values to distinguish between differences that are too large to be due to sampling error and

those that are small enough to be assumed to be due to sampling error. There is little doubt that microarrays/NGS have made a large contribution to our understanding of how cells respond under a variety of contexts, for example in environmental, developmental, and the medical sciences [1–3].

High throughput methods have, however, made interpretation of hypothesis testing more difficult because of the large number of comparisons that are done in each experiment [4]. That is, researchers will examine the effects of one or more treatment on the abundance of 1000s of transcripts. For each gene, there will be replication and a null hypothesis test of whether there is a statistically significant difference in relative expression levels among treatments. In most cases, the statistical threshold for rejecting the null hypothesis (i.e. α) is

---

* Correspondence: jeffhoul@unb.ca
[1]Department of Biology, Canadian Rivers Institute, University of New Brunswick, Saint John, NB E2L 4L5, Canada
Full list of author information is available at the end of the article

α = 0.05 although it may occasionally be set at a lower value such as 0.01 or 0.001. Thus, for any individual comparison, the probability of rejecting the null hypothesis when it is true is 5% (if the threshold is set at 0.05). When multiple tests are conducted on 1000's of transcripts, this creates the potential for hundreds of false positives (i.e. Type I error) at the experiment-wide scale, with the expected number of false positives depending on both the number of tests conducted (known) and the number of those tests where the treatment has no effect on gene expression (unknown). Researchers identified this problem early on and have used a variety of post-hoc approaches to controlling for false positives [5–9].

Approaches for adjusting *p*-values and reducing false 'positives' when testing for changes in gene expression, such as Bonferroni or Benjamini-Hochberg procedures, are designed to control experiment-wide error probabilities when many comparisons are being made. Typically they reduce the α for each test to a value much smaller than the default value of 0.05, so that the experiment-wide error is not as inflated due to the large number of comparisons being made. They all share the characteristic that they only explicitly address probabilities of Type I errors [4]. This has the effect of increasing the probability of false negatives (i.e. Type II errors) to varying degrees. This focus on Type I errors implies that it is much worse to conclude that gene expression is affected by a treatment when it is not than to conclude that expression is not affected by a treatment when, in reality, it is. Although there has been some focus on methods designed to balance Type I and Type II error rates [10], researchers rarely discuss the Type II implications of controlling Type I errors, and we believe this suggests that most researchers simply are not considering the effect of post-hoc adjustments on Type II error rates. Krzywinski and Altman [4] note the problem and offer practical advice, "we recommend always performing a quick visual check of the distribution of P values from your experiment before applying any of these methods". Our position is that this does not go far enough; we assert that post-hoc corrections to control Type I errors don't make sense unless (1) the researcher knows their Type II error probability (i.e. power) and (2) has explicitly identified the relative costs of Type I and II errors. We have recently developed a solution, optimal α, that balances α (the acceptable threshold for Type I errors – usually 0.05) and β (the acceptable threshold for Type II errors – often 0.20 but the standard practice is more variable than for α), minimizing the combined error rates and eliminating the need for any post-hoc adjustment [11–13]. In the context of transcriptomics, this reduces the overall error rate in identifying differentially expressed genes by finding the best trade-off between minimizing false detections of differential expression and minimizing nondetection of true differential expression.

While we have demonstrated this approach in the context of detecting environmental impacts of pulp and paper mills [13], it is of particular value in fields such as transcriptomics where many tests are conducted simultaneously. While microarrays and RNA-seq have been tremendous technological advances for transcriptomics, when coupled with low sample sizes, it magnifies multiple comparisons problems. The objectives of this paper were to apply optimal α to a set of published microarray data to demonstrate that using the optimal α approach reduces the probabilities of making errors and eliminates the need for any post-hoc adjustments. In addition, we discuss modifications to the experimental design of microarray data that directly address the problem of multiple comparisons.

## Methods
### Data collection
We collected data on microarray experiments conducted in teleost fishes spanning a period of 10 years (see Additional file 1: Data S1). Environmental toxicology is the research focus of one of the authors, however we point out here that this approach is not confined to aquatic toxicology and is applicable across disciplines. The search for microarray fish studies was conducted from January 2011–August 2011 using the search engines *Web of Science, Science Direct, PubMed (National Center for Biotechnology Information)*, and *Google Scholar*. Keywords and combinations of key words used in the search engine included "microarray", "gene expression", "DNA chip", "transcriptomics", "arrays", "fish", "teleost", and "aquatic". In addition, references from papers were reviewed for information on manuscripts not identified by the search engines. This intensive search resulted in representation of studies encompassing a wide range of teleost fishes and scientific disciplines (e.g. physiology, toxicology, endocrinology, and immunology). There were a total of 242 studies surveyed for information (Additional file 1: Data S1).

The extracted data from microarray experiments included fish species, family, sex, analyzed entity (e.g. cell, tissue), experimental treatment, concentration (if applicable), duration, exposure type, microarray platform, type of normalization, number of biological replicates, endpoints assessed, number of differentially expressed genes (DEGs) identified by the researchers, total gene probes on the array, average fold change of DEGs, and the method of post hoc analysis. Approximately 50% of these studies applied an FDR threshold as the method of choice for detecting differentially expressed genes. All microarray data were normalized by the authors of the original studies using the method of their choice (there are different methods but they differ only slightly).

## Calculating optimal A

For each study, we calculated optimal $\alpha$ levels [11] that minimized the combination of Type I and Type II error probabilities, and compared the Type I and II error probabilities resulting from this approach to those associated with using $\alpha = 0.05$. Data are summarized on a per-paper, not on a per test basis.

The calculation of an optimal $\alpha$ level requires information concerning the test type, the number of replicates, the critical effect size, the relative costs of Type I vs. Type II errors, and the relative prior probabilities of null vs. alternate hypotheses. Optimal $\alpha$ calculations are based on minimizing the combined probability or cost of Type I and II errors by examining the mean probability of making an error over the entire range of possible $\alpha$ levels (i.e. from 0 to 1). This is a 5 step process. Step 1 – Choose an $\alpha$ level between 0 and 1. Step 2 – Calculate $\beta$ for the chosen $\alpha$, sample size, critical effect size and variability of the data (this can be achieved using a standard calculation of statistical power for the statistical test being used, beta is 1 – statistical power), Step 3 – Calculate the mean of $\alpha$ and $\beta$, Step 4 – Choose a new $\alpha$ slightly smaller than the previous $\alpha$ and compare the mean error probability with the previous iteration. If it is larger choose a new $\alpha$ slightly larger than the previous $\alpha$. If it is smaller choose a new $\alpha$ slightly smaller than the current $\alpha$, Step 5 – Keep repeating until the improvement in mean error probability fails to exceed the chosen threshold – at this stopping point you have identified optimal $\alpha$. Several assumptions or constraints were made to enable consistent optimal $\alpha$ analysis of studies with a wide degree of technical and statistical methodologies:

## Assumptions

(1) We used the number of biological replicates in each group as the level of replication in each study. Microarrays were sometimes repeated on the same biological replicates but this was not treated as true replication, regardless of whether it was treated as replication within the study. Similarly, spot replicates of each gene on a microarray were not treated as replication, regardless of whether it was treated as replication within the study There were 39 studies which had levels of biological replication of $n = 1$, or $n = 2$. These studies were omitted from further statistical analysis, leaving 203 studies with biological replication of $n \geq 3$.

(2) Two hundred and three of the 242 studies identified were suitable for analysis and to ensure that the optimal $\alpha$ value in all 203 studies were calculated on the same test and are comparable, we analyzed each study as an independent, two-tailed, two-sample t-test even though some studies used confidence interval, randomization or Bayesian analyses instead of t-tests. ANOVA was also occasionally used instead of t-tests but even in cases where ANOVA was used, it was the post-hoc pairwise comparisons between each of the experimental groups and a control group that were the main focus. These post-hoc pairwise comparisons are typically t-tests with some form of multiple comparison adjustment. One-tailed or paired t-tests were sometimes used instead of two-tailed independent tests. Although these tests do increase power to detect effects, they do so by placing restrictions on the research question being asked.

(3) Critical effect size, in the context of t-tests, is the difference in the endpoint (in this case, gene expression) between treatment and control samples that you want to detect. In traditional null hypothesis testing, $\beta$ is ignored and critical effect sizes are not explicitly considered. We calculated optimal $\alpha$ levels at three potential critical effect sizes, defined in terms relative to the standard deviation of each gene (1 SD, 2 SD and 4 SD). Fold-changes or percent changes from the control group were occasionally used as critical effect sizes, but we avoided these effect sizes to maintain consistency across optimal $\alpha$ calculations and because we believe the difference in expression relative to the variability in the gene is more important than the size of effect relative to the control mean of the gene. A two-fold change in a gene may be well within the natural variability in expression of one gene and far outside the natural variability in expression of another gene. However, there may be contexts where fold-changes are more appropriate than standard deviations and optimal alpha can accommodate this by setting separate optimal alphas for each gene. Separate thresholds would be required because detecting a 2-fold change in a highly variable gene would result in a larger optimal alpha than detecting a 2-fold change a gene with little variability.

(4) We assumed the relative costs of Type I and Type II errors to be equal, representing a situation where researchers simply want to avoid errors, regardless of type. However, optimal alpha can accommodate any estimates of the relative costs of Type I and II errors (See code in Additional file 2: Appendix S1). So, where there is clear evidence of different relative Type I and II error costs they should be integrated into optimal alpha estimates. Multiple comparison adjustments that reduce Type I error rates without (1) estimating Type II error probability and (2) the relative costs of Type I and II errors are ill-advised.

(5) We assumed that the prior probabilities for the meta-analysis and the required within and among-study replication to be - $H_A$ prior probability = 0.50 and $H_o$ prior probability = 0.50. For the simulations comnparing optimal alpha error rates relative to traditional multiple comparisons approaches we used three prior probability scenarios - Scenario 1: $H_A$ prior probability = 0.50 and

$H_o$ prior probability = 0.50, Scenario 2: $H_A$ prior probability = 0.25 and $H_o$ prior probability = 0.75, and Scenario 3: $H_A$ prior probability = 0.10 and $H_o$ prior probability = 0.90. There has been relatively little empirical work done describing the proportion of genes that are affected by treatments in microarray studies but [14] examined the effects of mutations in different subunits of the transcriptional machinery on the percent of genes that showed differential expression and concluded that the percent of genes ranged from 3 to 100% with a mean of 47.5%. Another estimate by Pounds and Morris [15] suggested that slightly more than half the genes in a study examining two strains of mice showed differential gene expression. In addition, accurate estimation of global gene expression has been complicated by inappropriate assumptions about gene expression data [16, 17] and further research in this area is critical. There is no way of being certain of how many true positives and true negatives there are in each study but in the absence of any prior knowledge the rational assumption is that the probabilities are equal (Laplace's principle of indifference). However, in the context of gene expression a differential expression prior probability of 0.50 is at the high end and so we also examined $H_A$ prior probabilities of 0.25 and 0.10. Prior probabilities other than equal can be accommodated by optimal $\alpha$ and using other prior probabilities would result in quantitative differences in the results. However, the general conclusion that optimal alpha error rates will always be as low or lower than traditional approaches does not depend on the assumed prior probabilities.

## Analyses

*Minimum average of $\alpha$ and $\beta$ for each of the 203 studies at 3 different critical effect sizes (1, 2, and 4 SD's):* We calculated the average of $\alpha$ and $\beta$ using optimal $\alpha$ and the traditional approach of setting $\alpha = 0.05$. To do this we calculated optimal $\alpha$ for each of the 203 studies as described above, extracted the $\beta$ associated with optimal $\alpha$ for each of the 203 studies, and calculated the average of $\alpha$ and $\beta$. Similarly, we extracted the $\beta$ associated with $\alpha = 0.05$ for each of the 203 studies when and then calculated the average of $\alpha$ and $\beta$. We could then compare the average of $\alpha$ and $\beta$ for optimal $\alpha$ and $\alpha = 0.05$.

### Effect of post-hoc corrections on error rates (see Additional file 3: Data S2)
We simulated 15,000 tests of the effect of a treatment for each of 3 prior probability scenarios and 3 effect size scenarios. The prior probability scenarios were Scenario 1: $H_A$ prior probability = 0.50 and $H_o$ prior probability = 0.50, Scenario 2: $H_A$ prior probability = 0.25 and $H_o$ prior probability = 0.75, and Scenario 3: $H_A$ prior probability = 0.10 and $H_o$ prior probability = 0.90. The

effect size scenarios were Scenario 1: 1 SD, Scenario 2: 2 SD, and Scenario 3: 4 SD. All comparisons were made using two-tailed, two-sample t-tests. Based upon experience and the literature, gene expression studies vary widely in the proportion of genes that are differentially expressed and usually show a small effect (1 SD). We only select larger values (2 and 4 SD) above to illustrate the application of the optimal $\alpha$ compared to other post-hoc tests. All differences between treatment and control were chosen from normal distributions that reflected the 'true' differences (i.e. 0, 1, 2 or 4 SD's). We calculated error rates using optimal $\alpha$, $\alpha = 0.05$, $\alpha = 0.05$ with a Bonferroni correction and $\alpha = 0.05$ with a Benjamini-Hochberg False Discovery Rate correction. We then compared the total number of errors across all 15,000 tests for the four different approaches. For example, using Scenario 1 for both effect size and prior probability we compared the 4 approaches under the assumption that half the genes were affected by the treatment and the size of the effect for those 7500 genes was 1 SD. By contrast, using Scenario 3 for both prior probability and effect size we assumed that 1500 of the genes were differentially expressed and the size of the effect was 4 SD.

### Minimum number of within-study replicates needed to meet desired error rate
The same iterative process that can be used to calculate minimum average error rate for a specific sample size can be used to calculate minimum sample size for a specific average error rate. Here we identified a range of minimum acceptable average error rates from 0.00001 to 0.125 (reflecting the common practice of $\alpha = 0.05$ and $\beta = 0.2$) and calculated the minimum sample size required to achieve the desired error rates for 3 different effect sizes (i.e. 1, 2, and 4 SD's).

### Minimum among-study replication needed to meet desired error rate
An alternative to large within-study replication is to synthesize similar studies that have been replicated several times. Here we simply identified how often a study would have to be repeated at a specific optimal $\alpha$ to achieve a desired error rate. For example, to detect a 1 SD difference between treatment and control using a 2-sample 2-tailed t-test with a sample size of 4 the optimal $\alpha$ is 0.29. If optimal $\alpha$ is 0.29 but the desired error rate is 0.00001 we solve for x in $0.00001 = 0.29^x$ and conclude that 10 studies showing a significant difference between treatment and control expression would be necessary to meet our desired threshold. Similarly, $\beta$ at this optimal $\alpha$ is 0.38 and we would need 12 studies showing no significant difference between treatment and control to meet our desired error rate.

## Results

### Meta-analysis

Across all studies, the median number of genes tested with ≥3 replicates was 14,900 and the median number of replicates ≥3 was 4 (Fig. 1). Using optimal α instead of α = 0.05 resulted in a reduced probability of the combination of Type I and II errors of 19–29% (Table 1). One important conclusion is that under current practices, tests intended to detect effect sizes of 1 SD will make errors in 5% of tests if there are no treatment effects on any of the genes but the median level of replication (3 replicates per treatment) will make errors in more than 77% of tests if all the genes are affected by the treatment(s) and will make errors in more than 41% of the tests if half the genes are affected by treatment(s). That is, they will maintain the probability of making Type I errors at 0.05 but have highly inflated Type II error probabilities (i.e. low power). For tests intended to detect a 2 SD effect size, again the overall error rate will be 5% if none of the genes are affected by treatment(s) but will be more than 34% with median replication if all the genes are affected by the treatment(s) and almost 20% if half the genes are affected by treatment(s). So, current experimental design practices for microarrays are inadequate, especially with respect to Type II errors, and post-hoc corrections are not mitigating this problem (see below). It is important to note that we do not know the true error rates – for that we would have to know how many and which genes were actually differentially expressed. These are estimated error rates under the assumptions that (1) prior probabilities for $H_A$ and $H_0$ are equal and (2) critical effect sizes are SD =1,2, or 4.

### Sample size estimates (within-study replication)

Many microarray and RNA-seq studies ($n = 3$ per treatment) are only appropriate for detecting effects sizes at least of 4 SD at Type I and II error rates of 0.05 or greater. Traditionally, the least conservative acceptable error rates have been set at 0.05 for Type I errors and, when they consider Type II errors, at 0.20 for Type II errors. This implies an average error rate of 0.125 (i.e. $[\alpha + \beta] / 2 \le 0.125$, the average of α and β associated with using α = 0.05 and achieving 80% statistical power). To detect an effect of 2 SD at an error rate of 0.125 would require sample sizes greater than 5 per treatment, and detecting an effect of 1 SD would require at least 16 samples per treatment (Table 2).

### Repeating the experiment (among- study replication)

Using the optimal α minimize the combined probabilities of Type I and II errors, to reduce the probability of making a Type I error for any particular gene to 0.10 for an effect size of 1 SD using a sample size at the high end of what is usually used in microarray studies (i.e. 10 replicates per treatment), an experiment would have to show a statistically significant effect for a gene in two consecutive experiments. To reduce the probability to 0.001, the experiment would have to show a statistically significant effect for a gene in 4 consecutive experiments. Similarly, to reduce the probability of missing a real effect to 0.10 for an effect size of 1 SD, an experiment with 10 replicates per treatment would have to show no statistically significant effect for a gene in two consecutive experiments. To reduce this probability to 0.001, there would have to be no statistically significant results in 5 consecutive experiments. On the other hand, if the critical effect size is 4 SD, one experiment is all that would be needed for most traditional sample sizes and error rates (Tables 3a and b)

### Optimal α versus no post-hoc and traditional post-hoc analyses

We used three sets of simulated scenarios of 15,000 tests with 4 replicates per group. The scenarios differed in the

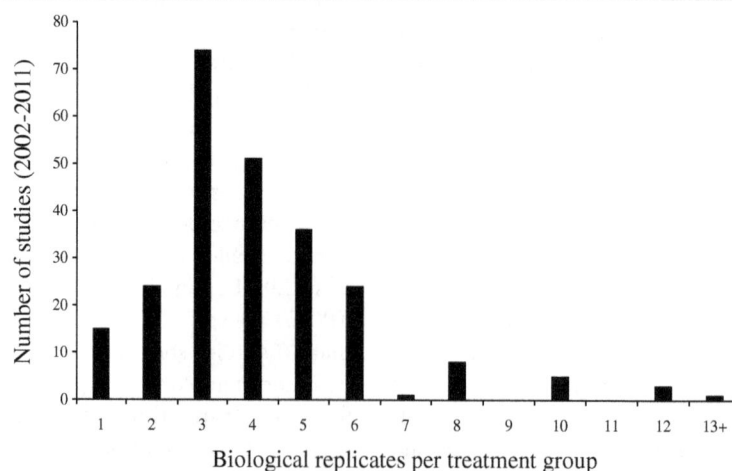

**Fig. 1** Distribution of the number of biological replicates per treatment group over 203 fish microarray papers published between 2002 and 2011

**Table 1** Type I and II error rates: *Median, 1st and 3rd quartiles, minimum and maximum α, β, average of α and β, and implied costs of Type I/II errors, evaluated for the standard α = 0.05 and for the optimal α approach, at 3 critical effect sizes (1, 2, and 4 SD), for 203 fish microarray papers with tests that have at least 3 replicates, published between 2002 and 2011 (assuming two-tailed, two-sample t-tests)*

| Critical effect size | Decision threshold | Statistical parameter | Minimum | 1st quartile | Median | 3rd quartile | Maximum |
|---|---|---|---|---|---|---|---|
| 1 standard deviation | standard α | α | 0.05 | 0.05 | 0.05 | 0.05 | 0.05 |
| | | β | 0.088 | 0.71 | 0.78 | 0.84 | 0.84 |
| | | (α + β)/2 | 0.069 | 0.38 | 0.41 | 0.45 | 0.45 |
| | | Implied Type I/II error cost ratio | 1.5 | 2.6 | 3.1 | 3.4 | 3.9 |
| | optimal α | α | 0.064 | 0.26 | 0.29 | 0.32 | 0.32 |
| | | β | 0.070 | 0.34 | 0.38 | 0.42 | 0.42 |
| | | (α + β)/2 | 0.067 | 0.30 | 0.33 | 0.37 | 0.37 |
| | | Implied Type I/II error cost ratio | 1 | 1 | 1 | 1 | 1 |
| 2 standard deviations | standard α | α | 0.05 | 0.05 | 0.05 | 0.05 | 0.05 |
| | | β | 0.0000015 | 0.21 | 0.34 | 0.54 | 0.54 |
| | | (α + β)/2 | 0.025 | 0.13 | 0.20 | 0.29 | 0.29 |
| | | Implied Type I/II error cost ratio | 0.00011 | 3.4 | 4.5 | 5.1 | 5.1 |
| | optimal α | α | 0.0011 | 0.11 | 0.15 | 0.21 | 0.21 |
| | | β | 0.00094 | 0.10 | 0.13 | 0.18 | 0.18 |
| | | (α + β)/2 | 0.0010 | 0.10 | 0.14 | 0.19 | 0.19 |
| | | Implied Type I/II error cost ratio | 1 | 1 | 1 | 1 | 1 |
| 4 standard deviations | standard α | α | 0.05 | 0.05 | 0.05 | 0.05 | 0.05 |
| | | β | 0 | 0.00023 | 0.0038 | 0.052 | 0.052 |
| | | (α + β)/2 | 0.025 | 0.025 | 0.027 | 0.051 | 0.051 |
| | | Implied Type I/II error cost ratio | 0.00011 | 0.014 | 0.19 | 1.9 | 1.9 |
| | optimal α | α | 0.000000031 | 0.013 | 0.028 | 0.065 | 0.065 |
| | | β | 0.000000017 | 0.0065 | 0.014 | 0.031 | 0.031 |
| | | (α + β)/2 | 0.000000024 | 0.0096 | 0.021 | 0.048 | 0.048 |
| | | Implied Type I/II error cost ratio | 1 | 1 | 1 | 1 | 1 |

assumed prior probability of the null and alternate hypotheses with Scenario 1 assuming a 50% probability of the alternate being true, Scenario 2 a 25% probability and Scenario 3 a 10% probability. Each scenario examined 3 different critical effect sizes, 1, 2, and 4 SD.

**Table 2** Replicate estimates: *Number of replicates per treatment needed to achieve maximum acceptable averages of α and β of 0.00001, 0.0001, 0.001, 0.01, 0.05, 0.1, and 0.125, at critical effects sizes of 1, 2, and 4 SD, for an independent two-tailed, two sample t-test*

| Maximum acceptable average of α and β | Number of samples required | | |
|---|---|---|---|
| | CES = 1SD | CES = 2SD | CES = 4SD |
| 0.00001 | 156 | 43 | 15 |
| 0.0001 | 120 | 33 | 12 |
| 0.001 | 85 | 24 | 9 |
| 0.01 | 50 | 14 | 5 |
| 0.05 | 27 | 8 | 3 |
| 0.1 | 18 | 6 | 3 |
| 0.125 | 16 | 5 | 3 |

Optimal α consistently resulted in fewer or the same overall errors when compared to any of the following approaches; no post-hoc test, Bonferroni correction, or an FDR (Table 4A-C).

Optimal α reduced the number of overall errors (α and β) relative to other approaches by as much as 96%. When the assumed prior probability of $H_A$ is low (i.e. 10%) and the critical effect size is small (i.e. 1 SD) Bonferroni and FDR adjustments do as well as optimal alpha because the threshold is so stringent that they find no significant results. Thus, the only error that is made is a Type II error and these approaches miss all 1500 true effects. Optimal alpha makes slightly fewer Type II error, at 1495. In addition, half of the 10 significant results found using the optimal alpha threshold are false positives resulting in the same number of errors which was 1500 and the same number as for Bonferroni or FDR adjustments. No post-hoc adjustments under these circumstances result in many more true effects being detected but also many more type I errors – more than half of the statistically significant results are false positives. The most

**Table 3** A and B. Required number of replicates: *A) Number of times a study would have to be repeated with the same conclusion to achieve an α of 0.00001, 0.0001, 0.001, 0.01, 0.05, 0.1, and 0.2, at critical effects sizes of 1, 2, and 4 SD, for an independent two-tailed, two sample t-test. (B) Number of times a study would have to be repeated with the same conclusion to achieve a β of 0.00001, 0.0001, 0.001, 0.01, 0.05, 0.1, and 0.2, at critical effects sizes of 1, 2, and 4 SD, for an independent two-tailed, two sample t-test*

A.

| Critical effect size | Within-study replication | Replication of the experiment needed to achieve | | | | | | |
|---|---|---|---|---|---|---|---|---|
| | | α = 0.00001 | α = 0.0001 | α = 0.001 | α = 0.01 | α = 0.05 | α = 0.1 | α = 0.2 |
| 1 SD | 4 | 10 | 8 | 6 | 4 | 3 | 2 | 2 |
| | 6 | 9 | 7 | 5 | 4 | 3 | 2 | 2 |
| | 8 | 8 | 6 | 5 | 3 | 3 | 2 | 2 |
| | 10 | 7 | 6 | 4 | 3 | 2 | 2 | 1 |
| 2 SD | 4 | 7 | 5 | 4 | 3 | 2 | 2 | 1 |
| | 6 | 5 | 4 | 3 | 2 | 2 | 1 | 1 |
| | 8 | 4 | 4 | 3 | 2 | 1 | 1 | 1 |
| | 10 | 4 | 3 | 2 | 2 | 1 | 1 | 1 |
| 4 SD | 4 | 4 | 3 | 2 | 2 | 1 | 1 | 1 |
| | 6 | 3 | 2 | 2 | 1 | 1 | 1 | 1 |
| | 8 | 2 | 2 | 2 | 1 | 1 | 1 | 1 |
| | 10 | 2 | 2 | 1 | 1 | 1 | 1 | 1 |

B.

| Critical effect size | Within-study replication | Replication of the experiment needed to achieve | | | | | | |
|---|---|---|---|---|---|---|---|---|
| | | β = 0.00001 | β = 0.0001 | β = 0.001 | β = 0.01 | β = 0.05 | β = 0.1 | β = 0.2 |
| 1 SD | 4 | 12 | 10 | 8 | 5 | 4 | 3 | 2 |
| | 6 | 10 | 8 | 6 | 4 | 3 | 2 | 2 |
| | 8 | 9 | 7 | 6 | 4 | 3 | 2 | 2 |
| | 10 | 8 | 6 | 5 | 3 | 2 | 2 | 2 |
| 2 SD | 4 | 6 | 5 | 4 | 3 | 2 | 2 | 1 |
| | 6 | 5 | 4 | 3 | 2 | 2 | 1 | 1 |
| | 8 | 4 | 3 | 3 | 2 | 1 | 1 | 1 |
| | 10 | 4 | 3 | 2 | 2 | 1 | 1 | 1 |
| 4 SD | 4 | 3 | 3 | 2 | 2 | 1 | 1 | 1 |
| | 6 | 2 | 2 | 2 | 1 | 1 | 1 | 1 |
| | 8 | 2 | 2 | 1 | 1 | 1 | 1 | 1 |
| | 10 | 2 | 2 | 1 | 1 | 1 | 1 | 1 |

conservative post-hoc adjustment, Bonferroni, routinely resulted in the largest overall error rate when the critical effect size was large while not using a post-hoc analysis resulted in fewer errors than either a Bonferroni or FDR except when the prior probability and critical effect size were small. Of course, the distribution of Type I and II errors varies among approaches, with no post hoc adjustment and the FDR adjustment resulting in a relatively large number of Type II errors when the critical effect size was 1 or 2 SD. However, no post-hoc adjustment produced relatively large number of Type I errors when the critical effect size was 4 SD while the FDR approach still resulted in more type II errors. Bonferroni resulted in zero Type I errors but

a large number of Type II errors at all effect sizes. Optimal alpha resulted in a much more even distribution of Type I and II errors except when the prior probability and critical effect size was small.

## Discussion

Researchers using high throughput expression techniques enjoy the benefits of global analyses, but must acknowledge the statistical issues associated with an extremely large number of comparisons. Problems may become exacerbated as even higher throughput techniques such as RNA-Seq become more common and genome projects continue to increase the capacity of

**Table 4** A-C. A comparison of the mean number of significant results among four different procedures for evaluating significance of multiple comparisons: *Type I errors, and Type II errors for 100 iterations of 15,000 simulated differential gene expression test using (1) α = 0.05 for all tests, (2) a Bonferroni correction to adjust the family-wise error rate (FWER) to 0.05, (3) the Benjamini-Hochberg procedure to adjust the false-discovery rate (FDR) to 0.05, and (4) optimal α*

| Critical effect size (CES) | Average of 100 iterations of 15,000 tests | α = 0.05 | Bonferroni FWER = 0.05 | Benjamini-Hochberg FDR = 0.05 | Optimal α |
|---|---|---|---|---|---|
| **A.** | | | | | |
| CES = 1SD | # of significant results | 2046 | 0 | 1 | 6776 |
| | # of Type I errors | 376 | 0 | 0 | 2143 |
| | # of Type II errors ≥ CES | 5829 | 7500 | 7499 | 2867 |
| | # of Type I and II errors | 6205 | 7500 | 7499 | 5010 |
| | % error reduction by using optimal α | 19.3% | 33% | 33% | - |
| CES = 2SD | # of significant results | 5298 | 3 | 1709 | 7659 |
| | # of Type I errors | 379 | 0 | 43 | 1130 |
| | # of Type II errors ≥ CES | 2581 | 7497 | 5834 | 970 |
| | # of Type I and II errors | 2960 | 7497 | 5876 | 2100 |
| | % error reduction by using optimal α | 29% | 72% | 64% | - |
| CES = 4SD | # of significant results | 7848 | 61 | 7560 | 7608 |
| | # of Type I errors | 378 | 0 | 190 | 212 |
| | # of Type II errors ≥ CES | 30 | 7439 | 130 | 105 |
| | # of Type I and II errors | 408 | 7439 | 320 | 317 |
| | % error reduction by using optimal α | 22% | 96% | 1% | - |
| **B.** | | | | | |
| CES = 1SD | # of significant results | 1400 | 0 | 0 | 1456 |
| | # of Type I errors | 562 | 0 | 0 | 590 |
| | # of Type II errors ≥ CES | 2912 | 3750 | 3750 | 2883 |
| | # of Type I and II errors | 3474 | 3750 | 3750 | 3473 |
| | % error reduction by using optimal α | 0.02% | 7% | 7% | - |
| CES = 2SD | # of significant results | 3032 | 1 | 119 | 3537 |
| | # of Type I errors | 562 | 0 | 5 | 791 |
| | # of Type II errors ≥ CES | 1280 | 3749 | 3636 | 1004 |
| | # of Type I and II errors | 1842 | 3749 | 3641 | 1795 |
| | % error reduction by using optimal α | 3% | 52% | 51% | - |
| CES = 4SD | # of significant results | 4295 | 31 | 3665 | 3826 |
| | # of Type I errors | 560 | 0 | 136 | 200 |
| | # of Type II errors ≥ CES | 15 | 3719 | 221 | 124 |
| | # of Type I and II errors | 575 | 3719 | 358 | 324 |
| | % error reduction by using optimal α | 44% | 91% | 9% | - |
| **C.** | | | | | |
| CES = 1SD | # of significant results | 1012 | 0 | 0 | 10 |
| | # of Type I errors | 680 | 0 | 0 | 5 |
| | # of Type II errors ≥ CES | 1167 | 1500 | 1500 | 1495 |
| | # of Type I and II errors | 1847 | 1500 | 1500 | 1500 |
| | % error reduction by using optimal α | 19% | 0% | 0% | - |

**Table 4** A-C. A comparison of the mean number of significant results among four different procedures for evaluating significance of multiple comparisons: *Type I errors, and Type II errors for 100 iterations of 15,000 simulated differential gene expression test using (1) α = 0.05 for all tests, (2) a Bonferroni correction to adjust the family-wise error rate (FWER) to 0.05, (3) the Benjamini-Hochberg procedure to adjust the false-discovery rate (FDR) to 0.05, and (4) optimal α (Continued)*

| | | | | | |
|---|---|---|---|---|---|
| CES = 2SD | # of significant results | 1662 | 1 | 3 | 1083 |
| | # of Type I errors | 677 | 0 | 0 | 334 |
| | # of Type II errors ≥ CES | 515 | 1499 | 1497 | 752 |
| | # of Type I and II errors | 1192 | 1499 | 1498 | 1086 |
| | % error reduction by using optimal α | 9% | 28% | 27% | - |
| CES = 4SD | # of significant results | 2169 | 12 | 1261 | 1539 |
| | # of Type I errors | 675 | 0 | 56 | 143 |
| | # of Type II errors ≥ CES | 6 | 1488 | 295 | 105 |
| | # of Type I and II errors | 681 | 1488 | 350 | 248 |
| | % error reduction by using optimal α | 64% | 83% | 29% | - |

Type II error rates and optimal α levels were evaluated using three different critical effect sizes (CES), representing effects as large as 1, 2, and 4 standard deviations (SD) of the data. The 15,000 simulated tests had 4 replicates in the experimental and control groups, and were constructed such that (A) $H_A$ prior probability = 0.50, $H_o$ prior probability = 0.50; (B) $H_A$ prior probability = 0.25, $H_o$ prior probability = 0.75; (C) $H_A$ prior probability = 0.10, $H_o$ prior probability = 0.90

microarray platforms. RNA-seq experiments are currently restricted due to cost to small sample sizes for each comparison which further exacerbates the error rates. Researchers have generally dealt with the issue of multiple comparisons by using one or more post-hoc adjustments designed to control Type I error rates [18, 19] and it is unlikely that one can publish transcriptomic datasets without using some form of post-hoc correction (e.g. FDR [20], Bonferroni, Tukey's range test, Fisher's least significant difference and Bayesian algorithms). Techniques are more conservative (i.e. less likely to result in a Type I error) or less conservative (more likely to result in a Type I error) and implicit in choosing one technique over another is a concern about making a Type II error. That is, the only reason to use a less conservative post-hoc adjustment is if one is concerned about the increasing Type II error rate associated with lowering the probability of making a Type I error. This has, inevitably, led to a large-scale debate that has been relatively unproductive because it is rarely focused on the fundamental issue, that all post-hoc adjustments are designed to reduce Type I error rates (i.e. concluding gene expression has been affected by a treatment when it has not) with little or no explicit regard for the inevitable increase in Type II error rates (i.e. concluding that the treatment has had no effect on gene expression when it has) [21]. Any informed decision about post-hoc adjustments requires a quantitative understanding of both α and β probabilities [22, 23] and a clear assessment of the relative costs of Type I and II errors. However, no post-hoc test currently attempts to explicitly and quantitatively integrate control of Type I and II errors simultaneously and the result is that none of them minimize either the overall error rates or costs of making an error.

One proposed solution to balancing concerns is to set Type I and II error thresholds to be equal [24]. However, the threshold that minimizes the probability of making an error may not be where the Type I and II error probabilities are equal and if Type I and II errors have equal costs, then we should seek to minimize their average probability with no concern for whether the individual probabilities are equal. This is a critical and underemphasized problem in bioinformatics. Our results demonstrate that using optimal α results in reduced error rates compared to using $p = 0.05$ with or without post-hoc corrections. However, it is unlikely that the improvement in error rates attributed to using optimal α will be the same as those estimated here. These results were calculated based on the assumption that the prior probabilities of the alternate being true were 0.5, 0.25 and 0.10, that the costs of Type I and II errors are equal, that the targeted critical effect sizes are 1, 2, or 4 SD and that the results in these 203 studies are representative of all disciplines. But optimal α can accommodate different assumptions about prior probabilities, relative error costs and critical effect sizes and, though the degree to which optimal alpha is superior to traditional approaches may vary, the fundamental conclusion that optimal α error probabilities are as good or better than traditional approaches holds under different assumptions about prior probabilities or critical effect sizes. That said, this is only certain to hold true when we make the assumption implied in null hypothesis testing, that there is either no effect ($H_0$) or there is an effect as large as the critical effect size ($H_A$).

**Multiple comparisons problem**

One particular advantage of optimal α is that it makes post-hoc corrections unnecessary and, in fact, undesirable (correction implies that something desirable has occurred when that isn't necessarily so – we were tempted to call it a post-hoc distortion). This should dispel some of the complexity and confusion surrounding the analysis of transcriptomic data. However, while optimal α can minimize the errors associated with the large number of comparisons made using microarrays for example, neither optimal α nor any form of post-hoc correction can eliminate the problems associated with multiple comparisons. Any post-hoc test that is done to lower the probability of a Type I error will increase the probability of making a Type II error. While optimal α minimizes the probability of making an error, there will still be an enormous number of unavoidable errors made simply because we are doing a large number of comparisons. There is no simple solution to solving the effects of multiple comparisons of error rates. However, progress can be made by developing new standards for the experimental design of microarray data including large increases in within-study replication, increased among study replication and/or use of 'network' approaches which broaden hypotheses to include suites of genes and reduce the total number of hypotheses being tested.

## Experimental design solutions
Our results suggest that standard within-study replication (i.e. 3–8 replicates per treatment) is adequate for critical effect sizes of 2 or 4 SD's at a target overall error rate of 0.05. Thus, increased replication would only be warranted if detecting smaller effect sizes were desirable. However, we question whether error rates = 0.05 are appropriate when 44 thousand comparisons are being made because a threshold error rate of 0.05 still results in thousands of errors. This is a particular problem when even a handful of statistically significant results might be considered reason enough for publication. We suggest that where thousands of comparisons are being made, the standard for statistical significance must be higher, say, 0.0001 or 0.00001 but not through the use of post-hoc corrections that will increase the probability of Type II errors. To meet these standards and retain high statistical power, within-study replication would require dramatic increases in the number of biological replicates used in experiments. Currently, the standard appears to be 3–8 biological replicates per treatment. Replication is limited by a variety of factors including financial costs, available person hours, sample availability, and physical space. However, where possible, it would often be preferable to test fewer genes using much larger samples sizes, especially because the price of microarrays/NGS will likely to continue to drop, making replication of 50, 100 or 200 possible and, in some cases, warranted. These may seem like drastic replication recommendations but the problems associated with high throughput of molecular

data are unusual and perhaps unprecedented and it is not surprising that when the number of comparisons that can be made at one time is large the number of replicates per comparison will also need to be large. Our results suggest that the number of replicates per treatment required to maintain acceptable error rates is large, and estimated to be 15–150, depending on the critical effect size.

An alternative approach is to replicate experiments rather than increasing the number of replicates within an experiment. This would involve identifying the number of times one would have to see the same result repeated, given a particular experimental design, before we would be willing to accept the result. Our results suggest that if 8–10 replicates per treatment are used and the critical effect size is 2 or 4 SD's then an experiment would not need to be repeated to meet at error rate = 0.05. However, the argument for a more stringent acceptable error rate applies here as well and so at an error rate of 0.0001 experiments would need to be repeated 2–10 times. It's not clear whether increased within-experiment or among experiment replication would be more efficient and may depend on the limiting factors in particular labs. However, it is clear that among-experiment replication adds an additional layer of inferential complexity because it would require interpretation of cases where a subset, but not all, of the experiments was consistent.

It's not clear whether within- or among-study replication is preferable – which is preferable may depend on context – but it is clear that one or both are necessary if conclusions of microarray studies are to be rigorous and reliable.

## Relative costs of type I and II errors
All of our analyses assumed that the costs of Type I and II errors were equal but we do not preclude the possibility that Type I and II errors should be weighted differently [25–27] and an additional advantage of optimal α is that the cost of error can be minimised rather than the probability of error. Thus, if the relative costs of Type I and II error can be estimated they can be integrated into the selection of appropriate statistical thresholds. The question of relative costs of Type I and II errors is a difficult and relatively unexplored one but the objectives of a study can often guide setting relative costs of Type I and II error. For example, preliminary work 'fishing' for genes that may respond to a specific treatment might be more concerned about missing genes that were actually affected than identifying genes as affected when they really weren't and would choose to set the cost of Type II errors greater than Type I errors. By contrast, a researcher attempting to identify a single gene (i.e. biomarker) that is regulated by a specific treatment or drug might decide that Type I error is a larger concern.

Post-hoc multiple comparison adjustments to reduce Type I errors at the expense of increased Type II errors imply that Type I errors are more serious than Type II errors. Although we don't believe Type I errors are actually more serious than Type II errors under all circumstances in which multiple comparison $\alpha$ adjustments have been used, similar outcomes can be obtained through the optimal $\alpha$ approach by selecting a large Type I / Type II error cost ratio a priori.

## SD versus k-fold effect size

One potential complication is that the convention in microarray analysis is often to use effect sizes expressed as multiples of the control mean (e.g. 1-fold, 2-fold or 4-fold change). Unlike effect sizes measured in standard deviations, which combine the raw effect size with the variability of the data, k-fold effect sizes, even when combined with a statistical significance threshold, do not quantitatively incorporate variability in the data. This implies that if a constant k-fold effect size was set as the critical effect size, say 2-fold, that there would be a different optimal $\alpha$ for each comparison, depending on the variability in expression of each gene. That is, if there were 44,000 comparisons, it would require 44,000 different optimal $\alpha$ levels. However, if k-fold effect sizes are used, solutions include (i) setting a single optimal $\alpha$ based on the 'average' variability across genes, (ii) grouping genes into variability categories and using a single optimal $\alpha$ for each category based on the average variability in each category, and (iii) setting an individual optimal $\alpha$ for each gene. While both –k-fold and SD effect sizes are reasonable indices of effects, using SD simplifies the application of optimal alpha. However, for microarray and RNA-seq, there are approaches such as Voom [28] that would allow using both k-fold changes and a single optimal alpha for all genes by making the variance homoscedastic.

## A priori $H_0$ and $H_A$ probabilities

The accuracy of estimates of optimal alpha is a function of assumptions about the a priori probabilities of $H_0$ and $H_A$. It is inarguably true that estimates of optimal alpha will be less reliable if estimates of a priori $H_0$ and $H_A$ probabilities are inaccurate. Thus, research into prior probabilities of global gene expression is critical. One key advantage of optimal alpha is that it has the explicit objective of identifying the threshold that will result in the lowest probability of cost of making a mistake. There is, indisputably, a true optimal alpha – if we know the true a priori probabilities of $H_0$ and $H_A$ and the true relative costs of Type I and II errors then the statistical threshold that minimizes the probability or cost of making an error can be calculated for any target critical effect size. By contrast, the approach of alpha = 0.05 with

post hoc corrections (i.e. Bonferroni or FDR) doesn't explicitly address Type II errors, it only explicitly addresses Type I errors. Of course, there is an implicit concern about Type II error – if there wasn't we would simply always set alpha = 0 and never make a Type I error. We don't set alpha = 0.00 because the goal is to make as few type I errors as possible *while* also detecting true effects. This implies that we should explicitly address (1) Type II errors, (2) the balance between Type I and II errors and (3) the assumptions about a priori $H_0$ and $H_A$ probabilities and relative costs of Type I and II errors. Optimal alpha does this and alpha = 0.05 with or without post-hoc corrections does not.

## Application

We have included R code (Additional file 2: Appendix S1) that can be used to calculate optimal $\alpha$ for t-tests ANOVA's and regressions. To apply optimal $\alpha$ to microarray analysis, you must choose the type of test, sample size, critical effect size, a priori null/alternate probabilities and relative costs of Type I/II errors.

(1) Type of test: We have written code for t-tests, ANOVA's and regressions. Here we have focused on t-tests. The code allows the choice of 1 or 2-tailed tests and one-, two- or paired-sample tests.
(2) Sample size: This is usually limited by time and/or money. Most microarray studies test many genes with relatively few replicates (3–6) and the result is a large number of false positives and negatives due to insufficient power. A different experimental design may be desired, testing fewer genes with a higher number of replicates (12–24) to reduce the number of errors. This includes a priori hypotheses for specific pathways perturbed by a stressor.
(3) Critical effect size type: This can be measured in standard deviations (e.g. 1, 2, and 4 SD's), absolute differences (e.g. difference in signal intensity) or relative differences (e.g. fold differences between treatment and control samples – 0.5, 1, 2× more/less gene expression). SD's are less labour intensive than absolute r relative differences because you set a single optimal $\alpha$ for all genes. If you target absolute or relative critical effects, it will require a separate optimal $\alpha$ for each gene tested in the microarray. This is because the variability among replicates for different genes will vary and therefore the power to detect the same absolute or fold difference will vary among genes.
(4) Critical effect size value: A single critical effect size or a range of critical effect sizes can be chosen. Unless there is a defensible reason for choosing a single critical effect size, we suggest selecting a range

of critical effect sizes that span small, moderate and large effects. Each critical effect will have a different optimal $\alpha$ (i.e. optimal $\alpha$ for large effects will be smaller than for moderate effects which will be smaller than for small effects) and this allows for different conclusions about small versus moderate versus large effects. For example, it would be reasonable to conclude that there is evidence for small effects but no evidence for moderate or large effects.

(5) A priori probability of null and alternate hypotheses: There must be explicit quantitative probabilities for these values. We suggest assuming these to be equal (i.e. 0.5 for both) unless you have theoretical or empirical reasons to believe otherwise.

(6) Relative costs of Type I and II errors: It is not unreasonable to expect that the costs of missing a real effect could be different than detecting a false effect and such difference can be assigned in the code for calculating optimal $\alpha$. However, unless there is a clear, explicit reason for estimating the costs to be different, we recommend assuming equal costs of Type I and II errors.

## Conclusions

While we don't have empirical estimates of the true error rates associated with the studies used in the meta-analysis, both the meta-analysis and simulations estimated error rates under simple and reasonable assumptions and suggest that optimal $\alpha$ provides a simple and superior approach to setting statistical thresholds for transcriptome analysis than the traditional $\alpha = 0.05$ with or without post-hoc adjustments. Using optimal $\alpha$ will provide significantly lower probabilities of making errors and will eliminate the need to use complex and controversial post-hoc adjustments. However, optimal $\alpha$ cannot eliminate the problems associated with the large number of tests that are traditionally carried out in transcriptome analysis. This problem will only become exacerbated as new high-throughput techniques such as RNA-Seq become more commonly used and increasing amounts of information are generated. Thus, moving forward, researchers should consider setting new standards for within and among-study replication and exploring novel approaches to evaluating gene expression data such as in the case of gene enrichment analyses.

## Abbreviations
ANOVA: Analysis of Variance; DEG: Differentially Expressed Genes; FDR: False Discovery Rate; NGS: Next Generation Sequencing; RNA-seq: RNA sequencing; SD: Standard Deviation

## Acknowledgements
Lillian Fanjoy for compiling studies.

## Funding
Natural Sciences and Engineering Research Council of Canada (JH and CJM). NSERC was not involved in the design or conclusions of this research.

## Authors' contributions
Concept and experimental design (JFM and JEH); Data collection (CM); Statistical analysis (JFM); First draft (JFM); Further drafts (JFM, JEH, CM). All authors approved the final version.

## Competing interests
The authors declare that they have no competing interests.

## Author details
[1]Department of Biology, Canadian Rivers Institute, University of New Brunswick, Saint John, NB E2L 4L5, Canada. [2]Center for Environmental and Human Toxicology & Department of Physiological Sciences, UF Genetics Institute, University of Florida, Gainesville, Florida 32611, USA.

## References
1. Pantalacci S, Sémon M. Transcriptomics of developing embryos and organs: a raising tool for evo-devo. J Exp Zool B Mol Dev Evol. 2015;324:363–71.
2. Xu J, Gong B, Wu L, Thakkar S, Hong H, Tong W. Comprehensive assessments of RNA-seq by the SEQC consortium: FDA-led efforts advance precision medicine. Pharmaceutics. 2016;8:E8.
3. Evans TG. Considerations for the use of transcriptomics in identifying the'genes that matter' for environmental adaptation. J Exp Biol. 2015;218: 1925–35.
4. Krzywinski M, Altman N. Points of significance: comparing samples – part II. Nat Methods. 2014;11:355–6.
5. Kooperberg C, Sipione S, LeBlanc M, Strand AD, Cattaneo E, Olson JM. Evaluating test statistics to select interesting genes in microarray experiments. Hum Mol Genet. 2002;19:2223–32.
6. Reiner A, Yekutieli D, Benjamini Y. Identifying differentially expressed genes using false discovery rate controlling procedures. Bioinformatics. 2003;19: 368–75.
7. Pounds S, Cheng C. Improving false discovery rate estimation. Bioinformatics. 2004;20:1737–45.
8. Qian HR, Huang S. Comparison of false discovery rate methods in identifying genes with differential expression. Genomics. 2005;86:495–503.
9. Brand JPL, Chen L, Cui X, Bartolucci AA, Page GP, Kim K, et al. An adaptive α spending algorithm improves the power of statistical inference in microarray data analysis. Bioinformation. 2007;10:384–9.
10. Gordon A, Chen L, Glazko G, Yakovlev A. Balancing type one and two errors in multiple testing for differential expression of genes. Comput Stat Data Anal. 2009;53:1622–9.
11. Mudge JF, Baker LF, Edge CB, Houlahan JE. Setting an optimal α that minimizes errors in null hypothesis significance tests. PLoS One. 2012;7:e32734.
12. Mudge JF, Edge CB, Baker LF, Houlahan JE. If all of your friends used α=0.05 would you do it too? Int. environ. Assess Manage. 2012;8:563–4.
13. Mudge JF, Barrett TJ, Munkittrick KR, Houlahan JE. Negative consequences of using α = 0.05 for environmental monitoring decisions: a case study from a decade of Canada's environmental effects monitoring program. Environ Sci Technol. 2012;46:9249–55.
14. Holstege FCP, Jenings EG, Wyrick JJ, Lee TI, Hengartner CJ, Green MR, et al. Dissecting the regulatory circuitry of a eukaryotic genome. Cell. 1998;95:717–28.
15. Pounds S, Morris SW. Estimating the occurrence of false positives and false negatives in microarray studies by approximating and partitioning the empirical distribution of p-values. Bioinformatics. 2003;19:1236–42.
16. Loven J, Orlando DA, Sigova AA, Lin CY, Rahl PB, Burge CB, et al. Revisiting global gene expression analysis. Cell. 2012;151:476–82.

17. Chen K, Hu Z, Xia Z, Zhao D, Li W, Tyler JK. The overlooked fact: fundamental need of spike-in controls for virtually all genome-wide analyses. Mol Cell Biol. 2016;36:662–7.

18. Cheng C, Pounds S. False discovery rate paradigms for statistical analyses of microarray gene expression data. Bioinformation. 2007;10:436–46.

19. Martin LJ, Woo JG, Avery CL, Chen H-S, North KE. Multiple testing in the genomics era: findings from genetic analysis workshop 15, group 15. Genet Epidemiol. 2007;31:S124–31.

20. Benjamini Y, Hochberg Y. Controlling the false discovery rate: a practical and powerful approach to multiple testing. J R Statist Soc B. 1995;57:289–300.

21. Mapstone BD. Scalable decision rules for environmental impact studies: effect size, type I and type II errors. Ecol Appl. 1995;5:401–10.

22. Kikuchi T, Pezeshk H, Gittins J. (2008) Bayesian cost-benefit approach to the determination of sample size in clinical trials. Stat. Med. 2008;27:68–82.

23. O'Brien C, van Riper C III, Myers DE. Making reliable decisions in the study of wildlife diseases: using hypothesis tests, statistical power, and observed effects. J Wild Dis. 2009;45:700–12.

24. Munkittrick KR, McGeachy SA, McMaster ME, Courtenay SC. Overview of freshwater fish studies from the pulp and paper environmental effects monitoring program. Water Qual Res J Canada. 2002;37:49–77.

25. Yang MCK, Yang JJ, McIndoe RA, She JX. Microarray experimental design: power and sample size considerations. Physiol Genomics. 2003;16:24–8.

26. Lin W-J, Hsueh H-M, Chen JJ. Power and sample size estimation in microarray studies. BMC Bioinformatics. 2010;11:48.

27. Matsui S, Noma H. Estimating effect sizes of differentially expressed genes for power and sample size assessments in microarray experiments. Biometrics. 2011;67:1225–35.

28. Law CW, Chen Y, Shi W, Smyth GK. Voom: precision weights unlock linear model analysis tools for RNA-seq read counts. Genome Biol. 2014;15(2):R29.

# 11

# Meta-analysis approach as a gene selection method in class prediction: does it improve model performance? A case study in acute myeloid leukemia

Putri W. Novianti[1,2,3*], Victor L. Jong[1,4], Kit C. B. Roes[1] and Marinus J. C. Eijkemans[1]

## Abstract

**Background:** Aggregating gene expression data across experiments via meta-analysis is expected to increase the precision of the effect estimates and to increase the statistical power to detect a certain fold change. This study evaluates the potential benefit of using a meta-analysis approach as a gene selection method prior to predictive modeling in gene expression data.

**Results:** Six raw datasets from different gene expression experiments in acute myeloid leukemia (AML) and 11 different classification methods were used to build classification models to classify samples as either AML or healthy control. First, the classification models were trained on gene expression data from single experiments using conventional supervised variable selection and externally validated with the other five gene expression datasets (referred to as the individual-classification approach). Next, gene selection was performed through meta-analysis on four datasets, and predictive models were trained with the selected genes on the fifth dataset and validated on the sixth dataset. For some datasets, gene selection through meta-analysis helped classification models to achieve higher performance as compared to predictive modeling based on a single dataset; but for others, there was no major improvement. Synthetic datasets were generated from nine simulation scenarios. The effect of sample size, fold change and pairwise correlation between differentially expressed (DE) genes on the difference between MA- and individual-classification model was evaluated. The fold change and pairwise correlation significantly contributed to the difference in performance between the two methods. The gene selection via meta-analysis approach was more effective when it was conducted using a set of data with low fold change and high pairwise correlation on the DE genes.

**Conclusion:** Gene selection through meta-analysis on previously published studies potentially improves the performance of a predictive model on a given gene expression data.

**Keywords:** Meta-analysis, Gene expression, Predictive modeling, Acute myeloid leukemia

## Background

The ability of microarray technology to simultaneously measure expression values of thousands of genes has brought major advances. The measurement of gene expression may be done within a relatively short time to quantify genome-wide expression levels. On the other hand, statistical analyses to extract useful information from such high dimensional data face well known challenges. Common mistakes in conducting statistical analyses were reported [1]. Particularly class prediction studies are subject to concerns about reliability of results [2], where genes involved in predictive models depend heavily on the subset of samples used to train the models. This is related to the likelihood of false positive findings due to the curse of dimensionality in microarray gene expressions datasets [3].

* Correspondence: p.novianti@vumc.nl
[1]Biostatistics & Research Support, Julius Center for Health Sciences and Primary Care, University Medical Center Utrecht, 3508, GA, Utrecht, The Netherlands
[2]Department of Epidemiology and Biostatistics, VU University medical center, Amsterdam, The Netherlands
Full list of author information is available at the end of the article

Methods for aggregating gene expression data across experiments exist [4, 5]. Data standardization is proposed as a preliminary step in cross-platform gene expression data analyses [6–8], as raw gene expression datasets are recommended to be used [9] and gene expression values may be incomparable across different experiments. Meta-analysis is known to increase the precision of the effect estimate and to increase the statistical power to detect a certain effect size (or fold change). In class prediction, meta-analysis methods can have different objectives, ranging from methods for combining effect sizes [10] or combining P values [11, 12] to rank-based methods [13]. However, there is no meta-analysis method known to be generally superior to others [14, 15].

In this study, we compared the performance of classification models on a given gene expression dataset between gene selection through meta-analysis on other studies and conventional supervised gene selection. A single gene expression dataset with less than a hundred samples is likely not enough to determine whether a particular gene is an informative gene [16]. Thus, gene selection based on multiple microarray studies may yield a more generalizable gene list for predictive modeling. We used raw gene expression datasets from six published studies in acute myeloid leukemia (AML) to develop predictive models using 11 different classification functions to classify patients with AML versus normal healthy controls. In addition, a simulation study was conducted to more generally assess the added value of meta-analysis for predictive modeling in gene expression data.

## Methods

As a starting point, we assume $D$ gene expression datasets are available for analysis. First, the $D$ raw datasets are individually preprocessed. Next, 11 classifiers are trained on expression values from the $j^{th}$ study ($j = 1, ..., D$) by incorporating variable selection procedure via limma method and externally validated on the remaining $D-1$ gene expression datasets. We refer to these models as *individual-classification* models.

To aggregate gene expression datasets across experiments, $D$ gene expression datasets are divided into three major sets, namely (i) a set for selecting probesets (SET1, consists of $D-2$ datasets), (ii) for predictive modeling using the selected probesets from SET1 (SET2, consists of one dataset) and (iii) for externally validating the resulting predictive models (SET3, consists of one dataset). The data division is visualized in Fig. 1. We next describe the predictive modeling with gene selection via meta-analysis (refer to as MA(meta-analysis)-classification model). First, significant genes from a meta-analysis on SET1 are selected. Next, classification models are constructed on SET2 using the selected genes from SET1. The models are then externally validated using the independent data in SET3. The MA-classification approach is briefly described in Table 1 and is elaborated in the next subsections.

### Data extraction

Raw gene expression datasets from six different studies were used in this study, as previously described elsewhere [16, 17], i.e. E-GEOD-12662 [18] (Data1), E-GEOD-14924 [19] (Data2), E-GEOD-17054 [20] (Data3), E-MTAB-220 [21] (Data4), E-GEOD-33223 [22] (Data5) and E-GEOD-37307 [23] (Data6). Five studies were conducted on Affymetrix Human Genome U133 Plus 2 array and one study was performed on U133A (Additional file 1: Table S1). The raw datasets were pre-processed by quantile normalization, background correction according to manufacturer's platform recommendation, $\log_2$ transformation

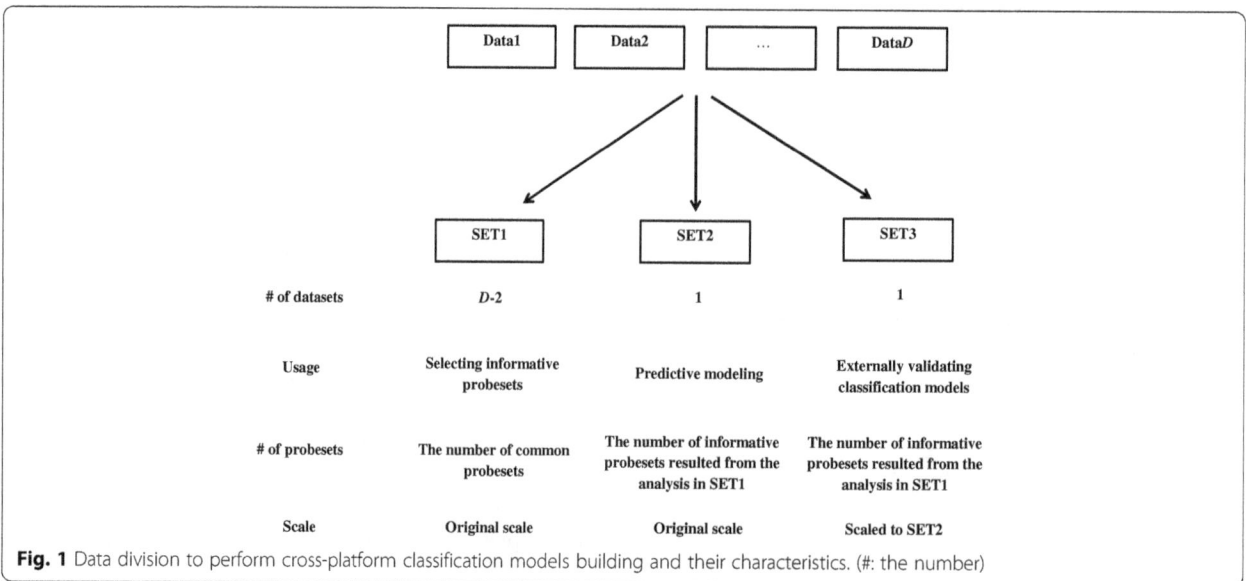

**Fig. 1** Data division to perform cross-platform classification models building and their characteristics. (#: the number)

**Table 1** An approach in building and validating classification models by using meta-analysis as gene selection technique

1. Data collection

Collect raw gene expression datasets, which possibly come from previous experiments and/or systematic search from online repositories.

2. Data preparation

(i) Individually preprocess raw gene expression datasets (i.e. normalization, background correction, log2 transformation).

(ii) Divide D available gene expression datasets into three sets, i.e. D-2 gene expression datasets to get a gene signature list (SET1), a gene expression set to train classification models (SET2) and a dataset to validate the models (SET3).

3. Meta-analysis for gene selection

(i) For each probesets, aggregate expression values from SET1 to get a signature list via random effect meta-analysis.

(ii) Record significant probesets (also refer to as informative probesets)

4. Predictive modeling

(i) In SET2, include informative probesets resulted from Step 3.

(ii) Divide samples in SET2 to a learning set and a testing set.

(iii) Perform cross validation in classification model modeling.

(iv) Evaluate optimum predictive models in the testing set.

5. External validation

(i) In SET3, include probesets that are informative from Step 3.

(ii) Scale gene expression values in SET3 with SET2 as a reference.

(iii) Validate classification models from Step 4 to the scaled gene expressions data in SET3.

and summarization of probes into probesets by median polish to deal with outlying probes. We limited analyses to 22,277 common probesets that appeared in all studies.

## Meta-analysis for gene selection

We aggregated $D$-$2$ gene expression datasets to extract informative genes by performing a random effects meta-analysis. This means meta-analysis acts as a dimensionality reduction technique prior to predictive modeling. For each probeset, we pooled the expression values across datasets in SET1 to estimate its overall effect size. Let $Y_{ij}$ and $\theta_{ij}$ denote the observed and the true study-specific effect size of probeset $i$ in an experiment $j$, respectively. The random effects model of a probeset $i$ is written as:

$$Y_{ij} = \theta_{ij} + \varepsilon_{ij}, \quad \text{where } \theta_{ij} = \theta_i + \delta_{ij} \text{ for } i = 1, ..., p \text{ and } j = 1, ..., (D-2),$$

where $p$ is the number of tested probesets, $\theta_i$ is the overall effect size of probeset $i$, $\varepsilon_{ij} \sim N(0; \sigma_{ij}^2)$ with $\sigma_{ij}^2$ as the within-study variance and $\delta_{ij} \sim N(0; \tau_i^2)$ with $\tau_i^2$ as the between-study or random effects variance of probeset $i$. The study-specific effect size $\theta_{ij}$ is defined as the corrected standardized mean different (SMD) between two groups, estimated by:

$$\hat{\theta_{ij}} = \left( \frac{\overline{x}_{ij0} - \overline{x}_{ij1}}{s_{ij}} \right) \left( 1 - \frac{3}{4(n_{j0} + n_{j1}) - 9} \right), \quad (1)$$

where $\overline{x}_{ij0} (\overline{x}_{ij1})$ is the mean of base-2 logarithmically transformed expression values of probeset $i$ in Group 0 (Group 1). $s_{ij}$ is originally defined as the square root of the pooled variance estimate of the within-group variances [24]. This estimation of $\sigma_{ij}$, however, is rather unstable in a small sample size study. We utilized the empirical Bayes approach implemented in limma to shrink extreme variances towards the overall mean variance. Thus, we define $s_{ij}$ as the square root of the variance estimate from the empirical Bayes t-statistics [25]. The second component in Eq.(1) is the Hedges' g correction for SMD [26]. The estimation of between-study variance $(\hat{\tau_i}2)$ was performed by Paule-Mandel (PM) method [27] as suggested by [28, 29]

For each probeset, a z-statistic was calculated to test the null hypothesis that the overall effect size in the random effects meta-analysis model is equal to zero (or a probeset is not differentially expressed). To adjust for multiple testing, P-values based on z-statistics were corrected at a false discovery rate (FDR) of $\alpha = 5\%$, using the Benjamini-Hochberg (BH) procedure [30]. We considered probesets that had a significant overall effect size as informative probesets. For each informative probeset $i$, the estimated overall effect size $\theta_i (\hat{\theta_i})$ is:

$$\hat{\theta_i} = \frac{\sum_j w_{ij} \theta_{ij}}{\sum_j w_{ij}}, \quad (2)$$

Where $w_{ij} = 1/(\hat{\tau_i}2 + s_{ij}^2)$.

## Classification model building

The following classification methods were used to construct predictive models: linear discriminant analysis (LDA), diagonal linear discriminant analysis (DLDA) [31], shrunken centroid discriminant analysis (SCDA) [32], random forest (RF) [33], tree-based boosting (TBB) [34], L2-penalized logistic regression (RIDGE), L1-penalized logistic regression (LASSO) [35], elastic net [36], feed forward neural networks (NNET) [37], support vector machines (SVM) [38] and k-nearest neighbors (kNN) [39]. A detailed description of the classification methods, model building procedure as well as the tuning -parameter(s) was presented in our previous study [40]. The class prediction modeling process for both individual- and MA-classification models was done by splitting the dataset in SET2 into a learning set $\mathscr{L}$ and a testing set $\mathscr{T}$. The learning set $\mathscr{L}$ was further split by cross validation into an inner-learning set and inner-testing set, to optimize the parameters in each classification model. The optimal models were then internally validated on the out-of-bag testing set $\mathcal{T}$. Henceforth, we referred to the testing set $\mathcal{T}$ as an internal-validation set $\mathcal{V}_0$.

For MA-classification models on SET2, we used all the probesets identified as differentially expressed by meta-analysis procedure in SET1, except for LDA, DLDA and NNET methods, which cannot handle a larger number of parameters than samples. For these methods, we incorporated top-X probesets to the predictive modeling, where X was less than or equal to the sample size minus 1. The top lists of probesets were determined by ranking all significant probesets on their absolute estimated pooled effect sizes ($\hat{\theta}_i$) from Eq.(2). As the number of probesets to be included was itself a tuning parameter, we varied the number of included probesets from 5 to the minimum number of within group samples. For other classification functions, we used the same values of tuning parameter(s) as described in our previous study [40].

For the individual-classification approach, we optimized the classification models based on a single gene expression dataset (SET2). Here, we applied the limma procedure [41] to determine top-X relevant probesets, controlling the false discovery rate at 5% using the BH procedure [30]. The optimum top-X was selected among{50, 100, 150, 200} for classification methods other than LDA, DLDA and NNET. We used the same number of selected probesets for the three aforementioned classification methods as in the MA-classification approach. In each case, we evaluated the classification models by the proportion of correctly classified samples to the number of total samples, known as a classification model accuracy.

## Model validation

The optimal classification models obtained from the previous step were externally validated on SET3. The $\log_2$ expression values of the data in SET3 for the probesets used in the classification models were scaled to the $\log_2$ expression values of the data in SET2, so that the learning and the validation sets had comparable range. For each probeset $i$, we assumed the expression values were in the interval $[a_i, b_i]$ in SET2 and $[c_i, d_i]$ in SET3. A $\log_2$ expression value $x_{is}$ of probeset $i$ in sample $s$ from SET3, was scaled to the scale of SET2 by the following transformation formula:

$$f(x_{is}) = a_i + \frac{(b_i - a_i)(x_{is} - c_i)}{(d_i - c_i)}, \quad d_i \neq c_i. \tag{3}$$

Predictive models were then applied to the scaled $\log_2$ gene expression data in SET3.

For individual-classification, we rotated the single learning dataset and validated the models on the other

*D-1* datasets. For MA-classification, we rotated the datasets used for selecting informative probesets (SET1) as well as learning (SET2) and validating (SET3) classification models. For each possible combination of *D-2* datasets, we repeated step 3–5 of our approach (Fig. 1). Due to a small number of samples in Data3, we omitted the predictive modeling process when it was selected as SET2. Hence, the possible gene expression datasets in SET2 were Data1, Data2, Data4, Data5 and Data6; and gene expression datasets in SET3 were Data1, Data2, Data3, Data4, Data5 and Data6, rendering thirty possible combinations to divide $D = 6$ datasets to three distinct sets.

## Simulation study

We generated synthetic datasets by conducting simulations similar to that described by Jong *et al* [42]. We refer to the publication for more detail description of each and every parameter stated in this sub-section. Among parameters to simulate gene expression data (Table 2, in [42]), we applied these following parameters for all simulation scenarios, i.e. (i) the number of genes per data set ($p = 1000$); (ii) the pairwise correlations of noisy genes were set equal to zero (implying $\Sigma_{33}$ in Fig. 1. reference [42] was equal to 0), (iii) the proportion of differentially expressed genes ($\pi = 10\%$) and; (iv) the parameter of an exponential distribution to draw the variances of the genes ($\lambda = 0.5$). Further, the number of samples per dataset ($n$), the $\log_2$ fold changes of differentially expressed (DE) genes ($\Delta$) and pairwise correlations of DE genes ($\rho$) were varied as follows: $n = 50, 100, 150$; $\Delta = 0.1, 0.5, 0.75$; and $\rho = 0.25, 0.5, 0.75$, respectively. We define pairwise correlation of noisy (*DE*) genes as the correlation between any and every two pairs of noisy

**Table 2** Parameters to generate simulated gene expression datasets

| Simulation ID | $n$ | $\Delta$ | $\rho$ | $DEG_{MA}^a$ | $DEG_{IND}^b$ |
|---|---|---|---|---|---|
| 1 | 50 | 0.1 | 0.75 | 12 | 72 |
| 2 | 50 | 0.5 | 0.5 | 57 | 34 |
| 3 | 50 | 0.75 | 0.25 | 70 | 62 |
| 4 | 100 | 0.1 | 0.75 | 12 | 14 |
| 5 | 100 | 0.5 | 0.5 | 53 | 56 |
| 6 | 100 | 0.75 | 0.25 | 67 | 50 |
| 7 | 150 | 0.1 | 0.75 | 15 | 23 |
| 8 | 150 | 0.5 | 0.5 | 52 | 26 |
| 9 | 150 | 0.75 | 0.25 | 58 | 57 |

Symbols. $n$: the number of samples in each generated dataset; $\Delta$: the $\log_2$ fold changes of differentially expressed (DE) genes. $\rho$: pairwise correlation of DE genes
[a]The number of genes that were stated as differentially expressed (DE) genes by MA approach from 50 cumulative studies. All the selected genes are true positives
[b]The number of true DE genes among the top-100 DE genes selected by limma procedure

(*DE*) genes. Table 2 shows nine combinations from these parameters, which reflect the amount of information in each simulated gene expression dataset. In the first block (simulation #1 to #3) for instance, the dataset generated by parameters in simulation #1 contains less information than the dataset generated by parameters in simulation #2, which is caused by the low degree of $\log_2$ fold changes and high correlation of DE genes.

For each scenario mentioned in Table 2, we simulated data that consisted of $n*52$ samples from the same population. The data was then randomly divided into 52 different sub-datasets of $n$ samples each (proportional to the classes). Next, the sub-datasets were randomly chosen to be considered as (i) SET1: a set of fifty datasets for selecting probes via meta-analysis; (ii) SET2: a dataset for predictive modeling; (iii) SET3: a dataset for validation. In the MA-predictive modeling, we estimated classification model accuracies when the number of studies for variable selection were ranging from 5 to 50 studies.

### Random effects linear regression

We quantified the difference in performance between classification models that were optimized with and without incorporating information from other studies in the simulation study by a random effects linear regression model. The difference of model accuracy between MA- and individual-classification procedure for a classification model $C$ based on a simulation scenario $S$ is denoted as $d_{CSM}$. Such differences were calculated when MA-classification procedure incorporated $M$ studies (where $M = 5{:}50$ by 5) to select features. Having rescaled the $d_{CSM}$ to be in the range of 0 and 1 by $\frac{1+d_{CSM}}{2}$, we then transformed $d_{CSM}$ using the logit function to get unbounded and more approximately normally distributed outcome values. Given in each simulation setting we calculated $d_{CSM}$ for different number of $M$ studies for feature selection in MA approach, we used a fully crossed random effects model, where simulation setting $S$ and the number of studies for MA-approach $M$ acted as clustering factors or random effects. Additionally, since the same classification methods were applied to build prediction models, classifier $C$ was added as a random effect term.

We then tested three determinants ($X_k$, $k = 1, 2, 3$) that might contribute to the difference in performance of classification models that were trained by two approaches ($d_{CSM}$), namely the number of samples per dataset ($n$), the $\log_2$ fold changes of differentially expressed (DE) genes ($\Delta$) and pairwise correlations of DE genes ($\rho$). Each of the determinant was individually evaluated in the random effects model. More formally, the random effects model for the $k^{th}$ determinant is written as:

$$d'_{CSM} = \beta_0 + \vartheta_{0C} + \vartheta_{0S} + \vartheta_{0M(S)} + \beta_1 X_k,$$

where $d'_{CSM}$ is the logit transformation of the scaled $d_{CSM}$; $\vartheta_{0S}$, $\vartheta_{0M(S)}$ and $\vartheta_{0C}$ are the random intercepts with respect to the simulation setting $S$ ($\vartheta_{0S} \sim N(0, \sigma^2_{0S})$), the number of studies for meta-analysis $M$ ($\vartheta_{0M(S)} \sim N(0, \sigma^2_{0M})$) and classification model $C$ ($\vartheta_{0C} \sim N(0, \sigma^2_{0C})$) respectively.

### Software

All analyses were performed in R statistical software using these packages: *affy* for preprocessing procedures [43]; *meta* for meta-analysis [44], *CMA* for predictive modeling [45], *lme4* for the random effects linear model [46] and *ggplot2* for data visualization [47].

### Results

We first present the performance of classification models when each individual study was used to optimize the classification functions (individual-classification procedure) in AML datasets. As the first illustration, we considered the case for which Data1 was used for optimization. To start with, we compared the distribution of expression values in the validation sets Data2 to Data6 to the expression values in Data1. There seemed to be a considerable difference in the distributions of expression values between studies, with Data6 having a lower range than other experiments, indicating that data standardization across studies was necessary (Fig. 2). Gene expression values in Data2 to Data6 were effectively scaled by using Eq.(3) so that they had comparable ranges as in Data1 (Additional file 1: Figure S1). The classification models optimized in Data1, were validated with Data2 to Data6. The classification models performed poorly in all 5 validation sets, notably worst in Data2 and Data4 (Additional file 1: Table S2). When Data2, Data4, Data5 and Data6 were used to optimize the classifiers, we found similar results (Additional file 1: Table S3-S6).

The comparison of the accuracies of classification models that were trained by MA- with individual-classification procedures based on optimization with Data1 is shown in Fig. 3. In most cases, MA-classification models outperformed individual-classification models. The difference of model accuracies between MA- and individual-classification approach was considerably larger when Data2 was used as a validation set. On average, classification methods that require the number of features to be smaller than the number of samples (i.e. NNET, LDA and DLDA), seemed to improve with the MA-classification approach. When validated against Data4, all models seemed to benefit from the MA-classification approach.

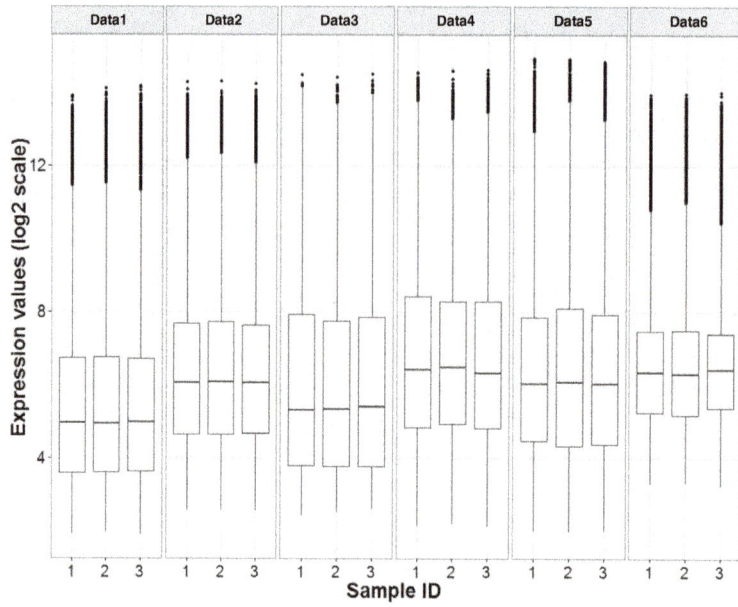

**Fig. 2** The distribution of expression values after pre-processing step from the first three samples in six experiments. The expression values are in $\log_2$ scale

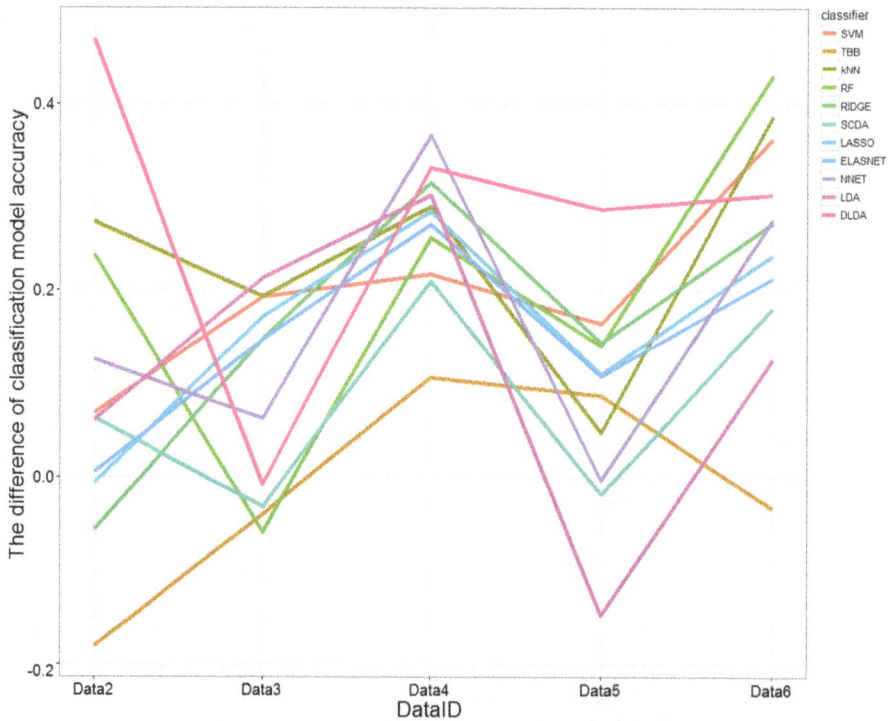

**Fig. 3** Plot of the difference of classification model accuracies between MA- and individual-classification approach, when Data1 was used as a training data

In the other cases (i.e. when Data4, Data5 and Data6 acted as a learning set), we noticed that MA-classification approach did not outperform the individual-classification models when the models were validated on Data2. The MA-classification approach reduced the classification model accuracies by up to 50%, as compared to individual-classification models. As the MA-classification approach mostly resulted in a lower number of genes used in the predictive models than individual-classification approach, it might be hard for MA-classification models to outperform individual-classification models when validated on Data2, as DE genes in this dataset (on average) had a low degree of $\log_2$ fold change (i.e. 0.471). On the other hand, most of MA-classification models outperformed individual-classification models when they were validated on Data3 (Additional file 1: Figure S2-S5). Given that (i) the MA-approach was better in selecting the "true" DE genes (results from the simulation study) and more importantly (ii) the average $\log_2$ fold change of the DE genes in Data3 was considerably high, i.e. 2.025, in most cases the classifiers benefited from the MA-approach. Incorporating information from other experiments in these datasets did not consistently improve the predictive ability of classification models when externally-validated. The simulation study was conducted to evaluate the difference of classification model accuracies between the MA- and individual-classification approach more generally. The results showed that the MA-classification approach was more likely to improve the classification model accuracy when it was conducted in a set of less informative datasets (Fig. 4). We defined a less informative dataset as a dataset with a small number of samples, a low degree of $\log_2$ fold changes of the DE genes and a high level of pairwise correlation of DE genes. In this type of dataset, feature selection via limma method hardly selected the true DE genes in the individual-classification approach. Among the true

100 DE genes in each simulated dataset, the limma procedure could select 14 to 72 DE genes. Meanwhile, all selected genes by MA approach were truly DE genes (Table 2). As we observed in the AML data, classification methods that require the number of features less than the number of samples (i.e. NNET, LDA and DLDA) performed better with the feature selection prior to predictive modeling via meta-analysis.

Factors that might contribute to the difference of classification model accuracy between the MA- and individual-classification approach, were individually evaluated by random effect models. This resulted in the $\log_2$ fold changes and pairwise correlation between DE genes as the significant factors. Both factors were consistent with the finding that a set of less informative datasets benefited from the MA-classification approach (shown by negative coefficient on $\Delta$ and positive coefficient on $\rho$). Further, there was no additional variation in the difference in performance between MA- and individual-classification approach that was associated with the number of datasets used to select features in meta-analysis approach ($\sigma^2_{0M} = 0$). A possible explanation of this finding could be that five datasets used in MA-classification approach were enough to select relevant variables so that the quality of the variable selection was not further increased by the increasing the number of datasets. This might also explain all the true positive genes selected by MA-approach in the simulation study. (Table 3)

## Discussion

This study applied a meta-analysis approach for feature selection in predictive modeling on gene expression data. Selecting informative genes among massive noisy genes in predictive modeling faces a great challenge in microarray gene expression data. Dimensionality reduction is applied to reduce the number of noisy genes as

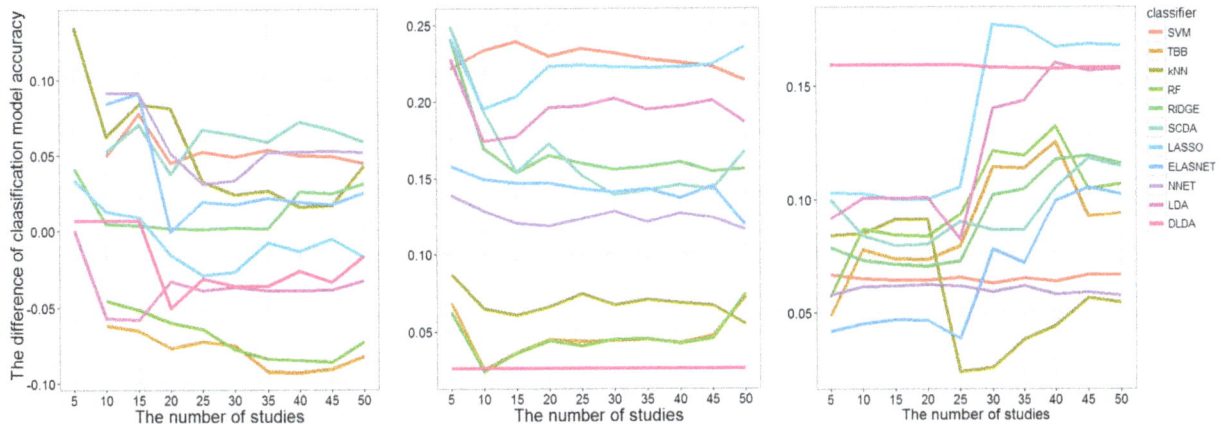

**Fig. 4** Plot of the difference of classification model accuracies between MA- and individual-classification approach in the simulated datasets, when $\Delta = 0.1$, $\gamma = 0.75$ and (a) $n = 50$ (Simulation 1) (b) $n = 100$ (Simulation 4) (c) $n = 150$ (Simulation 7). The aforementioned simulation parameters resulted in the less informative datasets

**Table 3** Results of the random effects models

| Factors | Coefficient | Confidence interval | | $\sigma_{OC}$ | Confidence interval | | $\sigma_{OS}$ | Confidence interval | | $\sigma_{OM(S)}$ | Confidence interval | |
|---|---|---|---|---|---|---|---|---|---|---|---|---|
| | | LL | UL | | LL | UL | | LL | UL | | LL | UL |
| $n$ | 0.0005 | -0.0005 | 0.0009 | 0.0244 | 0.0165 | 0.0404 | 0.0489 | 0.0289 | 0.0759 | 0.000 | 0.000 | 0.0039 |
| $\Delta$ | -0.1169 | -0.2041 | -0.0285 | 0.0245 | 0.0163 | 0.0402 | 0.0359 | 0.0159 | 0.0405 | 0.000 | 0.000 | 0.0039 |
| $\rho$ | 0.1489 | 0.0295 | 0.2636 | 0.0245 | 0.0165 | 0.0405 | 0.0369 | 0.0022 | 0.0579 | 0.000 | 0.000 | 0.0039 |

Each factor was evaluated individually in the random effects linear regression model. The coefficients were inverse transformed to the original scale of the difference of classification model accuracy between MA- and individual classification approach

*Abbreviations*: *LL* lower limit, *UL* upper limit

Symbols: $n$: the number of samples in each generated dataset; $\Delta$: the log2 fold change of differentially expressed (DE) genes. $\rho$: pairwise correlation of DE genes. $\sigma_{OC}$, $\sigma_{OS}$ and $\sigma_{OM(S)}$ are the standard deviation of the random intercepts with respect to classification model, scenario in the simulation study and the number of studies used for selecting relevant features via meta-analysis approach. See Method section for more details regarding the random effect models

well as to reduce the possibility of predictive models choosing clinically irrelevant biomarkers. An extra step to generate a gene signature list is usually applied in practice (e.g. by [48–53]), including predictive modeling via embedded classification methods (e.g. SCDA and LASSO). Selected informative genes may depend on the sub-samples used in the analysis [2], which may lead to the lack of direct clinical application [54].

Previous research on the application of meta-analysis in differential gene expression analysis showed that a single study might not contain enough samples to make a conclusion whether a particular gene is an informative gene. Among 12,211 common genes from 271 combined samples, 70 to 90% of the genes needed more samples in order to draw a conclusion [16]. A very low sample size as compared to the number of genes can cause false positive finding [3]. Involving thousands of samples is a straight forward solution but it can be very costly and time consuming. A possible solution to increase the sample size is by combining gene expression datasets with a similar research question through meta-analysis.

Meta-analysis is known as an efficient tool to increase statistical power and to obtain more generalizable results. Although a number of meta-analysis methods have been used as a feature selection technique in class prediction, no method has been shown to perform better than others [14, 17]. In this study, we combined the corrected standardized effect size for each gene by random effects models, similar to a study conducted by Choi *et al* [10]. However, we estimated the between-study variance by Paule-Mandel method, which outperforms the DerSimonian-Laird method in continuous outcome data [28]. We used a broad selection of classification functions to build predictive models in order to evaluate the added value of meta-analysis in aggregating information from gene expression across studies.

Six raw gene expression datasets resulting from a systematic search in a previous study in acute myeloid leukemia (AML) [16] were preprocessed, 22,277 common probesets were extracted and used for further analyses. We assessed the performance of classification models that were trained by each single gene expression

dataset. The models were then validated on datasets obtained from other studies. Classification models that were externally validated might suffer from heterogeneity between datasets, due to, for instance, different sample characteristics and experimental set-up.

For some datasets, gene selection through meta-analysis yielded better predictive performance as compared to predictive modeling on a single dataset, but for others, there was no major improvement. Evaluating factors that might account for the difference in performance of the two predictive modeling approaches on real-life datasets could be confounded by uncontrolled variables in each dataset. As such, we empirically evaluated the effects of fold change, pairwise correlation between DE genes and sample size on the added value of meta-analysis as a gene selection method in class prediction with gene expression data.

The simulation study was performed to evaluate the effect of the level of information contained in a gene expression dataset. For a given number of samples, we defined an informative gene expression data as a dataset with large $\log_2$ fold changes and low pairwise correlation of DE genes. The simulation study shows that the less informative datasets (i.e. Simulation 1, 4 and 6) benefited from MA-classification approach more clearly, than the more informative datasets. The limma feature selection method on a single dataset had a higher false positive rate of DE genes compared to feature selection via meta-analysis. Incorporating redundant genes in the predictive model may weaken the performance of a classification model on independent datasets. While conventional procedures use the same experimental data, meta-analysis uses a number of datasets to select features. Thus, the chances of sub-samples-dependent features to be included in a predictive model are reduced in MA- than in individual-classification approachand the gene signature may be widely applied.

For MA, we defined the effect size as a standardized mean difference between two groups. Although we individually selected differentially expressed probesets (i.e. ignoring correlation among probesets), we incorporated information from all probesets by applying limma procedure in estimating the within-group variances

(Eq.(1)). This empirical Bayes moderated t-statistics produces stable variances and it is proven to outperform ordinary t-statistics [55]. Marot *et al* implemented a similar approach in estimating unbiased effect sizes (Eq.(13) in [56]) and they suggested to apply such approach to estimate the study-specific effect size in meta-analysis of gene expression data.

We analyzed gene expression data at the probeset level. When more heterogeneous gene expression data from different platforms are used, mapping probesets to the gene level is a good alternative. Annotation packages from Bioconductor [57] and methods to deal with multiple probesets referring to the same gene may be considered, if such mapping is applied in a cross-platform gene expression study. A point to consider in cross-platform analysis of microarray experiments is data standardization. The same genes may have different signal in different experiments, due to e.g. different array technology and scanning process. We investigated the distributions of expression values across experiments and found incomparable ranges of expression values across experiments. Despite its simple nature, the scaling formula in Eq.(3) produces common ranges of gene expression values across experiments. Some methods to scale gene expression across experiments were proposed [7, 8, 10]. We do not expect that different scaling methods give significantly different findings as presented here, although it may be interesting to study.

We individually pre-processed the selected gene expression datasets, adjusted by the microarray platform in each and every study. A different preprocessing method may lead to different results of the prediction models, but it is not covered in this study. The predictive ability of a classification model may depend on a set of samples that is used in the preprocessing and normalization step. The rank-based genes is preferred over raw expression values to generate gene expression data [57]. Although we do not expect the present conclusions to change, it could be interesting to investigate this procedure further in this context.

## Conclusions

A meta-analysis (MA) approach was applied to select relevant features from multiple studies. Based on the simulation study, the MA approach was better in terms of variable selection than the predictive modeling by using a single dataset. In particular, a less informative dataset (which contains low $\log_2$ fold changes and highly correlated differentially expressed genes) was likely to benefit from feature selection via meta-analysis for class prediction. This also held for classification methods that require a smaller number of features than samples. Given the present public availability of omics datasets, meta-analysis approach can be used more often as an alternative gene selection method in class prediction.

## Abbreviations

AML: Acute myeloid leukemia; DE: Differentially expressed; DLDA: Diagonal linear discriminant analysis; kNN: k-nearest neighbors; LASSO: L1-penalized logistic regression; LDA: Linear discriminant analysis; MA: Meta-analysis; NNET: Feed forward neural networks; RF: Random forest; RIDGE: L2-penalized logistic regression; SCDA: Shrunken centroid discriminant analysis; SMD: Standardized mean difference; SVM: Support vector machines; TBB: Tree-based boosting

## Acknowledgements

The authors would like to thank the Biostatistics and Research Support group (Julius Center for Health Sciences and Primary Care, UMC Utrecht) for their inputs and comments during the study. The authors would also acknowledge the anonymous reviewers for their comments and constructive suggestions.

## Funding

Funding for publication charge: Biostatistics and Research support, Julius Center for Health Sciences and Primary Care, University Medical Center Utrecht, the Netherlands

## Authors' contributions

PWN, KCBR and MJCE designed the study. PWN then did systematic search, performed statistical analyses, and drafted the manuscript. VLJ contributed in doing simulation study and in interpreting the results. All authors critically reviewed and approved the final manuscript.

## Competing interests

The authors declare that they have no competing interests.

## Author details

[1]Biostatistics & Research Support, Julius Center for Health Sciences and Primary Care, University Medical Center Utrecht, 3508, GA, Utrecht, The Netherlands. [2]Department of Epidemiology and Biostatistics, VU University medical center, Amsterdam, The Netherlands. [3]Department of Pathology, VU University medical center, Amsterdam, The Netherlands. [4]Viroscience Laboratory, Erasmus Medical Center Rotterdam, 3015, CE, Rotterdam, The Netherlands.

## References

1. Dupuy A, Simon RM. Critical review of published microarray studies for cancer outcome and guidelines on statistical analysis and reporting. J Natl Cancer Inst. 2007;99(2):147–57.
2. Ein-Dor L, Kela I, Getz G, Givol D, Domany E. Outcome signature genes in breast cancer: is there a unique set? Bioinformatics (Oxford, England). 2005; 21(2):171–8.
3. Ein-Dor L, Zuk O, Domany E. Thousands of samples are needed to generate a robust gene list for predicting outcome in cancer. Proc Natl Acad Sci USA. 2006;103(15):5923–8.
4. Gormley M, Dampier W, Ertel A, Karacali B, Tozeren A. Prediction potential of candidate biomarker sets identified and validated on gene expression data from multiple datasets. BMC bioinformatics. 2007;8:415.
5. Miller JA, Cai C, Langfelder P, Geschwind DH, Kurian SM, Salomon DR, Horvath S. Strategies for aggregating gene expression data: the collapseRows R function. BMC bioinformatics. 2011;12:322.

6.  Heider A, Alt R. virtualArray: a R/bioconductor package to merge raw data from different microarray platforms. BMC bioinformatics. 2013;14:75.

7.  Autio R, Kilpinen S, Saarela M, Kallioniemi O, Hautaniemi S, Astola J. Comparison of Affymetrix data normalization methods using 6,926 experiments across five array generations. BMC bioinformatics. 2009;10 Suppl 1:S24.

8.  Warnat P, Eils R, Brors B. Cross-platform analysis of cancer microarray data improves gene expression based classification of phenotypes. BMC bioinformatics. 2005;6:265.

9.  Ramasamy A, Mondry A, Holmes CC, Altman DG. Key issues in conducting a meta-analysis of gene expression microarray datasets. PLoS Med. 2008;5(9):e184.

10. Choi JK, Yu U, Kim S, Yoo OJ. Combining multiple microarray studies and modeling interstudy variation. Bioinformatics (Oxford, England). 2003;19 Suppl 1:i84–90.

11. Lu TP, Hsu YY, Lai LC, Tsai MH, Chuang EY. Identification of gene expression biomarkers for predicting radiation exposure. Sci Rep. 2014;4:6293.

12. Rhodes DR, Yu J, Shanker K, Deshpande N, Varambally R, Ghosh D, Barrette T, Pandey A, Chinnaiyan AM. Large-scale meta-analysis of cancer microarray data identifies common transcriptional profiles of neoplastic transformation and progression. Proc Natl Acad Sci U S A. 2004;101(25):9309–14.

13. Fishel I, Kaufman A, Ruppin E. Meta-analysis of gene expression data: a predictor-based approach. Bioinformatics (Oxford, England). 2007;23(13): 1599–606.

14. Phan JH, Young AN, Wang MD. Robust microarray meta-analysis identifies differentially expressed genes for clinical prediction. Sci World J. 2012;2012:989637.

15. Campain A, Yang YH. Comparison study of microarray meta-analysis methods. BMCBioinformatics. 2010;11:408.

16. Novianti PW, van der Tweel I, Jong VL, Roes KC, Eijkemans MJ. An Application of Sequential Meta-Analysis to Gene Expression Studies. Cancer Inform. 2015;14 Suppl 5:1–10.

17. Jong VL, Novianti PW, Roes KC, Eijkemans MJ. Exploring homogeneity of correlation structures of gene expression datasets within and between etiological disease categories. Stat Appl Genet Mol Biol. 2014;13(6):717–32.

18. Payton JE, Grieselhuber NR, Chang LW, Murakami M, Geiss GK, Link DC, Nagarajan R, Watson MA, Ley TJ. High throughput digital quantification of mRNA abundance in primary human acute myeloid leukemia samples. J Clin Invest. 2009;119(6):1714–26.

19. Le DR, Taussig DC, Ramsay AG, Mitter R, Miraki-Moud F, Fatah R, Lee AM, Lister TA, Gribben JG. Peripheral blood T cells in acute myeloid leukemia (AML) patients at diagnosis have abnormal phenotype and genotype and form defective immune synapses with AML blasts. Blood. 2009;114(18):3909–16.

20. Majeti R, Becker MW, Tian Q, Lee TL, Yan X, Liu R, Chiang JH, Hood L, Clarke MF, Weissman IL. Dysregulated gene expression networks in human acute myelogenous leukemia stem cells. Proc Natl Acad Sci USA. 2009;106(9): 3396–401.

21. Beghini A, Corlazzoli F, Del GL, Re M, Lazzaroni F, Brioschi M, Valentini G, Ferrazzi F, Ghilardi A, Righi M, et al. Regeneration-associated WNT signaling is activated in long-term reconstituting AC133bright acute myeloid leukemia cells. Neoplasia. 2012;14(12):1236–48.

22. Bacher U, Schnittger S, Macijewski K, Grossmann V, Kohlmann A, Alpermann T, Kowarsch A, Nadarajah N, Kern W, Haferlach C, et al. Multilineage dysplasia does not influence prognosis in CEBPA-mutated AML, supporting the WHO proposal to classify these patients as a unique entity. Blood. 2012;119(20):4719–22.

23. Stirewalt DL, Pogosova-Agadjanyan EL, Ochsenreither S. Aberrant expressed genes in AML. ArrayExpress Archive of Functional Genomics Data. 2012. https://www.ebi.ac.uk/arrayexpress/experiments/E-GEOD-37307/.

24. Whitehead A. Estimating the Treatment Difference in an Individual Trial. In: Meta-Analysis Of Controlled Clinical Trials. Sussex: John Wiley & Sons, Ltd; 2002. p. 23–55.

25. Smyth GK. Linear models and empirical bayes methods for assessing differential expression in microarray experiments. Stat Appl Genet Mol Biol. 2004;3:Article3.

26. Borenstein M, Hedges LV, Higgins JPT, Rothstein HR. Effect Sizes Based on Means. In: Introduction toMeta-Analysis. Sussex: John Wiley & Sons, Ltd; 2009. p. 21–32.

27. Paule RM J. Consensus Values and Weighting Factors. J Res Natl Bur Stand. 1982;87(5):377.

28. Novianti PW, Roes KC, van der Tweel I. Estimation of between-trial variance in sequential meta-analyses: a simulation study. Contemp Clin Trials. 2014; 37(1):129–38.

29. van der Tweel I, Bollen C. Sequential meta-analysis: an efficient decision-making tool. Clin Trials. 2010;7(2):136–46.

30. Benjamini Y, Hochberg Y. Controlling the False Discovery Rate: A Practical and Powerful Approach to Multiple Testing. J R Stat Soc Ser B Methodol. 1995;57(1):289–300.

31. McLachlan G. Discriminant Analysis and Statistical Pattern Recognition (Wiley Series in Probability and Statistics). New Jersey: Wiley-Interscience; 2004.

32. Tibshirani R, Hastie T, Narasimhan B, Chu G. Class Prediction by Nearest Shrunken Centroids, with Applications to DNA Microarrays. Stat Sci. 2003; 18(1):104–17.

33. Breiman L. Random Forests. Mach Learn. 2001;45(1):5–32.

34. Friedman J. Greedy Function Approximation: A Gradient Boosting Machine. In: Annals of Statistics. 2000. p. 1189–232.

35. Hastie T, Tibshirani R, Friedman J. The Elements of Statistical Learning: Data Mining, Inference, andPrediction. 2nd ed. New York: Springer; 2009.

36. Zou H, Hastie T. Regularization and variable selection via the elastic net. J R Stat Soc Series B Stat Methodology. 2005;67(2):301–20.

37. Bishop CM. Pattern Recognition and Machine Learning (Information Science and Statistics). New Jersey: Springer-Verlag New York, Inc; 2006.

38. Boser BE, Guyon IM, Vapnik VN. A training algorithm for optimal margin classifiers. In: Proceedings of the fifth annual workshop on Computational learning theory; Pittsburgh, Pennsylvania, USA. New York: ACM; 1992. p. 144–52.

39. Ripley BD, Hjort NL. Pattern Recognition and Neural Networks. New York: Cambridge University Press; 1995.

40. Novianti PW, Jong VL, Roes KC, Eijkemans MJ. Factors affecting the accuracy of a class prediction model in gene expression data. BMC bioinformatics. 2015;16:199.

41. Smyth GK. limma: Linear Models for Microarray Data Bioinformatics and Computational Biology Solutions Using R and Bioconductor. In: Bioinformatics and Computational Biology Solutions Using R and Bioconductor. Edited by Gentleman R, Carey V, Huber W, Irizarry R, Dudoit S. New York: Springer New York; 2005. p. 397–420.

42. Jong VL, Novianti PW, Roes KC, Eijkemans MJ. Selecting a classification function for class prediction with gene expression data. Bioinformatics (Oxford, England). 2016;32(12):1814–22.

43. Gautier L, Cope L, Bolstad BM, Irizarry RA. affy–analysis of Affymetrix GeneChip data at the probe level. Bioinformatics. 2004;20(3):307–15.

44. Schwarzer G. meta: General Package for Meta-Analysis. R News. 2007;7(3):40–5.

45. Slawski M, Daumer M, Boulesteix AL. CMA: a comprehensive Bioconductor package for supervised classification with high dimensional data. BMC Bioinformatics. 2008;9:439.

46. Bates D, Maechler M, Bolker B, Walker S, Christensen RHB, Singmann H, Dai B, Grothendieck G, Green P. Fitting Linear Mixed-Effects Models Using lme4. J Stat Softw. 2015;67(1):1–48.

47. Wickham H. ggplot2: Elegant Graphics for Data Analysis. New York: Springer-Verlag; 2009.

48. Arijs I, Li K, Toedter G, Quintens R, Van LL, Van SK, Leemans P, De HG, Lemaire K, Ferrante M, et al. Mucosal gene signatures to predict response to infliximab in patients with ulcerative colitis. Gut. 2009;58(12):1612–9.

49. Kabakchiev B, Turner D, Hyams J, Mack D, Leleiko N, Crandall W, Markowitz J, Otley AR, Xu W, Hu P, et al. Gene expression changes associated with resistance to intravenous corticosteroid therapy in children with severe ulcerative colitis. PLoS One. 2010;5(9). doi:10.1371/journal.pone.0013085.

50. Scian MJ, Maluf DG, Archer KJ, Suh JL, Massey D, Fassnacht RC, Whitehill B, Sharma A, King A, Gehr T, et al. Gene expression changes are associated with loss of kidney graft function and interstitial fibrosis and tubular atrophy: diagnosis versus prediction. Transplantation. 2011;91(6):657–65.

51. Menke A, Arloth J, Putz B, Weber P, Klengel T, Mehta D, Gonik M, Rex-Haffner M, Rubel J, Uhr M, et al. Dexamethasone stimulated gene expression in peripheral blood is a sensitive marker for glucocorticoid receptor resistance in depressed patients. Neuropsychopharmacology. 2012; 37(6):1455–64.

52. Rasimas J, Katsounas A, Raza H, Murphy AA, Yang J, Lempicki RA, Osinusi A, Masur H, Polis M, Kottilil S, et al. Gene expression profiles predict emergence of psychiatric adverse events in HIV/HCV-coinfected patients on interferon-based HCV therapy. J Acquir Immune Defic Syndr. 2012;60(3):273–81.

53. Lunnon K, Sattlecker M, Furney SJ, Coppola G, Simmons A, Proitsi P, Lupton MK, Lourdusamy A, Johnston C, Soininen H, et al. A blood gene expression marker of early Alzheimer's disease. J Alzheimers Dis. 2013;33(3):737–53.

54. Ransohoff DF. Promises and limitations of biomarkers. Recent results in cancer research Fortschritte der Krebsforschung Progres dans les recherches sur le cancer. 2009;181:55–9.

55. Jeffery IB, Higgins DG, Culhane AC. Comparison and evaluation of methods for generating differentially expressed gene lists from microarray data. BMC Bioinformatics. 2006;7:359.

56. Marot G, Foulley JL, Mayer CD, Jaffrezic F. Moderated effect size and P-value combinations for microarray meta-analyses. Bioinformatics (Oxford, England). 2009;25(20):2692–9.

57. Gentleman RC, Carey VJ, Bates DM, Bolstad B, Dettling M, Dudoit S, Ellis B, Gautier L, Ge Y, Gentry J, et al. Bioconductor: open software development for computational biology and bioinformatics. Genome Biol. 2004;5(10):R80.

# 12

# Evaluation of the impact of Illumina error correction tools on de novo genome assembly

Mahdi Heydari[1,4], Giles Miclotte[1,4], Piet Demeester[1,4], Yves Van de Peer[2,3,4,5] and Jan Fostier[1,4*] ⓘ

## Abstract

**Background:** Recently, many standalone applications have been proposed to correct sequencing errors in Illumina data. The key idea is that downstream analysis tools such as *de novo* genome assemblers benefit from a reduced error rate in the input data. Surprisingly, a systematic validation of this assumption using state-of-the-art assembly methods is lacking, even for recently published methods.

**Results:** For twelve recent Illumina error correction tools (EC tools) we evaluated both their ability to correct sequencing errors and their ability to improve *de novo* genome assembly in terms of contig size and accuracy.

**Conclusions:** We confirm that most EC tools reduce the number of errors in sequencing data without introducing many new errors. However, we found that many EC tools suffer from poor performance in certain sequence contexts such as regions with low coverage or regions that contain short repeated or low-complexity sequences. Reads overlapping such regions are often ill-corrected in an inconsistent manner, leading to breakpoints in the resulting assemblies that are not present in assemblies obtained from uncorrected data. Resolving this systematic flaw in future EC tools could greatly improve the applicability of such tools.

**Keywords:** Next-generation sequencing, Error correction, Illumina, Genome assembly

## Background

Modern Illumina systems generate sequencing data with very high throughput and low financial cost. Illumina estimates that over 90% of sequencing data worldwide are generated on Illumina platforms. This data is characterized by a relatively short read length (100–300 bp) and a high accuracy (1–2% errors, mostly substitutions) [1]. Data generated on Illumina platforms suffers from various sources of bias, most notably a higher number of sequencing errors towards the 3'-end of the reads and a non-uniform distribution of reads across the genome [2].

Despite its short read length, Illumina data is often used for *de novo* genome assembly, sometimes complemented by data generated through other platforms. Most short-read assemblers first generate a de Bruijn graph from the input reads [3]. This graph represents all $k$-mers that occur in the input reads and the overlap between them. As such, de Bruijn graphs are used to efficiently establish the overlap between individual reads. The original genomic sequence is then represented as some path through the de Bruijn graph.

The presence of sequencing errors significantly complicates this task: a single sequencing error in a read results in up to $k$ erroneous $k$-mers in the de Bruijn graph. These $k$-mers create artifacts in the de Bruijn graph such as spurious dead ends, parallel paths and chimeric connections [4]. Despite the low error rate, erroneous $k$-mers can vastly outnumber true $k$-mers, challenging the identification of the original sequence. To reduce the number of erroneous $k$-mers, trimming tools can be used as a primary solution to discard parts of each input read that have a per-base quality score below a user-defined threshold. However, this further reduces the read length and might aggravate the coverage bias.

Error correction tools (EC tools) on the other hand, try to identify and correct the sequencing errors. Often, this is achieved by generating a $k$-mer coverage spectrum from the input data and replacing poorly covered (and hence likely erroneous) $k$-mers by similar $k$-mers with a

*Correspondence: jan.fostier@ugent.be
[1] Department of Information Technology, Ghent University-imec, IDLab, B-9052 Ghent, Belgium
[4] Bioinformatics Institute Ghent, B-9052 Ghent, Belgium
Full list of author information is available at the end of the article

higher coverage. Sometimes, this process is further guided by using the per-base quality scores. Many standalone read error correction algorithms and implementations have been proposed for Illumina data, including ACE [5], BayesHammer [6], BFC [7], BLESS [8], BLESS 2 [9], Blue [10], EC [11], Fiona [12], Karect [13], Lighter [14], Musket [15], Pollux [16], Quake [17], QuorUM [18], RACER [19], SGA-EC [20] and Trowel [21]. For a comprehensive overview of the characteristics of these EC tools and those for other sequencing platforms, we refer to [22].

The key idea is that the prior application of EC tools on raw Illumina sequencing data provides assembly methods with cleaner input data and hence improves the quality of assembly both in terms of reduced fragmentation (i.e., longer contigs or scaffolds) and higher accuracy of the resulting assemblies. As a secondary goal, the prior use of EC tools may reduce the memory usage and the runtime of the assembly tool. This is useful when assembling larger genomes, a task that is typically quite resource-intensive.

Surprisingly, most EC tools are not evaluated on their ability to improve the quality of *de novo* genome assembly with modern assemblers, but rather directly on their ability to correct sequencing errors. Using simulated Illumina data, such an evaluation is straightforward as error-free data is known. In that case, the *error correction gain*, a metric that expresses to what degree the error rate is reduced, is used to describe the performance of EC tools. With real Illumina data, the error correction performance is typically assessed through the use of a read mapper: both corrected and uncorrected reads are aligned to their corresponding reference genome and various performance metrics are derived to express the reduction in mismatches in the respective alignments. EC tools that result in more aligned reads and/or alignments with fewer mismatches are assumed to be superior.

We argue that a lower average error-rate in the input data does not necessarily lead to better assembly results. First, the vast majority of sequencing errors are benign to the assembly process. For example, consider a sequencing error that gives rise to one or more erroneous $k$-mers that otherwise do not exist in the sequenced genome. In the de Bruijn graph, such sequencing error causes a spurious dead end or a short parallel path. These graph artifacts are easily detected and corrected for by many assembly tools assuming the corresponding true $k$-mers occur with sufficient coverage in the input reads. Only a relatively small fraction of sequencing errors is truly problematic, for example when they give rise to erroneous $k$-mers that do exist elsewhere in the genome. These errors thus give rise to spurious 'chimeric' connections between nodes in the de Bruijn graph that are otherwise distantly located in the original sequence. As such, they may result in mis-assemblies and/or shorter contig sizes. A second class of problematic errors are those that occur in regions with

very low coverage. Such errors may render the assembly tool unable to detect overlap between reads because no $k$-mers are shared. Overall, an EC tool that is able to correct all benign sequencing errors and not a single problematic sequencing error might exhibit a high error correction gain but will not substantially improve the assembly process. Second, EC tools might introduce new errors in the sequence data. If such events are rare and unbiased, they may not pose a great threat to the assembly process. However, if EC tools systematically make the same mistake in a given context, the genome assembler may not be able to recover from this error.

Most state-of-the-art genome assembly tools have built-in algorithms to detect and handle sequencing errors, either directly or implicitly through a correction procedure on the de Bruijn graph. The prior use of standalone EC tools thus only makes sense if they outperform these built-in error correction algorithms. Table 1 lists for every EC tool the accuracy analyses that were performed in the accompanying publication. Even though all tools were evaluated for their ability to reduce sequencing errors, their ability to improve the genome assembly process is either lacking or performed with older assembly tools. Also, recent review papers on EC tools [23, 24] did not contain such analyses.

In this paper, we review twelve recently published EC tools. We compiled a benchmark suite of eight public datasets sequenced from organisms with a genome size ranging from 2 to 116 Mbp and assessed the performance of the different EC tools both on their potential to correct the sequencing errors and on their ability to improve assembly results using four assemblers (DISCOVAR [25], IDBA [26], SPAdes [27] and Velvet [4]). We discuss the impact on the resulting assembly quality and investigate systematic errors in some of the EC tools. Finally, computational efficiency (memory usage and runtime) of the different EC tools is discussed. Note that the effect of error correction for other applications such as variant calling is beyond the scope of this paper.

## Methods

### Error correction tools

Twelve state-of-the-art (published in 2012 or later) EC tools for Illumina data were included in this review and listed in Table 1. We were unable to produce corrected reads with QuorUM and EC and hence these tools were excluded in this study.

EC tools have been classified according to their underlying algorithmic principles in several review papers [22, 23, 28]. In Table 1, tools were classified according to their main algorithmic approach: $k$-mer spectrum based or multiple sequence alignment (MSA) based. The $k$-mer spectrum based tools operate on the level of individual $k$-mers. First, the complete set of $k$-mers that occur

**Table 1** List of EC tools evaluated in this paper

| EC tool | Algorithm | Data structure | Indel support | Accuracy analysis | Assembly analysis | Year |
|---------|-----------|----------------|---------------|-------------------|-------------------|------|
| ACE | *k*-mer | *k*-mer trie | | Read level | - | 2015 |
| BayesHammer | *k*-mer | Hamming graph | | Read level | SPAdes | 2013 |
| BFC | *k*-mer | Bloom filter | | Read level | Velvet, ABySS [34] | 2015 |
| BLESS 2 | *k*-mer | Bloom filter | | Read level | Gossamer [35] | 2016 |
| Blue | *k*-mer | Hash table | ✓ | Read level | Velvet | 2014 |
| Fiona | MSA | Suffix tree | ✓ | Base level | - | 2014 |
| Karect | MSA | Partially-ordered graph | ✓ | Read, base level | Velvet, SGA, Celera [36] | 2015 |
| Lighter | *k*-mer | Bloom filter | | Read level | Velvet | 2013 |
| Musket | *k*-mer | Bloom filter | | Base level | SGA | 2013 |
| RACER | *k*-mer | Hash table | | Read level | - | 2013 |
| SGA-EC | MSA | Suffix array | | Read level | SGA | 2012 |
| Trowel | *k*-mer | Hash table | | Read, base level | Velvet, SOAPdenovo [37] | 2014 |

The algorithmic approach is either *k*-mer spectrum based ('*k*-mer') or multiple sequence alignment based ('MSA'). Tools can be further classified according to data structure and heuristics used. Some tools are able to correct insertions or deletions. In their accompanying publication, all tools were assessed directly on their ability to reduce error rate, either on the read or base level. Most tools did not use assembly analyses with modern assemblers in their evaluation. SPAdes was used for the evaluation of BayesHammer, but no comparison was made with assembly results from uncorrected data

in the input data and their corresponding frequency is determined. Second, reads that contain rarely occurring *k*-mers are assumed to contain sequencing errors and are modified, using a minimum edit distance strategy, such that these *k*-mers are replaced by similar, more frequently occurring *k*-mers. In contrast, MSA-based tools operate on the level of reads. First, reads that are assumed to represent overlapping genomic regions are clustered together and a consensus is obtained through multiple alignment. Second, reads are corrected according to the consensus alignment. While all EC tools considered in this review rely on either of these two approaches, there is still a great diversity in the specific implementation heuristics and data structures (bloom filter, hash table, suffix tree, ...).

Most tools require users to specify a *k*-mer length to be used during the error correction procedure. The optimal value can differ from one dataset to another, depending on the coverage, genome size and error distribution. This optimal value was empirically obtained by running the EC tool multiple times with different *k*-mer sizes and selecting the *k*-mer size that yields the most contiguous SPAdes assembly results as measured in terms of N50. This optimal value was used to produce the results of Table 4. For all other tables and figures, the default or recommended *k*-mer size was used for all datasets. Parameters and settings are provided in Additional file 1: Section 1. All tools support multithreading, and with the exception of ACE and RACER, the number of parallel threads can be specified. Those tools were run with 32 threads. Runtime and peak memory usage were measured with the GNU 'time -v' command. We recorded elapsed (wall clock) time and peak resident memory usage. All tools

were run on a machine with four Intel(R) Xeon(R) E5-2698 v3 @ 2.30 GHz CPUs (64 cores in total) and 256 GB of memory.

**Data**

Tools are benchmarked on eight datasets for which both a high quality reference genome and real Illumina data are publicly available (see Table 2). Genome sizes range from 2 Mbp (*Bifidobacterium dentium*) to 116 Mbp (*Drosophila melanogaster*) while read coverage varies from 29 X to 612 X. Data is produced by the Illumina HiSeq, MiSeq and GAII platforms with read lengths varying between 100 bp and 251 bp. Two of the datasets have a variable read length due to read trimming, all other datasets have fixed read lengths.

To assess the performance of tools on simulated data, synthetic Illumina reads for the same set of organisms were generated using ART [29]. The same coverage and read lengths were used as for the real data (Additional file 1: Section 2). ART also generates a corresponding set of error-free reads, which greatly facilitates the evaluation of EC tools on synthetic data.

**Error metrics**

The error rate is the ratio of the total number of sequencing errors (substitutions or indels) and the number of nucleotides in the input data. Error correction performance is measured as follows: true positives (TP) correspond to corrected errors; true negatives (TN) correspond to initially correct bases left untouched; false positives (FP) correspond to newly introduced errors; false negatives (FN) correspond to unidentified errors. The error correction gain (EC gain) is defined as:

**Table 2** Real datasets used for the evaluation of EC tools

| Abbr. | Organism | Reference ID | Genome size | Cov. | Sequencing platform | Read length | Trimmed reads | Dataset ID | Ref. |
|---|---|---|---|---|---|---|---|---|---|
| D1 | *Bifidobacterium dentium* | Nc013714.1 | 2.6 Mbp | 373 X | Illumina MiSeq | 251 bp | | SRR1151311 | [23] |
| D2 | *Escherichia coli K-12 DH10B* | NC010473 | 4.5 Mbp | 418 X | Illumina MiSeq | 150 bp | | Ill. Data library | [10] |
| D3 | *Escherichia coli K-12 MG1655* | NC000913 | 4.5 Mbp | 612 X | Illumina GAII | 100 bp | | ERA000206 | [10] |
| D4 | *Salmonella enterica* | NC011083.1 | 4.7 Mbp | 97 X | Illumina MiSeq | 239 bp | ✓ | SRR1206093 | [23] |
| D5 | *Pseudomonas aeruginosa* | ERR330008 | 6.1 Mbp | 169 X | Illumina MiSeq | 120 bp | ✓ | ERR330008 | [10] |
| D6 | *Homo sapiens* Chr. 21 | HG19 | 45.2 Mbp | 29 X | Illumina HiSeq | 100 bp | | Ill. Data library | [10] |
| D7 | *Caenorhabditis elegans* | WS222 | 97.6 Mbp | 58 X | Illumina HiSeq | 101 bp | | SRR543736 | [23] |
| D8 | *Drosophila melanogaster* | Release 5 | 116.4 Mbp | 52 X | Illumina HiSeq | 100 bp | | SRR823377 | [23] |

$$\text{EC gain} = \frac{\text{TP} - \text{FP}}{\text{TP} + \text{FN}}.$$

The EC gain measures the degree in which the error rate is reduced. A gain of 100% means all errors were corrected and no new errors were introduced. The sensitivity (true positive rate – TPR) is defined as follows:

$$\text{TPR} = \frac{\text{TP}}{\text{TP} + \text{FN}}.$$

### Evaluation of assembly results

To assess the impact of error correction on *de novo* assembly results, the following assemblers were used: DISCOVAR, IDBA, SPAdes and Velvet. All four assemblers have built-in error correction functionality. Velvet, IDBA and SPAdes remove erroneous $k$-mers through the identification of parallel paths ('bubbles' and 'tips') in the de Bruijn graph. SPAdes and IDBA iteratively increase the $k$-mer size. This way, they take advantage of shorter $k$-mers for a sensitive detection of overlap between reads and of longer $k$-mers for dealing with repeat resolution. DISCOVAR uses a different methodology: for each read, a group of 'true friends' is determined. These are reads that share a $k$-mer with the read and that do not have a high quality base difference with the read. DISCOVAR then corrects each read based on the consensus sequence obtained from the multiple sequence alignment of its true friends.

We investigated the underlying causes of suboptimal assembly results after error correction. MUMmer [30] was used to align contigs, and to check if the contig has no structural misassemblies. In order to determine the $k$-mer frequencies Jellyfish [31] was used.

## Results and discussion
### Ability of EC tools to correct sequencing errors

In order to estimate the reduction in error rate through the use of EC tools, both uncorrected and corrected data were aligned to the corresponding reference genome using BWA [32]. For all datasets D1-D8 and EC tools,

the fraction of reads that align with respectively $m = 0$ and $m > 9$ mismatches is reported in Additional file 1: Section 3.1. All EC tools are able to substantially reduce the number of mismatches required for read alignment. This is especially true for bacterial genomes, where often >95% of the corrected reads show perfect alignment with the reference. In contrast, for larger genomes, this is typically in the range of 60–80%. Error correction also reduces the fraction of highly erroneous reads (i.e., reads that require more than 9 mismatches to align), albeit to varying degrees. For the largest dataset D8 (*D. melanogaster*), Fig. 1 provides a more detailed breakdown of the number of mismatches $m$ required for read alignment. Initially, about 50% of the uncorrected reads perfectly align. ACE shows the highest increase of this figure to 60.14%. ACE also has the lowest percentage of highly erroneous reads.

After applying error correction to a read, there is no guarantee that BWA will again align that read to the same genomic location. Therefore, this evaluation metric might favor overly aggressive EC tools that transform reads into similar reads that do exist in the genome, but that do not

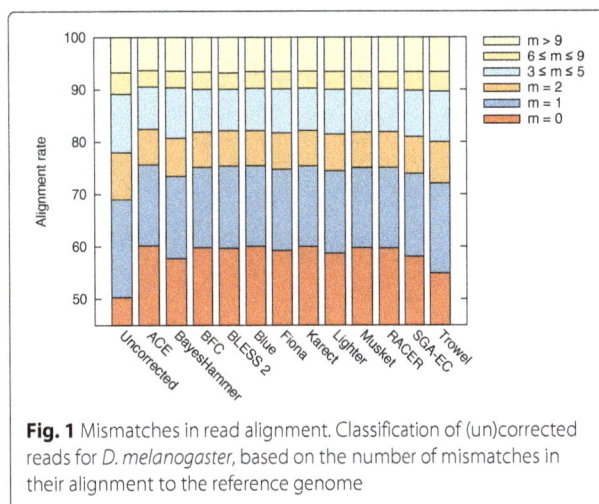

**Fig. 1** Mismatches in read alignment. Classification of (un)corrected reads for *D. melanogaster*, based on the number of mismatches in their alignment to the reference genome

represent the actual sequenced genomic region. Therefore, in an alternative evaluation metric, we assume that the error-free read is represented by the segment of the reference genome to which the uncorrected read aligns. Uncorrected reads that can not be mapped to the reference genome are excluded from this evaluation. As BayesHammer and BLESS 2 do not provide a one-to-one correspondence between input and output, they are not included in this evaluation.

Table 3 shows the EC gain, the percentage of corrected errors and the number of newly introduced errors per Mbp of read data for each of the eight datasets. Detailed confusion matrices are provided in Additional file 1: Section 3.2.2. Major differences in EC gain can now be observed between the different EC tools. All EC tools perform much better on the smaller bacterial genomes (D1-D5), than on the larger eukaryotes (D6-D8). For all datasets, Karect shows the highest number of

**Table 3** Accuracy comparison of EC tools in terms of EC gain, percentage of corrected errors, and number of newly introduced errors per Mbp of read data

|  | D1 | D2 | D3 | D4 | D5 | D6 | D7 | D8 |
|---|---|---|---|---|---|---|---|---|
| Error correction gain (%) | | | | | | | | |
| ACE | 96.3 | 97.9 | 98.7 | 96.2 | 91.1 | 41.7 | -3.3 | 25.9 |
| BFC | 78.7 | 84.3 | 80.2 | 81.4 | 78.6 | 52.8 | 63.3 | 24.1 |
| Blue | 98.5 | 98.8 | 98.7 | 96.7 | 95.4 | 51.1 | 65.2 | 28.8 |
| Fiona | 87.4 | 94.6 | 97.5 | 85.5 | 91.4 | 55.0 | 65.8 | 29.8 |
| Karect | 99.4 | 99.8 | 99.7 | 98.5 | 98.2 | 63.1 | 75.5 | 34.3 |
| Lighter | 85.4 | 93.8 | 92.5 | 80.1 | 84.6 | 45.7 | 50.3 | 21.7 |
| Musket | 91.3 | 93.6 | 93.4 | 88.0 | 87.1 | 49.5 | 59.2 | 23.5 |
| RACER | 92.3 | 94.4 | 97.0 | 88.3 | 94.0 | 17.4 | 32.6 | 22.3 |
| SGA-EC | 55.3 | 67.2 | 45.5 | 53.1 | 65.2 | 48.7 | 60.6 | 23.0 |
| Trowel | 38.4 | 49.4 | 38.8 | 40.5 | 46.8 | 13.2 | 1.1 | 10.5 |
| Percentage of corrected errors (sensitivity) | | | | | | | | |
| ACE | 97.7 | 98.5 | 99.2 | 98.0 | 97.0 | 61.3 | 73.8 | 34.5 |
| BFC | 78.8 | 84.4 | 80.2 | 81.4 | 78.7 | 54.1 | 63.8 | 24.7 |
| Blue | 98.7 | 99.3 | 99.1 | 97.0 | 95.7 | 59.9 | 70.6 | 31.4 |
| Fiona | 87.5 | 94.8 | 97.7 | 85.5 | 91.7 | 60.6 | 71.7 | 31.5 |
| Karect | 99.4 | 99.9 | 99.7 | 98.5 | 98.2 | 64.4 | 76.7 | 35.5 |
| Lighter | 85.5 | 94.0 | 92.7 | 80.2 | 86.3 | 48.9 | 59.1 | 24.3 |
| Musket | 91.3 | 93.6 | 93.4 | 88.1 | 87.3 | 52.9 | 65.3 | 26.4 |
| RACER | 92.9 | 95.8 | 98.2 | 89.0 | 94.8 | 59.2 | 68.2 | 34.0 |
| SGA-EC | 55.3 | 67.2 | 45.5 | 53.1 | 65.3 | 50.4 | 61.3 | 23.2 |
| Trowel | 39.0 | 49.9 | 43.4 | 40.9 | 47.6 | 23.6 | 31.2 | 11.8 |
| Number of errors introduced per Mbp | | | | | | | | |
| ACE | 44 | 23 | 40 | 151 | 194 | 1217 | 2375 | 1123 |
| BFC | 2 | 3 | 7 | 2 | 3 | 83 | 15 | 73 |
| Blue | 8 | 20 | 30 | 31 | 10 | 547 | 167 | 341 |
| Fiona | 2 | 7 | 14 | 6 | 9 | 347 | 183 | 218 |
| Karect | 0 | 1 | 3 | 1 | 1 | 80 | 36 | 157 |
| Lighter | 2 | 6 | 14 | 8 | 56 | 202 | 273 | 332 |
| Musket | 1 | 2 | 5 | 3 | 6 | 214 | 190 | 383 |
| RACER | 21 | 62 | 97 | 58 | 27 | 2603 | 1097 | 1524 |
| SGA-EC | 1 | 3 | 6 | 2 | 3 | 105 | 22 | 24 |
| Trowel | 21 | 26 | 376 | 41 | 25 | 647 | 930 | 172 |

true positives (errors that were successfully corrected) and the lowest number of false negatives (uncorrected errors). With the exception of dataset D7 (*C. elegans*) and D8 (*D. melanogaster*), Karect also has the lowest number of false positives (newly introduced errors). Overall, Karect has the highest error correction gain for all datasets.

For most datasets, BFC, SGA-EC and Trowel correct significantly fewer sequencing errors compared with other EC tools. BFC and SGA-EC appear to be conservative as they introduce only a small number of new errors. In contrast, ACE, Racer and Trowel often introduce a significant amount of new errors. Note that for dataset D7, the EC gain of ACE is negative, indicating a higher number of sequencing errors after error correction than in the uncorrected data: ACE successfully corrects about 10.8 million errors but introduces almost 11.3 million new errors.

For comparison, *artificial* data was generated for the eight genomes using the same read length and coverage as the corresponding real datasets. Data was corrected using identical settings as before. The confusion matrix and derived metrics can be unambiguously constructed for artificial data since the true, error-free read is known (see Additional file 1: Section 3.2.3). BFC now shows the highest gain for four datasets, while Karect and Fiona each have the highest gain for two datasets. The numbers indicate that EC tools perform much better on artificial data than on real data. This is due to the fact that simulated data are produced according to simplified models that may fail to capture the intricacies of real data.

### Ability of EC tools to improve genome assembly

To evaluate the effect of error correction on *de novo* genome assembly, both uncorrected and corrected reads were assembled using respectively DISCOVAR, IDBA, SPAdes and Velvet. The resulting assemblies were evaluated using QUAST [33] and detailed reports for all combinations of assemblers and EC tools are provided in Additional file 1: Section 4 for reference. We found that SPAdes and DISCOVAR consistently produced higher quality contigs than Velvet and IDBA. We were unable to produce assemblies with DISCOVAR using the reads that were corrected by Trowel and Fiona. Therefore, only SPAdes assemblies are discussed in detail in the remainder of this section.

Table 4 shows the contig and scaffold NGA50 values for all eight datasets and EC tools. For the EC tools that allow the *k*-mer size to be specified, the optimal value of *k* was used (see Additional file 1: Section 1). The NGA50 represents the characteristic length of the assembled contigs/scaffolds that can be contiguously aligned to the reference genome. These contigs/scaffolds thus contain no major structural assembly errors and a higher NGA50 hence implies a less fragmented assembly. For smaller

genome sizes (datasets D1-D5), the prior application of EC tools often does not significantly influence the scaffold NGA50. For dataset D3, many tools are able to improve the contig NGA50, sometimes significantly. Remarkably, for dataset D5 (*P. aeruginosa*) most EC tools lead to a somewhat lower scaffold NGA50 compared to the assembly result obtained from uncorrected data. However, the NGAx plot of this dataset reveals no major differences in assembly quality between corrected and uncorrected reads (see Additional file 1: Section 4.3.5). For the larger genomes, the use of EC tools does occasionally improve assembly results, especially on dataset D6 (Human, chr. 21) where eight out of twelve EC tools lead to a higher scaffold NGA50. On the largest datasets D7 and D8 however, error correction may significantly deteriorate the assembly quality. In some cases, the NGA50 obtained is less than half of the corresponding value on uncorrected data.

Especially for dataset D8 (*D. melanogaster*), the prior use of different EC tools results in a large variability in assembly quality (see Fig. 2). Only Blue, Karect and SGA-EC improve the NGA50 for this dataset. In contrast, error correction with ACE, BLESS 2, Fiona or RACER leads to significantly shorter scaffolds. Additionally, a lower percentage of the genome was found to be covered by scaffolds and a higher rate of insertions, deletions and mismatches was observed (see Additional file 1: Section 4).

At this point it should be stressed that error correction does consistently lead to substantially better assembly results for Velvet or IDBA. However, in our hands, the NGA50 values obtained with Velvet or IDBA were much lower than with SPAdes or DISCOVAR. Even after error correction, Velvet and IDBA yield significantly shorter contigs than SPAdes or DISCOVAR. From this we conclude that the built-in error correction procedures in Velvet and IDBA are less accurate than those in SPAdes and DISCOVAR.

### Error rate versus assembly quality

Even though EC tools almost always reduce the error rate in the input data, they do not necessarily lead to better assemblies. In order to better understand these contrasting observations, we investigated why the use of corrected data can lead to a more fragmented assembly. For the largest dataset (D8), the two largest contigs (> 400 kbp each) that were correctly assembled from uncorrected data were selected. The corresponding (shorter) contigs obtained from assemblies on corrected data were aligned to these contigs and visualized in Fig. 3. With the exception of Trowel, all error correction tools lead to a more fragmented assembly of at least one of these contigs. Breakpoints, i.e., endpoints of the shorter contigs, caused by error correction do not appear to occur at random positions. Rather, different EC tools often cause breakpoints at the same positions. For example, in

**Table 4** NGA50 of respectively contigs (top) and scaffolds (bottom) assembled by SPAdes before and after error correction

| Tools | D1 | D2 | D3 | D4 | D5 | D6 | D7 | D8 |
|---|---|---|---|---|---|---|---|---|
| Contig NGA50 | | | | | | | | |
| Uncorrected | 397 392 | 92 570 | 119 253 | 231 409 | 264 881 | 8 559 | 6 429 | 50 484 |
| ACE | 397 392 = | 92 570 = | 125 608 ↑ | 231 409 = | 264 881 = | 8 771 ↑ | 3 143 ⇊ | 28 679 ⇊ |
| BayesHammer | 397 392 = | 92 344 ↓ | 132 564 ⇈ | 231 409 = | 264 881 = | 9 075 ↑ | 6 540 ↑ | 53 534 ↑ |
| BFC | 397 392 = | 92 570 = | 132 876 ⇈ | 231 409 = | 264 881 = | 9 375 ↑ | 6 389 ↓ | 49 185 ↓ |
| BLESS 2 | 397 392 = | 92 570 = | 119 265 ↑ | 231 409 = | 264 881 = | 7 975 ↓ | 3 047 ⇊ | 23 814 ⇊ |
| Blue | 397 392 = | 92 708 ↑ | 132 876 ⇈ | 231 409 = | 289 353 ↑ | 7 628 ⇊ | 6 191 ↓ | 50 486 ↑ |
| Fiona | 397 392 = | 92 611 ↑ | 119 253 = | 231 409 = | 264 881 = | 9 224 ↑ | 5 346 ⇊ | 45 472 ↓ |
| Karect | 397 392 = | 92 611 ↑ | 132 876 ⇈ | 231 409 = | 264 881 = | 9 865 ⇈ | 6 392 ↓ | 54 132 ↑ |
| Lighter | 397 392 = | 92 570 = | 132 564 ⇈ | 231 409 = | 289 353 ↑ | 9 609 ⇈ | 6 423 ↓ | 50 440 ↓ |
| Musket | 397 392 = | 92 566 ↓ | 132 876 ⇈ | 231 409 = | 264 881 = | 9 293 ↑ | 6 170 ↓ | 46 377 ↓ |
| RACER | 397 392 = | 92 523 ↓ | 112 393 ↓ | 231 409 = | 264 881 = | 7 336 ⇊ | 3 244 ⇊ | 21 538 ⇊ |
| SGA-EC | 397 392 = | 92 344 ↓ | 119 255 ↑ | 231 409 = | 264 881 = | 9 296 ↑ | 6 435 ↑ | 52 105 ↑ |
| Trowel | 397 392 = | 92 344 ↓ | 119 335 ↑ | 231 409 = | 264 881 = | 7 808 ↓ | 6 389 ↓ | 48 357 ↓ |
| Scaffold NGA50 | | | | | | | | |
| Uncorrected | 397 392 | 97 353 | 132 876 | 231 409 | 289 353 | 8 829 | 6 472 | 60 554 |
| ACE | 397 392 = | 97 353 = | 133 713 ↑ | 231 409 = | 264 881 ↓ | 9 190 ↑ | 3 158 ⇊ | 35 392 ⇊ |
| BayesHammer | 397 392 = | 97 353 = | 133 309 ↑ | 231 409 = | 264 881 ↓ | 9 443 ↑ | 6 576 ↑ | 58 570 ↓ |
| BFC | 397 392 = | 97 353 = | 133 088 ↑ | 231 409 = | 264 881 ↓ | 9 664 ↑ | 6 419 ↓ | 59 613 ↓ |
| BLESS 2 | 397 392 = | 97 353 = | 132 876 = | 231 409 = | 264 881 ↓ | 8 441 ↓ | 3 073 ⇊ | 35 638 ⇊ |
| Blue | 397 392 = | 97 288 ↓ | 133 309 ↑ | 231 409 = | 289 353 = | 7 841 ⇊ | 6 183 ↓ | 61 289 ↑ |
| Fiona | 397 392 = | 97 353 = | 132 876 = | 231 409 = | 264 881 ↓ | 9 491 ↑ | 5 385 ⇊ | 54 188 ⇊ |
| Karect | 397 392 = | 97 353 = | 133 058 ↑ | 231 409 = | 264 881 ↓ | 10 302 ⇈ | 6 446 ↓ | 62 304 ↑ |
| Lighter | 397 392 = | 97 353 = | 133 309 ↑ | 231 409 = | 289 353 = | 9 955 ⇈ | 6 468 ↓ | 59 697 ↓ |
| Musket | 397 392 = | 97 353 = | 133 088 ↑ | 231 409 = | 264 881 ↓ | 9 502 ↑ | 6 219 ↓ | 55 842 ↓ |
| RACER | 397 392 = | 97 353 = | 132 876 = | 231 409 = | 264 881 ↓ | 7 603 ⇊ | 3 266 ⇊ | 23 783 ⇊ |
| SGA-EC | 397 392 = | 97 353 = | 132 876 = | 231 409 = | 264 881 ↓ | 9 640 ↑ | 6 483 ↑ | 60 636 ↑ |
| Trowel | 397 392 = | 97 353 = | 132 876 = | 231 409 = | 264 881 ↓ | 8 107 ↓ | 6 435 ↓ | 57 078 ↓ |

Arrows in the table are based on their value relative to the NGA50 value obtained from uncorrected data as follows: ⇊ < -10% < ↓ < 0% < ↑ < +10% < ⇈

Fig. 3, the breakpoints marked as 'A' and 'B' each occur in four cases.

In order to identify the mechanisms that cause breakpoints, the $k$-mer spectrum of both corrected and uncorrected data along the two contigs was examined. In this section, $k = 21$ is used throughout, as it corresponds to the smallest $k$-mer size that is used to establish overlap between individual reads by the multi-$k$ SPAdes assembler. In Fig. 3, black bars visualize the locations of 'lost true 21-mers', i.e., 21-mers that do exist in the reference sequence (hence 'true') and also do exist in the uncorrected data but that are no longer present in the corrected data (hence 'lost'). Lost true $k$-mers hence refer to those $k$-mers that were systematically, but erroneously removed during error correction. In many cases, lost true 21-mers

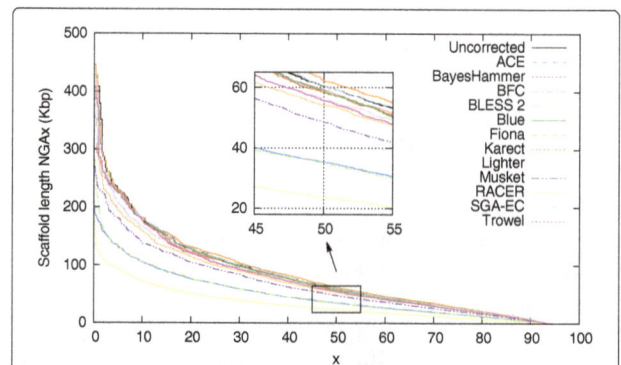

**Fig. 2** SPAdes assemblies. SPAdes assembly results for *D. melanogaster* for (un)corrected data. Scaffolds with length NGAx or larger contain x% of the genome

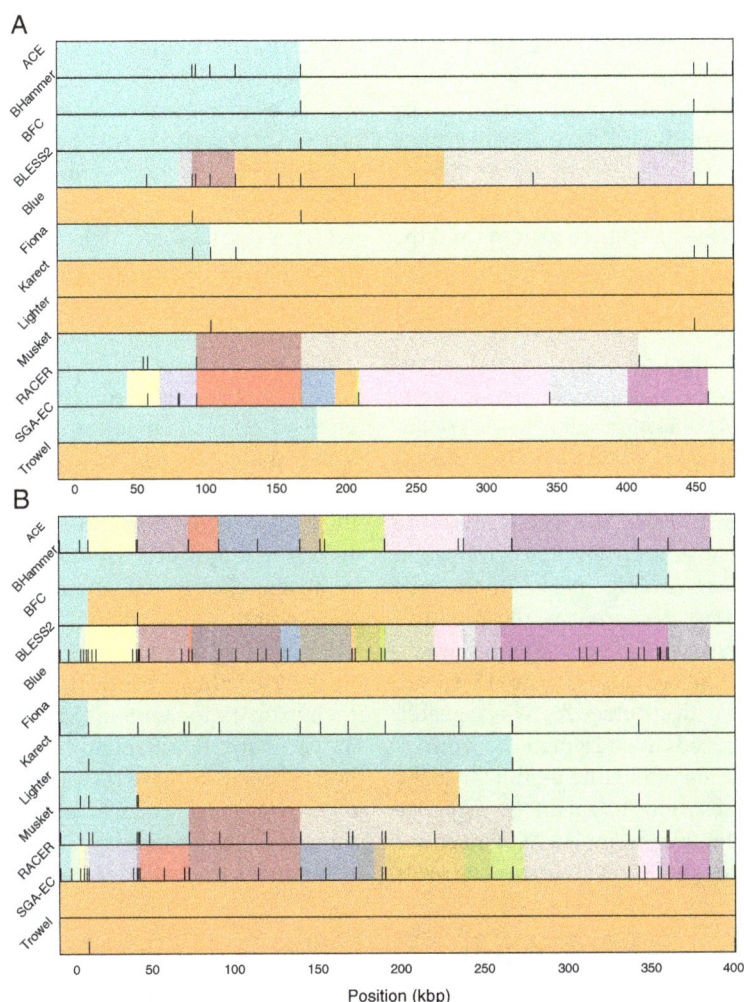

**Fig. 3** Fragmented assembly using corrected data. Contigs assembled from corrected data are aligned to the largest (*top*) and second largest (*bottom*) contig obtained from uncorrected data. Different colors denote different contigs. *Black bars* indicate the location of lost true *k*-mers in the contigs. This indicates a possible causal relationship between lost true *k*-mers and the breakpoints in the assemblies of corrected data

occur in the direct vicinity of breakpoints, indicating a possible causal relationship between lost true 21-mers and these breakpoints (see Fig. 3).

To varying degrees, all EC tools suffer from lost true *k*-mers. For dataset D8, Fig. 4 shows the 21-mer spectrum of the uncorrected data, along with the lost true 21-mer spectrum for the individual EC tools. Unsurprisingly, true *k*-mers are almost exclusively lost when their corresponding coverage in the uncorrected data is low. Indeed, a lower than expected coverage is an important feature for EC tools to select candidate errors. Trowel and SGA-EC appear most conservative in terms of lost true *k*-mers: almost no true 21-mers that occur > 2 times are removed. In contrast, ACE, BLESS 2, Musket and RACER remove a significant number of true 21-mers, some of which occur > 10 times in the initial data. These EC tools lead to a more fragmented assembly, which becomes especially evident for the second biggest contig (cfr. Fig. 3).

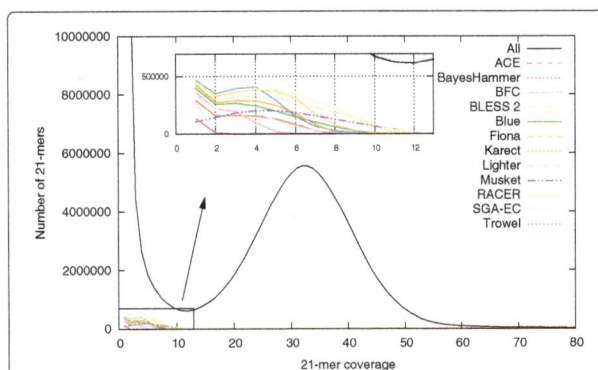

**Fig. 4** Lost true 21-mers spectrum. For dataset D8, this figure shows the 21-mer spectrum of the uncorrected data, along with the lost true 21-mer spectrum for all EC tools. EC tools erroneously remove low frequency true 21-mers during error correction

In principle, a lost true $k$-mer should not necessarily lead to a breakpoint. If all reads that initially contain the lost true $k$-mer(s) are modified in a consistent manner, the assembler will still be able to correctly identify the overlap between those reads and the lost true $k$-mers would appear as mismatches in the resulting assembly. In practice, the lost true $k$-mers will likely be replaced by $k$-mers that actually occur elsewhere in the genome and the genome assembler will be challenged by a spurious repeat that it may or may not be able to resolve. Vice versa, not all breakpoints due to error correction are directly related lost true $k$-mers. The ill-correction of reads could potentially only lead to a decrease in coverage without losing the true $k$-mer in all reads. This can still result in a breakpoint.

In practice however, we find that breakpoints due to error correction are often related to lost true $k$-mers (cfr. Fig. 3). Further inspection revealed that true $k$-mers are typically lost in regions that suffer from poor coverage in the direct vicinity of a local coverage peak. Often, such sudden increase in coverage is caused by the presence of a short repeated element. For example, Fig. 5 shows a genomic region with low $k$-mer coverage (around 7 X) that contains a repeated $k$-mer with coverage 35. This repeated $k$-mer also occurs in other reads that originate from different genomic locations. We can therefore assume that the EC tool makes erroneous decisions based on the sequence content of these reads. In this example, ACE makes a large number of substitutions in originally error-free reads

causing 75 consecutive lost true $k$-mers. Clearly, the error correction procedure is not performed in a consistent manner for all reads, rendering the assembler unable to detect overlap between these reads and ultimately leading to a breakpoint. For the same reasons, BLESS 2 and RACER also break at this specific location.

As a second example, Fig. 6 shows a short 22 bp long AT repeat with very high coverage (nearly 14 000 X), in a genomic region with otherwise low coverage. Musket introduces a new error in two out of four overlapping reads. Within this specific context, these substitutions cause a number of true $k$-mers to be lost. More importantly, because the error correction is not performed in an identical manner across all four reads overlapping this locus, the overlap is broken and a breakpoint is introduced. Similarly, due to the same AT repeat, Fiona introduces errors that result in a number of lost true $k$-mers. In this case however, the newly introduced errors result in mismatches in the assembled sequence rather than a breakpoint.

From these examples, the limitations of $k$-mer spectrum based error correction tools become evident. Due to their primary focus on individual $k$-mers, they do not take into account the surrounding context in which the $k$-mer occurs. Because these tools correct reads individually, different corrections may be applied to different reads even though the reads overlap the same genomic region. This may render de Bruijn graph assemblers unable to detect

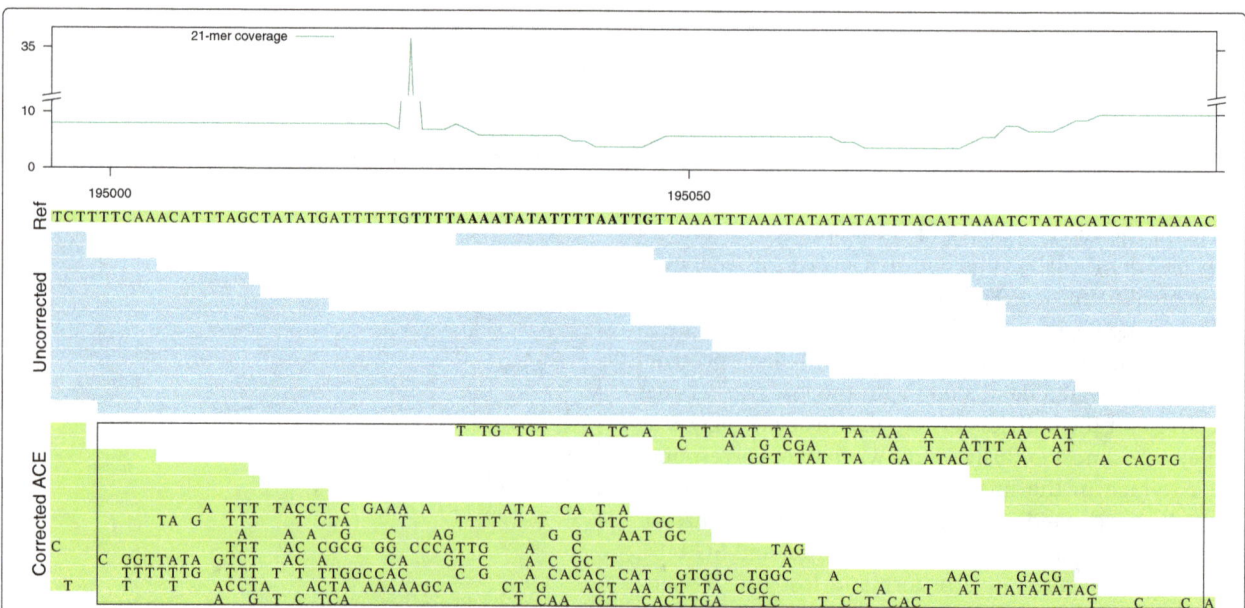

**Fig. 5** Alignment of uncorrected and ACE-corrected reads in the neighborhood of a contig breakpoint: The first track shows the 21-mer coverage of the uncorrected data. The second track (*Ref*) contains part of the reference genome, which is assembled into one contig from uncorrected data. A repeated 21-mer is indicated in *red*. The third track (*Uncorrected*) shows the alignment of the uncorrected, but error-free reads to the reference. The fourth track (*Corrected*) uses these same alignment positions, but with the sequence content of the corrected reads. Newly introduced errors are indicated by a character in the reads. The *rectangle* in the fourth track indicates 75 overlapping 21-mers that are lost as a result of erroneous error correction

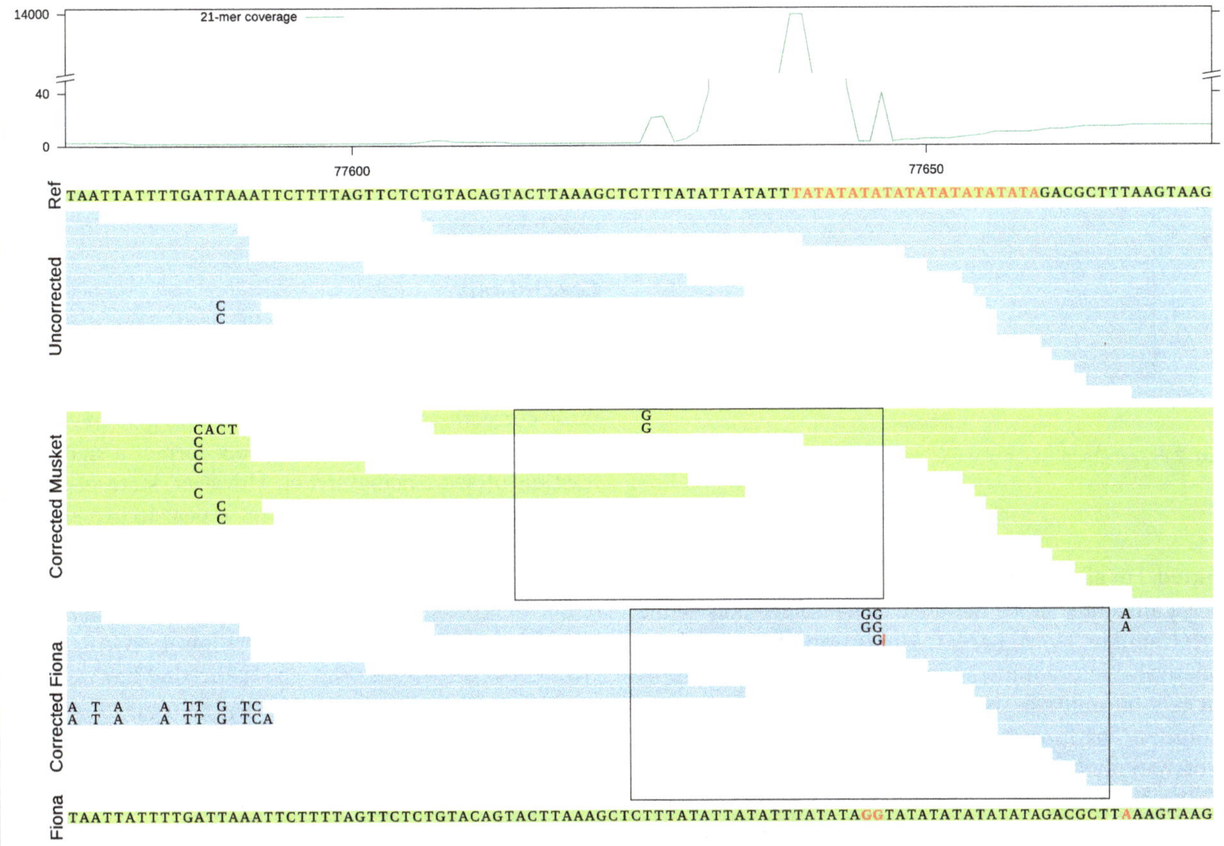

**Fig. 6** Alignment of uncorrected and corrected reads by Musket and Fiona in the neighborhood of a contig breakpoint: Lost true *k*-mer can result in two different scenarios. The first track shows the 21-mer coverage of the uncorrected data. The second track (*Ref*) shows a part of the reference genome, which is assembled into one contig from uncorrected data. A frequently occurring AT-repeat is indicated in *red*. The third track (*Uncorrected*) shows the alignment of the uncorrected reads to the reference. The fourth and the fifth tracks (*Corrected Musket* and *Corrected Fiona*) use these same alignment positions, but with the sequence content of corrected reads by Musket and Fiona. The sixth track is the assembled contig from corrected reads by Fiona. The *rectangles* indicate the regions in corrected reads by Musket and Fiona that no longer contain any true 21-mers. The coverage is low around an 'AT' repeat with coverage 13750x in the uncorrected data. Musket incorrectly changed two bases, breaking the connection between two groups of reads. In contrast, in the Fiona-corrected reads, the connection is not lost. Instead the lost true *k*-mers in Fiona appear as mismatches in the assembled contig

overlap between those reads. In that respect, error correction tools that rely on multiple sequence alignments (MSA) are in principle less susceptible to this kind of error. As overlapping reads are clustered and aligned, the error correction is systematic across those reads. MSA-based tools indeed yield higher NGA50 values on average.

These results demonstrate that evaluating error correction tools directly on their ability to reduce error rate has significant limitations as there is often no clear correlation between such metrics and the ability to improve assembly. For example, on datasets D8, ACE ranked fourth in terms of gain and showed the highest number of corrected reads that align error-free to the reference genome. Yet, ACE-corrected reads do not lead to good assembly results on this dataset.

We should emphasize that error correction is not always destructive: EC tools can improve the quality of assembly

in certain cases. For example, even though Karect also suffers from a significant number of 'lost true *k*-mers' (see Fig. 4), the tool leads to the highest NGA50 values in many cases (see Table 4). Again for dataset D8, we selected the longest contig (> 500 kbp) that was correctly assembled from corrected data by Karect and aligned the corresponding (shorter) contigs obtained from assemblies on uncorrected data. A specific case where Karect removes errors that subsequently lead to the correct connection between two contigs is shown in Additional file 1: Section 5.

**Time and space requirements**

Figures 7 and 8 show the memory usage and runtime of the EC tools (see Additional file 1: Section 6.1 for detailed tables). Since it is not possible to specify the number of threads for ACE and RACER, they were

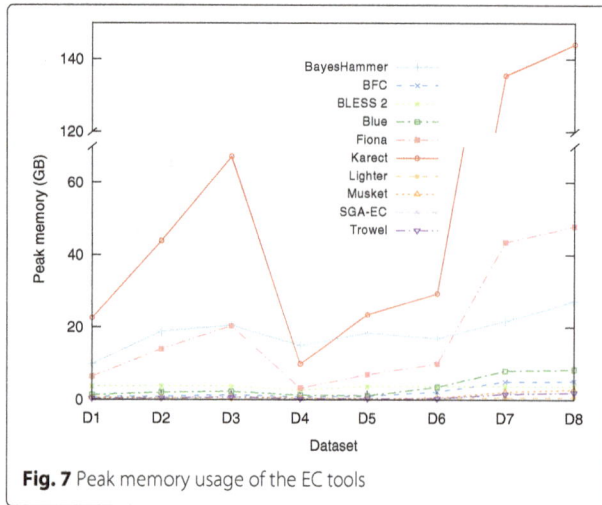

**Fig. 7** Peak memory usage of the EC tools

omitted. For all datasets, BayesHammer, Fiona and Karect use significantly more memory than other tools while BayesHammer, Fiona, Karect, Musket, and SGA-EC have a relatively high runtime. In general, we note that all tools that rely on multiple sequence alignments require more resources. The tools that rely on Bloom filters (BLESS 2, Lighter and BFC) are both memory efficient and fast.

Given the reduced error in the input data, we evaluate the potential of error correction tools to reduce the peak memory usage and/or runtime of the assembly process itself. Since error correction is computationally intensive, this may be an important aspect of error correction tools. Peak memory usage and runtime were measured for all assemblies with SPAdes and DISCOVAR (Additional file 1: Figures S3–S6). The runtime of DISCOVAR shows no decrease after error correction, while the peak memory usage decreases slightly. Conversely, the runtime of

SPAdes does decrease after error correction, but the peak memory usage does not.

The peak memory usage and runtime tables for artificial data show that Lighter and SGA-EC are again among the most memory-efficient tools, while Karect and Fiona consume more memory than any other tools. Lighter is the fastest tool followed by BLESS 2 in all the cases (Additional file 1: Section 6.2).

## Conclusions

The performance of different EC tools was compared using two approaches: the ability of EC tools to correct sequencing errors in Illumina data, and the effects of those corrections on the resulting *de novo* genome assembly quality. We found that EC tools correct a significant fraction of sequencing errors. However, state-of-the-art Illumina assemblers do not always appear to benefit from this. The assembly results for eight different datasets with SPAdes and DISCOVAR show that the prior application of EC tools often does not lead to a significant increase in NGA50, and in fact may result in a lower NGA50. Many erroneous corrections occur in regions that have low read coverage and in the vicinity of highly frequent repeats. Due to the low coverage, error correction tools incorrectly assume the presence of sequencing errors. The repeated elements on the other hand cause erroneous substitutions to be applied. A too aggressive and/or inconsistent transformation of such reads in such region may lead to loss of information from which no recovery is possible during the assembly process. This inevitably leads to an increased assembly fragmentation. Additionally, the prior use of EC tools does not lead to a major decrease in overall runtime and/or memory requirements compared with the assembly from uncorrected data.

From a methodological point of view, multiple sequence alignment (MSA) based methods might have an advantage over methods that operate on isolated *k*-mers. MSA-based methods take multiple reads into account when applying substitutions and hence appear to make more consistent corrections across overlapping reads.

We recommend future EC tools to be primarily evaluated on their ability to improve assembly results using state-of-the-art assemblers and sufficiently large datasets. Only a relatively small fraction of sequencing errors are truly impacting the assembly process. It is the behavior of the error correction tool on precisely these cases that will ultimately determine its degree of success.

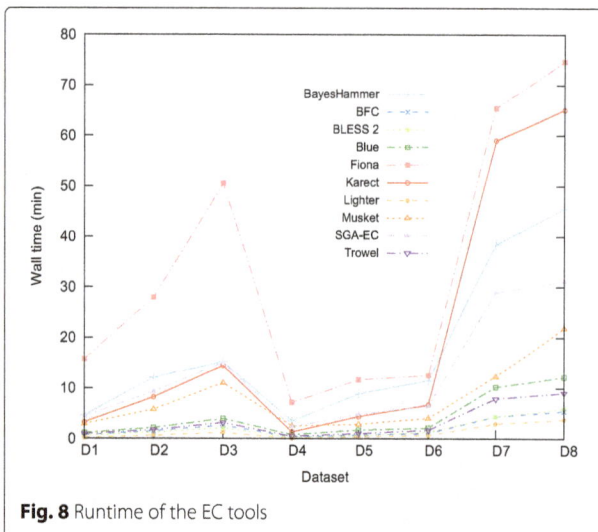

**Fig. 8** Runtime of the EC tools

## Abbreviations

bp: Base pair; D1…D8: Dataset 1…8; EC: Error correction; FN: False negative; FP: False positive; GB: Gigabyte; GHz: Gigahertz; indel: Insertion or deletion; kbp: Kilobase pair; Mbp: Megabase pair; MSA: Multiple sequence alignment; TN: True negative; TP: True positive; TPR: True positive rate

## Acknowledgments
Computational resources and services were provided by the Flemish Supercomputer Center, funded by Ghent University, the Hercules Foundation and the Flemish Government – EWI.

## Funding
This work was funded by The Research Foundation - Flanders (FWO) (G0C3914N).

## Authors' contributions
MH, GM, and JF designed the research study and developed the scripts. MH conducted the benchmark experiments. MH,GM, PD, YVP and JF analyzed and interpreted the data and wrote the manuscript. All authors read and approved the final manuscript.

## Competing interests
The authors declare that they have no competing interests.

## Author details
[1]Department of Information Technology, Ghent University-imec, IDLab, B-9052 Ghent, Belgium. [2]Center for Plant Systems Biology, VIB, B-9052 Ghent, Belgium. [3]Department of Plant Biotechnology and Bioinformatics, Ghent University, B-9052 Ghent, Belgium. [4]Bioinformatics Institute Ghent, B-9052 Ghent, Belgium. [5]Department of Genetics, Genome Research Institute, University of Pretoria, Pretoria, South Africa.

## References
1. Minoche AE, Dohm JC, Himmelbauer H. Evaluation of genomic high-throughput sequencing data generated on Illumina HiSeq and genome analyzer systems. Genome Biol. 2011;12(11):112. doi:10.1186/gb-2011-12-11-r112.
2. Ross MG, Russ C, Costello M, Hollinger A, Lennon NJ, Hegarty R, Nusbaum C, Jaffe DB. Characterizing and measuring bias in sequence data. Genome Biol. 2013;14(5):51. doi:10.1186/gb-2013-14-5-r51.
3. Compeau PE, Pevzner PA, Tesler G. How to apply de Bruijn graphs to genome assembly. Nat Biotechnol. 2011;29(11):987–91. doi:10.1038/nbt.2023.
4. Zerbino DR, Birney E. Velvet: algorithms for de novo short read assembly using de Bruijn graphs. Genome Res. 2008;18(5):821–9. doi:10.1101/gr.074492.107.
5. Sheikhizadeh S, de Ridder D. ACE: accurate correction of errors using K-mer tries. Bioinformatics. 2015;31(19):3216–8. doi:10.1093/bioinformatics/btv332.
6. Nikolenko SI, Korobeynikov AI, Alekseyev Ma. BayesHammer: Bayesian clustering for error correction in single-cell sequencing. BMC Genomics. 2013;14 Suppl 1(Suppl 1):7. doi:10.1186/1471-2164-14-S1-S7.
7. Li H. BFC: correcting Illumina sequencing errors. Bioinformatics. 2015;31(17):2885–7. doi:10.1093/bioinformatics/btv290.
8. Heo Y, et al. BLESS: bloom filter-based error correction solution for high-throughput sequencing reads. Bioinformatics. 2014;30(10):1354–62. doi:10.1093/bioinformatics/btu030.
9. Heo Y, Ramachandran A, Hwu WM, Ma J, Chen D. BLESS 2: accurate, memory-efficient and fast error correction method. Bioinformatics. 2016;32(15):2369–71. doi:10.1093/bioinformatics/btw146.
10. Greenfield, et al. Blue: correcting sequencing errors using consensus and context. Bioinformatics. 2014;30(19):2723–32. doi:10.1093/bioinformatics/btu368.
11. Saha S, Rajasekaran S. EC: an efficient error correction algorithm for short reads. BMC Bioinforma. 2015;16(Suppl 17):2. doi:10.1186/1471-2105-16-16-S17-S2.
12. Schulz MH, Weese D, Holtgrewe M, Dimitrova V, Niu S, Reinert K, Richard H. Fiona: a parallel and automatic strategy for read error correction. Bioinformatics. 2014;30(17):356–63. doi:10.1093/bioinformatics/btu440.
13. Allam A, et al. Karect: accurate correction of substitution, insertion and deletion errors for next-generation sequencing data. Bioinformatics. 2015;31(21):3421–28. doi:10.1093/bioinformatics/btv415.
14. Song L, Florea L, Langmead B. Lighter: fast and memory-efficient sequencing error correction without counting. Genome Biol. 2014;15(11):509. doi:10.1186/s13059-014-0509-9.
15. Liu Y, Schröder J, Schmidt B. Musket: a multistage k-mer spectrum-based error corrector for Illumina sequence data. Bioinformatics. 2013;29(3):308–15. doi:10.1093/bioinformatics/bts690.
16. Marinier E, Brown DG, McConkey BJ. Pollux: platform independent error correction of single and mixed genomes. BMC Bioinforma. 2015;16(1):10. doi:10.1186/s12859-014-0435-6.
17. Kelley DR, et al. Quake: quality-aware detection and correction of sequencing errors. Genome Biol. 2010;11(11):116. doi:10.1186/gb-2010-11-11-r116.
18. Marcais G, Yorke JA, Zimin A. QuorUM: An error corrector for Illumina reads. PLoS ONE. 2015;10(6):1–13. doi:10.1371/journal.pone.0130821. 1307.351v1.
19. Ilie L, Molnar M. RACER: Rapid and accurate correction of errors in reads. Bioinformatics. 2013;29(19):2490–3. doi:10.1093/bioinformatics/btt407.
20. Simpson J, Durbin R. Efficient de novo assembly of large genomes using compressed data structures. Genome Res. 2012549–56. doi:10.1101/gr.126953.111.Freely.
21. Lim EC, Müller J, Hagmann J, Henz SR, Kim ST, Weigel D. Trowel: a fast and accurate error correction module for Illumina sequencing reads. Bioinformatics. 2014;30(22):3264–5. doi:10.1093/bioinformatics/btu513.
22. Alic AS, Ruzafa D, Dopazo J, Blanquer I. Objective review of de novo stand-alone error correction methods for NGS data. Wiley Interdisc Rev Comput Mol Sci. 2016;6(April). doi:10.1002/wcms.1239. arXiv:1011.1669v3.
23. Yang X, Chockalingam SP, Aluru S. A survey of error-correction methods for next-generation sequencing. Brief Bioinform. 2013;14(1):56–66. doi:10.1093/bib/bbs015.
24. Molnar M, Ilie L. Correcting Illumina data. Brief. Bioinform. 2015;16(4):588–99. doi:10.1093/bib/bbu029.
25. Weisenfeld NI, Yin S, Sharpe T, Lau B, Hegarty R, Holmes L, Sogoloff B, Tabbaa D, Williams L, Russ C, Nusbaum C, Eric S, Maccallum I, Jaffe DB. Comprehensive variation discovery in single human genomes. 2015;46(12):1350–5. doi:10.1038/ng.3121.Comprehensive.
26. Peng Y, Leung HCM, Yiu SM, Chin FYL. In: Berger B, editor. IDBA – A Practical Iterative de Bruijn Graph De Novo Assembler. Berlin: Springer; 2010, pp. 426–40.
27. Bankevich A, Nurk S, Antipov D, Gurevich Aa, Dvorkin M, Kulikov AS, Lesin VM, Nikolenko SI, Pham S, Prjibelski AD, Pyshkin AV, Sirotkin AV, Vyahhi N, Tesler G, Alekseyev Ma, Pevzner Pa. SPAdes: a new genome assembly algorithm and its applications to single-cell sequencing. J Comput Biol. 2012;19(5):455–77. doi:10.1089/cmb.2012.0021.
28. Laehnemann D, Borkhardt A, McHardy AC. Denoising DNA deep sequencing data-high-throughput sequencing errors and their correction. Brief Bioinform. 2016;17(1):154–79. doi:10.1093/bib/bbv029.
29. Huang W, Li L, Myers JR, Marth GT. ART: a next-generation sequencing read simulator. Bioinformatics. 2012;28(4):593–4. doi:10.1093/bioinformatics/btr708.
30. Delcher AL, Kasif S, Fleischmann RD, Peterson J, White O, Salzberg SL. Alignment of whole genomes. Nucleic Acids Res. 1999;27(11):2369–76. doi:10.1093/nar/27.11.2369.
31. Marcais G, Kingsford C. A fast, lock-free approach for efficient parallel counting of occurrences of k-mers. Bioinformatics. 2011;27(6):764–0. doi:10.1093/bioinformatics/btr011.
32. Li H, Durbin R. Fast and accurate short read alignment with Burrows-Wheeler transform. Bioinformatics. 2009;25(14):1754–60. doi:10.1093/bioinformatics/btp324.
33. Gurevich A, et al. QUAST: quality assessment tool for genome assemblies. Bioinformatics. 2013;29(8):1072–5. doi:10.1093/bioinformatics/btt086.

34. Simpson JT, Wong K, Jackman SD, Schein JE, Jones SJM, Birol I. ABySS: A parallel assembler for short read sequence data. Genome Res. 2009;19(6): 1117–23. doi:10.1101/gr.089532.108.

35. Conway T, Wazny J, Bromage A, Zobel J, Beresford-smith B. Gossamer - A resource-efficient de novo assembler. Bioinformatics. 2012;28(14): 1937–8. doi:10.1093/bioinformatics/bts297.

36. Miller JR, Delcher AL, Koren S, Venter E, Walenz BP, Brownley A, Johnson J, Li K, Mobarry C, Sutton G. Aggressive assembly of pyrosequencing reads with mates. Bioinformatics. 2008;24(24):2818–24. doi:10.1093/bioinformatics/btn548.

37. Luo R, Liu B, Xie Y, Li Z, Huang W, Yuan J, He G, Chen Y, Pan Q, Liu Y, Tang J, Wu G, Zhang H, Shi Y, Liu Y, Yu C, Wang B, Lu Y, Han C, Cheung DW, Yiu SM, Peng S, Xiaoqian Z, Liu G, Liao X, Li Y, Yang H, Wang J, Lam TW, Wang J. Soapdenovo2: an empirically improved memory-efficient short-read de novo assembler. GigaScience. 2012;1(1): 18. doi:10.1186/2047-217X-1-18.

# sgnesR: An R package for simulating gene expression data from an underlying real gene network structure considering delay parameters

Shailesh Tripathi[1], Jason Lloyd-Price[2,3], Andre Ribeiro[3,5], Olli Yli-Harja[6,5], Matthias Dehmer[4] and Frank Emmert-Streib[1,5*]

## Abstract

**Background:** sgnesR (Stochastic Gene Network Expression Simulator in R) is an R package that provides an interface to simulate gene expression data from a given gene network using the stochastic simulation algorithm (SSA). The package allows various options for delay parameters and can easily included in reactions for promoter delay, RNA delay and Protein delay. A user can tune these parameters to model various types of reactions within a cell. As examples, we present two network models to generate expression profiles. We also demonstrated the inference of networks and the evaluation of association measure of edge and non-edge components from the generated expression profiles.

**Results:** The purpose of sgnesR is to enable an easy to use and a quick implementation for generating realistic gene expression data from biologically relevant networks that can be user selected.

**Conclusions:** sgnesR is freely available for academic use. The R package has been tested for R 3.2.0 under Linux, Windows and Mac OS X.

**Keywords:** Gene expression data, Gene network, Simulation

## Background

Networks provide a statistical and mathematical framework for the general understanding of the complex functioning of biological systems because the causal relationship between different entities, such as proteins, genes or metabolites, defines how a cellular system functions collectively. This leads to an emergent behavior, e.g., with respect to phenotypic aspects of organisms [1–4]. Unfortunately, understanding of the system's functioning of a cell is not an easy task and one reason for this is that the causal inference of gene network itself is a formidable problem [5, 6]. For this reason, we provide the R package sgnesR (Stochastic Gene Network Expression Simulator in R). Specifically, sgnesR can be used to generate biologically realistic gene expression data based on an underlying gene regulatory network that can be used to test network inference methods qualitatively. In this way an inferred network can be compared with the known *true* gene regulatory network, which is for most real biological systems unknown requiring the usage of approximations, e.g., by using transcriptional regulatory networks or protein interaction networks [7]. Overall, our package sgnesR enables the quantitative estimation of important statistical measures, e.g., the power, false discovery rate or AUROC values of such inferred networks. Furthermore, the resulting gene expression profiles can be itself of use for instance for comparison with real measurements of gene expression values for the identification of model parameters.

In general, the simulation of biologically realistic gene expression values is a challenging task because it requires the specification of transcription and translation mechanisms of biological cells, which are far from being understood in every detail. Specifically, there are two major components that need to be defined for the

*Correspondence: v@bio-complexity.com
[1]Predictive Medicine and Data Analytics Lab, Department of Signal Processing, Tampere University of Technology, Tampere, Finland
[5]Institute of Biosciences and Medical Technology, Tampere, Finland
Full list of author information is available at the end of the article

simulation of such a process. The first relates to the connection structure among the genes and the second to the parameter values of the modeling equations. The connection structure corresponds to the regulatory network which defines which genes control the expression of other genes. Our package sgnesR allows the usage of previously inferred biological networks or the usage of artificially simulated networks. For the identification of the parameters of the modeling equations of the transcription and translation processes values can be sampled from plausible distributional assumptions.

In the following, we discuss some existing methods that have been proposed and implemented for the simulation of gene expression data. An overview of these simulation methods for which software implementations are available is shown in Table 1. One of the most widely used methods is *syntren* [8]. Syntren uses an interaction kinetics model based on the equations of Michaelis-Menten and Hill kinetics. In contrast, *netsim* applies a fuzzy logic for the representation of interactions for a given topology of a gene regulatory network and differential equations to generate expression data [9]. Despite these differences, both simulation methods aim at emulating a biological model of transcription regulation and translation. A completely different approach is used by *GeneNet* [10]. This method samples network data from a Gaussian graphical model (GGM) for a given network structure. A similar approach is used in [11].

Our R package *sgnesR* provides an easy-to-use interface to simulate gene expression data generated by the stochastic simulation algorithm (SSA) [12, 13]. That means a gene regulatory network is modeled whose activation patterns are defined by the transcription and translation which are modeled as multiple time delayed events. The delays itself can be drawn from a variety of distributions and the reaction rates can be determined via complex functions or from physical parameters. The original implementation of the 'Stochastic Gene Networks Simulator' (*SGNSim*) algorithm [13] is available in C/C++. However, by providing the R interface *sgnesR*, it is possible to perform all relevant analysis steps, e.g., for testing network inference methods or for investigating pathway methods, within the R environment. This is not only convenient but leads to a natural integration of all parts making the overall analysis reproducible in the most straight forward way [14]. In addition, our package *sgnesR* allows selection capabilities for various biological and artificially simulated gene regulatory networks that can be used

**Table 1** A list of network sampling and simulation methods

| Methods ⇓ \ Features ⇒ | Method-based on | Input | Output |
|---|---|---|---|
| sgnesR (SGN sim [13]) | A set of biochemical reactions where transcription and translation of genes and proteins are modelled as multiple time delayed events and their activities are modelled by a stochastic simulation algorithm (SSA) [20] | S4 data object with a network of *igraph* class. | S4 data object which consists expression data matrix. |
| AGN [25] | Set of biochemical reactions in the form of a network, simulation of the kinetics of systems of biochemical reactions based on differential equations. | SMBL | Text file |
| GenGe [26] | Non linear differential equation system where degradation of biological molecules are modelled by a linear or Michalies-Menten kinetic and translation is described by a linear kinetic law by using several global and local perturbation parameters. | SMBL | Text file (numeric values). |
| GRENDEL [27] | A set of differential equation system uses hill kinetics based activation and repression functions for the transcription rate law. | SMBL | Text file (numeric values) |
| NetSim [9] | Differential equations are used to to model the dynamics of transcription and degradation along with the integration of fuzzy logic in order to define the complex regulatory mechanism | adjacency matrix with other parameters | list object in R |
| RENCO [28] | Uses pre defined network topology or generates topologies to model ordinary differential equations and use Copasi for simulating expression data. | Text file | Text file |
| SynTReN [8] | The interactions of a network uses non-linear functions based on Michaelis-Menten and hill enzyme kinetic equations to model gene regulation | Text file | Text file |

as realistic wiring diagrams for the interactions between genes.

The paper is organized as follows. In the next section we describe our gene expression simulator *sgnesR* in detail and present some working examples. These examples will demonstrate the capabilities of *sgnesR*. The paper finishes with a summary and conclusions.

## Implementation

In this section, we provide a description of the organizational structure corresponding to the workflow of the *sgnesR* package and its components. Schematically, the overview of the workflow is shown in Fig. 1. The first step consists in specifying the network topology. Here the user has two choices: A) use an external network or B) generate a simulated network. For B) we are using the *igraph* package in R. The *igraph* package provides a comprehensive set of functions that allows to generate or create several types of networks and compute several network related features; for the visualization of networks see [15]. A user can easily generate a network forming the connections for a set of reactions as the input of the SGNS algorithm [13]. Alternatively, a user can select biological networks as input as provided by public databases, e.g., [16, 17]. For convenience, we provide two biological networks in the sgnesR package. The first one is a transcription regulator network of E. coli [18] and the second a subnetwork of the human signaling network [19].

In addition to the specification of a graph topology, the assignment of initial populations of RNAs and proteins for each node and the activation or suppression indicator for each edge of the network are initialized in the first step of the *sgnesR* package. In the following, a brief description of the generation of the set of reactions from a network topology is provided.

Suppose, we have a network consisting four genes (nodes) A, B, C and D. Their interactions are described as follows:

```
B -[activates]-> A
C -[activates]-> A
D -[suppress]-> A
```

In order to represent the following network topology as a set of chemical reactions we assume that each node is represented by a promoter, an RNA and a protein product. For example the node A is represented as ProA (promoter), RA (RNA) and PA (protein produce). In the following example below, A interacts with three nodes so A has three different promoter sites where the protein products of different genes (B, C and D) bind to activate or suppress the expression of A. The set of reactions are divided into three sections as follows:

1. Reactions for translation and degradation for each gene: In this step, three steps of reactions describe the translation of RNAs of each node into the protein products and the respective decay of each RNA and protein product. The example is shown below.

```
RA --[ <translation rate> ]
    --> RA (<RNA-delay>)
        + PA (<protein-delay>);
RA --[ <rna degradation rate> ]--> ;
PA --[ <protein degradation rate> ]--> ;

RB --[ <translation rate> ]--> RB+ PB;
RB --[ <rna degradation rate> ]--> ;
PB--[ <protein degradation rate> ]--> ;

RC --[ <translation rate> ]--> RC + PC;
RC --[ <rna degradation rate> ]--> ;
PC--[ <protein degradation rate> ]--> ;

RD --[ <translation rate> ]--> RD+ PD;
RD --[ <rna degradation rate> ]--> ;
PD --[ <protein degradation rate> ]--> ;
```

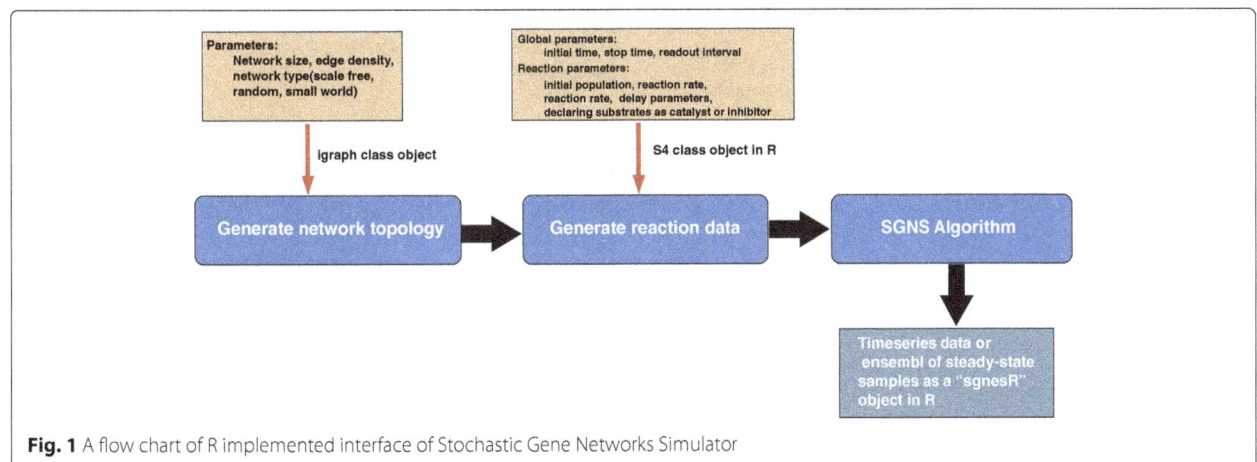

**Fig. 1** A flow chart of R implemented interface of Stochastic Gene Networks Simulator

2. Binding-unbinding reactions: This set of reactions describe the binding of protein products of interacting genes to the promoter sites of interacted gene. In the given example, genes B and C activate and gene D suppress the expression of gene A so the protein products of B, C, and D interact with their respective promoter sites ProA.NoB, ProA.NoC and ProA.NoD in gene A and form intermediary products ProA.B, ProA.C and ProA.D. These intermediary products take part in the transcription process of the gene A. The gene D suppresses the expression of gene A, in this process an intermediary product of suppressor gene (ProA.D) is formed by Protein product of D (PD) by binding to the promoter site of the gene A (ProA.NoD). The intermediary product of suppressor gene D (ProA.D) does not allow to express gene A, therefore avoids the transcription process and releases after sometime. The example of binding and the unbinding of proteins to promoters sites is shown below.

```
ProA.NoB + PB --[ <binding rate> ]
        --> ProA.B;
ProA.B --[ <unbinding rate> ]
        --> ProA.NoB + PB;
ProA.NoC + PC --[ <binding rate> ]
        --> ProA.C;
ProA.C --[ <unbinding rate> ]
        --> ProA.NoC + PC;
ProA.NoD + PD --[ <binding rate> ]
        --> ProA.D;
ProA.D --[ <unbinding rate> ]
        --> ProA.NoD + PD;
```

3. Transcription reactions: This is a set of reactions of the transcription process of the gene to which all possible combinations of the intermediary products of the activators of the genes contributes to the expression of gene A. In this example, the two activators B and C can have three possible choices to contribute to the expression of A in which the intermediary product of only B, intermediary product of only C and intermediary products of both B and C contribute to the expression of the RNA of gene A. The example reaction is shown below:

```
ProA.B + ProA.NoC + ProA.NoD
        --[ <transcription rate> ]
        --> ProA.B(<promoter-delay>)
+ ProA.NoC(<promoter-delay>)
+ ProA.NoD+ RA(<promoter-delay>) ;
ProA.NoB + ProA.C + ProA.NoD
        --[ <transcription rate> ]
        --> ProA.NoB(<promoter-delay>)
+ ProA.C(<promoter-delay>)
+ ProA.NoD(<promoter-delay>)
```

```
+ RA(<promoter-delay>) ;
ProA.B + ProA.C + ProA.NoD
        --[ <transcription rate> ]
        --> ProA.B(<promoter-delay>)
+ ProA.C(<promoter-delay>)
+ ProA.NoD(<promoter-delay>)
+ RA(<promoter-delay>)
```

These three sets of reactions along with other reaction parameters are passed to the SGNS algorithm to generate the expression profiles for the different genes. The additional reaction parameters needed are the initial population, reaction rates and delay parameters which are described in the following:

- Initial populations: The initial population of parameters assigns the initial values of promoters, RNAs and proteins for all the genes in the network.
- Reaction rates: The reaction rate parameter assigns values for *reaction-rate* to different reaction types for translation and degradation reactions as translation rate, RNA degradation rate and protein degradation rate. For binding and unbinding reactions it assigns binding and unbinding rates and for transcription rates it assigns transcription rate.
- Delay parameters: The delay parameter assigns a delay time for RNAs and proteins in translation and degradation reactions to be released at a certain time point. Also, the promoter delay is assigned to the products of transcriptions reactions to be released at a certain time point.

The sgnesR package provides two options to obtain the expression profiles of different genes as either time series data or steady-state values. The time series data is a set of expression values of different genes between the different time points of starting time and end time of reactions which are captured at fixed time intervals. The steady state values are final expression values of different genes at the end of the reaction. Furthermore the sgnesR packages allows to repeat the simulation of a input network *n* times and generates this way an ensemble of steady-state expression values of sample size *n*.

**Results and discussion**

In this section, we present some working examples for the usage of our package *sgnesR*. These examples demonstrate some of the available features of its capabilities. The *sgnesR* package provides options to apply various parameters using base R functions and a variety of network topologies, based on several network features as parameters for generating simulated data. Further parameters are assigned to each reaction by defining two data objects of the "rsgns.param" and "rsgns.data" class. These are defined as follows.

- "rsgns.param": This class defines the initial parameters which include "start time", "stop time" and "read-out interval" for time series data.
- "rsgns.data": The class defines a data object for the input which includes the network topology and other parameters such as the initial populations of RNA and protein molecules of each node/gene, rate constants, delay parameters and initial population parameters of different molecules.
- "rsgns.waitlist": This class defines the molecules placed in a waiting list and to be released a specific number of molecules at a particular time during the reaction. This class includes "nodes", "time", "mol" and "type" for time series data.

### R functions for generating data from a given network

- *getreactions* : This function generates an object of class "rsgns.reactions" which contains a set of reactions, their initial values and the wait-list of reactions. This object can be supplied to the SGNS API for generating gene expression data. The "rsgns.reactions" object is a list containing six components which are "population", "activation", "binding_unbinding", "trans_degradation" and "waitlist". Each component of the list is a matrix object and user can modify those reaction parameters depending on the requirements before passing it to "rsgns.rn" function as an input.
- *rsgns.rn*: This function is an interface to the SGNS API for simulating timeseries data. A user can either provide a "rsgns.reactions" class object directly to the function or the "rsgns.data" class object to receive the output. There are further options available to tune the reaction parameters. The function itself returns a "sgnesR" class object which contains the generated expression data, the input network and the reaction kinetics information.
- *plot.sgnesR*: This function provides different options to visualize the expression profiles. The function has two major options available. The first one is to visualize the expression values in terms of RNA numbers at different time points and the second option is to visualize the distribution of RNA numbers for different nodes/genes at different time points or the sample-distribution of an ensemble of steady state values.

### Generating time series data from a scale-free network

The first example we demonstrate how to use sgnesR package to generate time series data from a scale-free network. The code for this is presented in Example 1. For reasons of simplicity, in this example we do not consider delay parameters for the translation and transcription processes (see Example 2 for an extension). The visualization

of the network and the generated expression values are shown in Fig. 2.

### Generating time series data from a scale-free network with delay parameters

In Example 2 we provide a working example to generate time series data from a scale-free network with delay parameters. That means we are assigning delay parameters for the translation reactions of the RNA delay and the protein delay and in transcription reactions for a promoter delay. The user can assign delay parameters chosen from a Gaussian distribution with different mean values and variance. Further choices are delay functions such as a gamma distribution or an exponential function for the delays. However, for simulating real biological gene expression data it is preferable to use the "gamma" function to assign delays [20].

### Generating steady-state samples of expression values from an Erdos-Renyi network

Here 'steady-state samples' means 'asymptotic samples' in the sense that we run our simulations until the expression values of the genes reach constant values where a further continuation of the simulations lead to no further changes of expression values of the molecules. Example 3 provides a working example to demonstrate the usage of our package. The visualization of the results of the network and the distribution of the ensemble of generated expression profiles is shown in Fig. 3. We want to remark that the 'sample' option for the function 'rsgns.rn' means that the simulations are repeated n times, as defined by the value of 'sample=n', by using the same initial values of all parameters. In case the user wants to use different initial values, then 'sample=1' needs to be used and an explicit loop over 'rsgns.rn' needs to be carried out.

### Generating time series data from a known set of equations

In this example we demonstrate how to use sgnesR package to generate time series data from a user defined set of reactions. The code for this is presented in Example 4. This example is based on the toggle switch reactions without cooperative binding. The purpose of this example is to simulate a set of reactions when we know the information of promoter regions along with RNA and protein binding information. Suppose the equations are described as follows:

1. ProA + *Ind −[0.002]−> A + ProA
2. ProB + *Ind −[0.002]−> B + ProB
3. A −[0.005]−>
4. B −[0.005]−>
5. A + ProB + *ProA −[0.2]−> ProB.A
6. B + ProA + *ProB −[0.2]−> ProA.B

Example 1: Generation of time series data from a scale-free network without delay parameters

1: Generation of a random scale free network with 20 nodes using barabasi-game model [21].

```
g<-sample_pa(20)
```

2: Assigning random initial values for the RNAs and protein products for each node.

```
V(g)$Ppop <- (sample(100, vcount(g), rep=T))
V(g)$Rpop <- (sample(100, vcount(g), rep=T))
```

3: Assign -1 or +1 to each directed edge to represent that an interacting node is either acting as a activator, if +1, or as a suppressor, if -1

```
sm <- sample(c(1,-1), ecount(g), rep=T, p=c(.8,.2))
E(g)$op <- sm
```

4: Initiate global reaction parameters.

```
rp<-new(``rsgns.param'',time=0, stop_time=1000, readout_{i}nterval=.1)
```

5: Specify the reaction parameters.

6: Specifying the reaction rate constant vector for the following reactions: (1) Translation rate, (2) RNA degradation rate, (3) Protein degradation rate, (4) Protein binding rate, (5) unbinding rate, (6) transcription rate.

```
rc <- c(0.002, 0.005, 0.005, 0.005, 0.01, 0.02)
```

7: Specify the reaction rate function for the protein unbinding reactions

```
rn1 <- list(``invhill'', c(10,2), c(0,1))
rn2 <- list(``'',``'')
rn <- list(rn2,rn2,rn1)
```

8: Specifying the input data object

```
rsg <- new(``rsgns.data'',network=g, rn.rate.function=rn, rconst=rc)
```

9: Call the R function for the SGN simulator

```
xx <- rsgns.rn(rsg, rp)
```

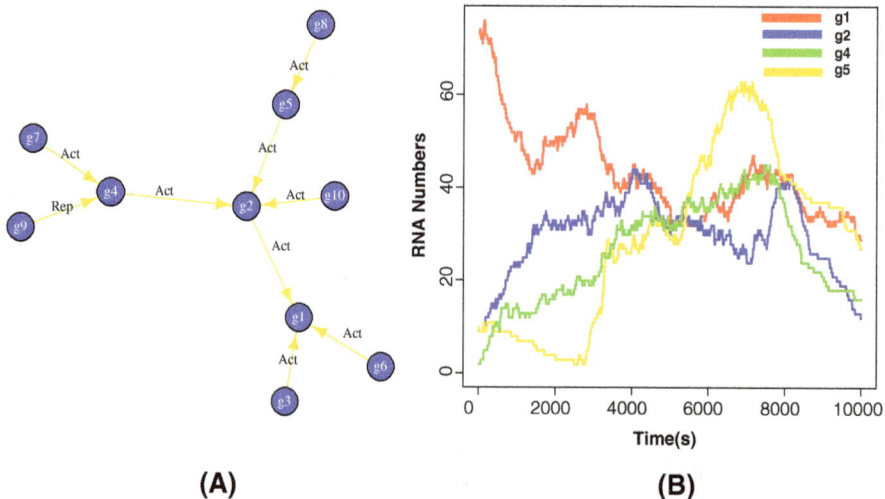

**(A)**                                                                                          **(B)**

**Fig. 2** A plot of sample network and the expression values at different time points of different nodes from the simulation. **a** The input network **b** Expression values of genes which show incoming edges

---

Example 2: Generation of time series data from a scale-free network by assigning delay parameters.

---

1: Generation of a random scale-free network with 20 nodes using barabasi-game model [21].

```
g<-sample_pa(20)
```

2: Assigning initial values to the RNAs and protein products to each node randomly.

```
V(g)$Ppop <- (sample(100,vcount(g), rep=T))
V(g)$Rpop <- (sample(100, vcount(g), rep=T))
```

3: Assign -1 or +1 to each directed edge to represent that an interacting node is either acting as a activator, if +1, or as a suppressor, if -1

```
sm <- sample(c(1,-1), ecount(g), rep=T, p=c(.8,.2))
E(g)$op <- sm
```

4: Specify global reaction parameters.

```
rp<-new(``rsgns.param'',time=0,stop_time=1000,readout_interval=.1)
```

5: Specify the reaction parameters.
6: Declaring reaction rate constant vector for following reactions: (1) Translation rate, (2) RNA degradation rate, (3) Protein degradation rate, (4) Protein binding rate, (5) unbinding rate, (6) transcription rate.

```
rc <- c(0.002, 0.005, 0.005, 0.005, 0.01, 0.02)
```

7: Specifying the reaction rate function for the protein unbinding reactions

```
rn1 <- list(``invhill'', c(10,2), c(0,1))
rn2 <- list(``'',``'')
rn <- list(rn2,rn2,rn1)
```

8: Defining the delay parameters for RNA and protein delay and promoter delay

```
dl1 <- list(``gamma'', c(5,15)) #promoter delay
dl2 <- list(``gamma'', c(3,12)) #RNA delay
dl3 <- list(``gamma'', c(4,12)) #protein delay
dlsmp <- list(dl1, dl2, dl3)
```

9: Specifying the input data object

```
rsg <- new(``rsgns.data'',network=g, rn.rate.function=rn, rconst=rc)
```

10: Call the R function for the SGN simulator

```
xx <- rsgns.rn(rsg, rp)
```

---

7. ProB.A −[0.01]−> ProB + A
8. ProA.B −[0.01]−> ProA + B
9. ProB.A −[0.005]−> ProB
10. ProA.B −[0.005]−> ProA

### Application in network inference
In this section, we present two examples to generate expression profiles and the inference of networks from the expression profiles using BC3NET [22]. BC3NET is a network inference method based on the ensemble of inferred networks by assigning an edge for a gene-pair if at least one of these two genes show maximal mutual information with respect to all other genes [23]. For simulation, we chose two types of networks the first one are the scale-free artificial networks with 50 nodes and edges of different edge densities. The second network is a subnetwork of *ecoli* transcription regulatory network [24] which contains 59 nodes and 60 edges. The subnetwork is shown in Fig. 5(a). The generated expression profiles of *ecoli* transcription subnetwork are based on

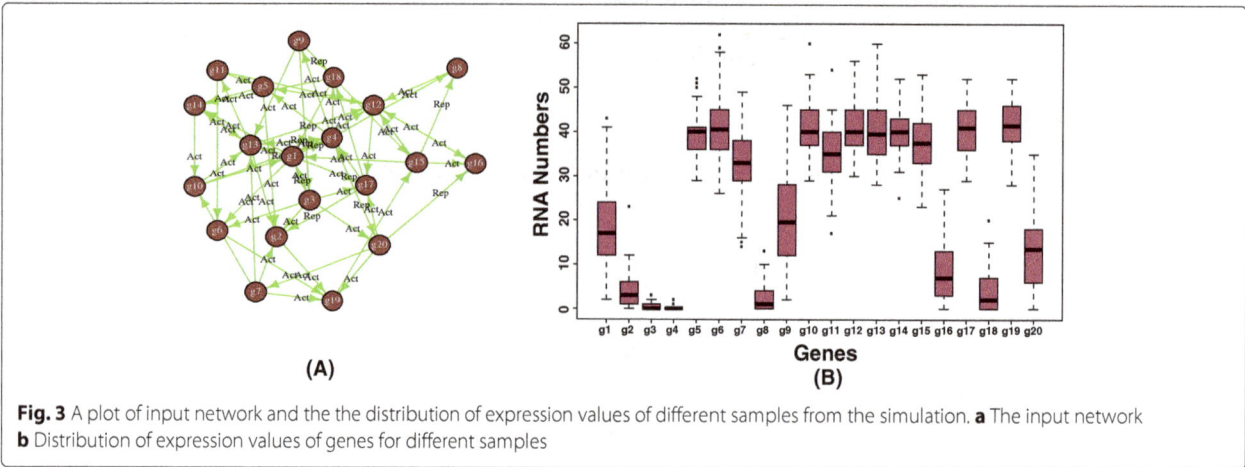

**Fig. 3** A plot of input network and the the distribution of expression values of different samples from the simulation. **a** The input network **b** Distribution of expression values of genes for different samples

hypothetical promoter regions where an RNA molecule of a gene binds to a hypothetical promoter region of another gene if there is an edge exist between them. The other parameters of the reactions are hypothetical assumptions for the reactions. The details of these parameters and generation of expression profiles are provided in the *supplementary R* file (ecolisim_script.R). In the first step, we generate expression profiles of artificial networks and *ecoli* subnetwork using *sgnesR*, in the second step we used expression profile for inferring networks using *BC3NET*. For all three types of artificial networks, we repeat simulation 20 times. For each simulation step, the mutual information is calculated between all pairs of nodes using *BC3NET* which assigns weights to all pair of nodes. In this simulation, we highlight the distribution of weights of gene-pairs which are connected by edges and gene-pairs which are not connected with each other (non-edge). The results are shown in Fig. 4. Similarly, we generate expression profiles using the *ecoli* network and inferred the network using *BC3NET*. The distribution of weights of gene-pairs which are connected by edges and gene-pairs which are not connected with each other (non-edge) are shown in Fig. 5(b). In these examples, we clearly see that the *BC3NET* assigns higher weights by

computing mutual information of expression profiles to the pairs of nodes for edge components compare to the non-edge components of simulated networks and *ecoli* subnetwork. Similarly, the other measures can be used to evaluate the performance of different network inference methods.

### Computational complexity

Overall, the computational complexity of the algorithm depends on the edge density of the used network and specifically on the in-degree of each node. However, for networks with up to $\sim$ 1000 genes the package generates rapid results. A practical overview of the run time of our sgnesR package is shown in Table 2. The average run time is shown in seconds for different network sizes. We repeated the analysis 10 times for each network size shown in the table.

We would like to remark that the theoretical computational complexity of the implementation of the SGNS algorithm has a formal time complexity of $O\left(TR*(D\log R + \log W)\right)$. Where T = simulation time, R = number of reactions, D = max degree in propensity update dependency graph between reactions, W = max wait list size. However, our sgnesR package contains an

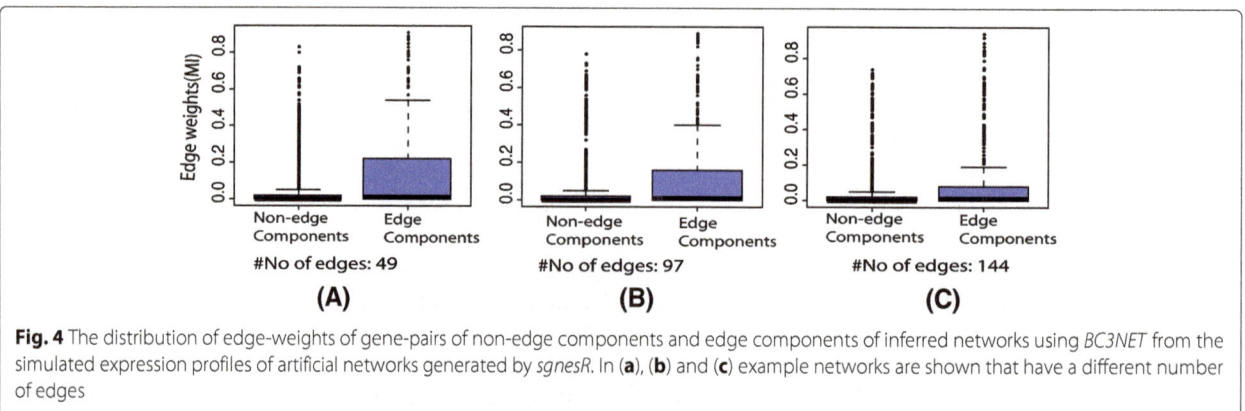

**Fig. 4** The distribution of edge-weights of gene-pairs of non-edge components and edge components of inferred networks using *BC3NET* from the simulated expression profiles of artificial networks generated by *sgnesR*. In (**a**), (**b**) and (**c**) example networks are shown that have a different number of edges

---

Example 3: Generation of steady-state samples of expression values from an Erdos-Renyi network without delays

1: Generation of a random scale-free network with 20 nodes using an Erdos-Renyi network model.

```
g <- erdos.renyi.game(20,.15, directed=T)
```

2: Assigning initial values to the RNAs and protein products to each node randomly.

```
V(g)$Ppop <- (sample(100,vcount(g), rep=T))
V(g)$Rpop <- (sample(100, vcount(g), rep=T))
```

3: Assign -1 or +1 to each directed edge to represent that an interacting node is acting either as a activator, if +1, or as a suppressor, if -1

```
sm <- sample(c(1,-1), ecount(g), rep=T, p=c(.8,.2))
E(g)$op <- sm
```

4: Specifying global reaction parameters.

```
rp<-new(``rsgns.param'',time=0,stop_time=1000,readout_interval=500)
```

5: Specifying the reaction rate constant vector for following reactions: (1) Translation rate, (2) RNA degradation rate, (3) Protein degradation rate, (4) Protein binding rate, (5) unbinding rate, (6) transcription rate.

```
rc <- c(0.002, 0.005, 0.005, 0.005, 0.01, 0.02)
```

6: Declaring input data object

```
rsg <- new(``rsgns.data'',network=g, rconst=rc)
```

7: Call the R function for SGN simulator

```
xx <- rsgns.rn(rsg, rp, timeseries=F, sample=50)
```

---

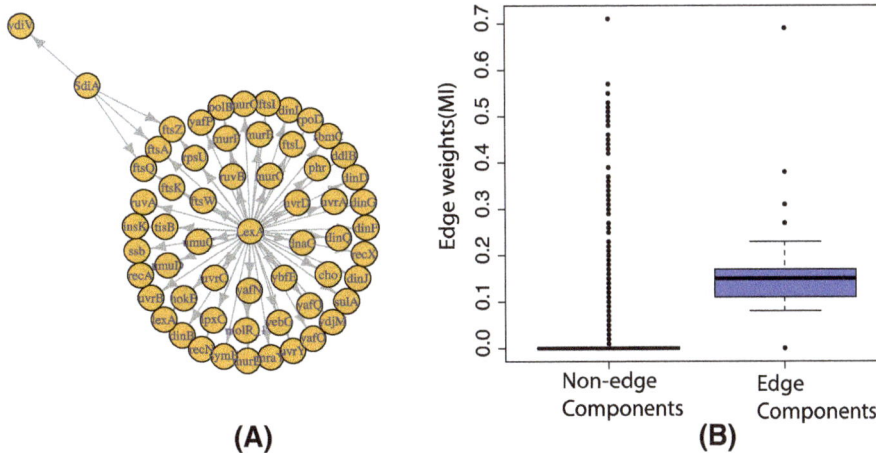

**(A)**          **(B)**

**Fig. 5 a** A subnetwork of transcription regulatory network of *ecoli* used to simulate expression profiles using *sgnesR*. **b** The distribution of edge-weights of gene-pairs of non-edge components and edge components of inferred network using *BC3NET* from the expression profiles of *ecoli* subnetwork generated by *sgnesR*

---

Example 4: Generation of expression values from a toggle switch reactions

---

1: Initialize a dataframe object

```
toggle <- getrndf()
```

2: Set different properties of molecules participating in the reactions and adding to the object "toggle".

```
setmolprop(''toggle'', rnindex=1, name=''ProA'', molcount=1,type=''s'',
           rc=.0002,pop=1)
setmolprop(''toggle'', rnindex=1, name=''Ind'', inhib=''*'', molcount=1,type=''s'',
           rc=.0002,pop=100)
setmolprop(''toggle'', rnindex=1, name=''A'', type=''p'', pop=1)
setmolprop(''toggle'', rnindex=1, name=''ProA'', type=''p'')

setmolprop(''toggle'', rnindex=2, name=''ProB'', molcount=1, type=''s'',
           rc=.0002,pop=1)
setmolprop(''toggle'', rnindex=2, name=''Ind'', inhib=''*'', molcount=1,type=''s'',
           rc=.0002)
setmolprop(''toggle'', rnindex=2, name=''B'', type=''p'', pop=1)
setmolprop(''toggle'', rnindex=2, name=''ProB'', type=''p'')

setmolprop(''toggle'', rnindex=3, name=''A'', type=''s'', rc=.005)
setmolprop(''toggle'', rnindex=4, name=''B'', type=''s'', rc=.005)

setmolprop(''toggle'', rnindex=5, name=''A'', molcount=1, type=''s'', rc=.2)
setmolprop(''toggle'', rnindex=5, name=''ProB'', molcount=1, type=''s'')
setmolprop(''toggle'', rnindex=5, name=''ProA'',inhib=''*'', molcount=1,type=''s'')
setmolprop(''toggle'', rnindex=5, name=''ProB.A'', molcount=1, type=''p'',pop=0)

setmolprop(''toggle'', rnindex=6, name=''B'', molcount=1, type=''s'', rc=.2)
setmolprop(''toggle'', rnindex=6, name=''ProA'', molcount=1, type=''s'')
setmolprop(''toggle'', rnindex=6, name=''ProB'',inhib=''*'', molcount=1,type=''s'')
setmolprop(''toggle'', rnindex=6, name=''ProA.B'', molcount=1, type=''p'',pop=0)

setmolprop(''toggle'', rnindex=7, name=''ProB.A'', type=''s'',rc=0.01)
setmolprop(''toggle'', rnindex=7, name=''ProB'', type=''p'')
setmolprop(''toggle'', rnindex=7, name=''A'',type=''p'')

setmolprop(''toggle'', rnindex=8, name=''ProA.B'', type=''s'',rc=0.01)
setmolprop(''toggle'', rnindex=8, name=''ProA'',type=''p'')
setmolprop(''toggle'', rnindex=8, name=''B'',type=''p'')

setmolprop(''toggle'', rnindex=9, name=''ProB.A'', type=''s'', rc=.005)
setmolprop(''toggle'', rnindex=9, name=''ProA'',type=''p'')
setmolprop(''toggle'', rnindex=10, name=''ProA.B'',type=''s'', rc=.005)
setmolprop(''toggle'', rnindex=10, name=''ProB.A'',type=''p'')

rw <- new(''rsgns.waitlist'', time=c(1000000), mol=c(100), type=c(''Ind''))
rp <- new(''rsgns.param'', time=0, stop_time=200000, readout_interval=50)
```

3: Obtaining the set of reactions and call the R function for the SGN simulator

```
xx <- getreactions(toggle, waitlist=rw)
rnsx <- rsgns.rn(xx, rp)
```

4: Specifying global reaction parameters.

```
rp<-new(''rsgns.param'',time=0, stop_time=1000, readout_interval=500)
```

**Table 2** Estimated time by *sgnesR*, in seconds for different type of networks

| Network size | Average edge size | Maximum degree (Average) | Average run time (seconds) |
|---|---|---|---|
| 20 | 21.9 | 6.4 | 0.25 |
| 50 | 55.4 | 9.0 | 0.42 |
| 100 | 114.0 | 10.7 | 1.92 |
| 150 | 165.2 | 12.4 | 7.77 |
| 200 | 227.1 | 12.5 | 14.10 |
| 500 | 560.9 | 15.4 | 116.31 |
| 1000 | 1110.8 | 17.8 | 391.04 |

additional layer of complexity consisting of the automatic generation of all reaction equations for a given network topology.

## Conclusions

In this paper, we described the R implementation of the *sgnesR* (Stochastic Gene Network Expression Simulator) package. The main objective of the sgnesR package is to utilize the applicability of gene expression simulations, e.g., for validating the performance of network inference methods [5, 6]. The *sgnesR* package allows an easy-to-use interface for the simulation of gene expression profiles from a given network structure. A user can easily either utilize a given biological network or generate a topological structure of different network types for which reaction parameters are specified in correspondence to given constraints. In our package the reaction parameters can be modeled and used in a very flexible manner, e.g., with respect to the underlying parameter distributions. The resulting gene expression data can be either obtained as time series data for user defined sampling time steps or as steady-steady data.

### Abbreviations
AUROC: Area under receiver operator characteristics (ROC) curve; sgnesR: Stochastic gene network expression simulator in R; SSA: Stochastic simulation algorithm

### Acknowledgement
Matthias Dehmer thanks the Austrian Science Funds for supporting this work (project P26142).

### Funding
Source of funding is not available.

### Authors' contributions
FES, OYH, MD, ST conceived and designed the analysis. ST, FES implemented the algorithms, and analyzed the data. FES, JLP, AR, OYH, MD, ST wrote the paper. All authors approved the final version.

### Competing interests
The authors declare that they have no competing interests.

### Author details
[1] Predictive Medicine and Data Analytics Lab, Department of Signal Processing, Tampere University of Technology, Tampere, Finland. [2] Department of Biostatistics, Harvard T.H. Chan School of Public Health, Harvard University, Boston, USA. [3] Laboratory of Biosystem Dynamics, Department of Signal Processing, Tampere University of Technology, Tampere, Finland. [4] Institute for Theoretical Informatics, Mathematics and Operations Research, Department of Computer Science, Universität der Bundeswehr München, Munich, Germany. [5] Institute of Biosciences and Medical Technology, Tampere, Finland. [6] Computational Systems Biology, Department of Signal Processing, Tampere University of Technology, Tampere, Finland.

### References
1. Kauffman SA. The origins of order: Self-organization and selection in evolution. Underst Origs. 1992;65:153–81.
2. Schadt EE. Molecular networks as sensors and drivers of common human diseases. Nature. 2009;461:218–23.
3. Emmert-Streib F, Glazko GV. Network Biology: A direct approach to study biological function. Wiley Interdiscip Rev Syst Biol Med. 2011;3(4):379–91.
4. Vidal M. A unifying view of 21st century systems biology. FEBS Lett. 2009;583(24):3891–4.
5. Emmert-Streib F, Glazko GV, Altay G, de Matos Simoes R. Statistical inference and reverse engineering of gene regulatory networks from observational expression data. Front Genet. 2012;3:8.
6. Markowetz F, Spang R. Inferring cellular networks–a review. BMC Bioinforma. 2007;8:5.
7. de Matos Simoes R, Dehmer M, Emmert-Streib F. B-cell lymphoma gene regulatory networks: Biological consistency among inference methods. Front Genet. 2013;4:281.
8. Van den Bulcke T, Van Leemput K, Naudts B, van Remortel P, Ma H, Verschoren A, De Moor B, Marchal K. Syntren: a generator of synthetic gene expression data for design and analysis of structure learning algorithms. BMC Bioinforma. 2006;7(1):43. doi:10.1186/1471-2105-7-43.
9. Di Camillo B, Toffolo G, Cobelli C. A gene network simulator to assess reverse engineering algorithms. Ann N Y Acad Sci. 2009;1158(1):125–42. doi:10.1111/j.1749-6632.2008.03756.x.
10. Castelo R, Roverato A. Reverse engineering molecular regulatory networks from microarray data with qp-graphs. J Comput Biol. 2009;16(2):213–7.
11. Opgen-Rhein R, Strimmer K. From correlation to causation networks: a simple approximate learning algorithm and its application to high-dimensional plant gene expression data. BMC Syst Biol. 2007;1(1):37. doi:10.1186/1752-0509-1-37.
12. Ribeiro AS, Zhu R, Kauffman SA. A general modeling strategy for gene regulatory networks with stochastic dynamics. J Comput Biol. 2006;13(9):1630–9.
13. Ribeiro AS, Lloyd-Price J. Sgn sim, a stochastic genetic networks simulator. Bioinformatics. 2007;23(6):777.
14. Peng RD. Reproducible research in computational science. Science. 2011;334(6060):1226–7.
15. Tripathi S, Dehmer M, Emmert-Streib F. NetBioV: An R package for visualizing large network data in biology and medicine. Bioinformatics. 2014;30(19):2834–6.
16. Breitkreutz BJ, Stark C, Reguly T, Boucher L, Breitkreutz A, Livstone M, Oughtred R, Lackner DH, Bähler J, Wood V, Dolinski K, Tyers M. The BioGRID Interaction Database: 2008 update. Nucl Acids Res. 2008;36(suppl_1):D637–40.
17. Aranda B, Achuthan P, Alam-Faruque Y, Armean I, Bridge A, Derow C, Feuermann M, Ghanbarian AT, Kerrien S, Khadake J, Kerssemakers J, Leroy C, Menden M, Michaut M, Montecchi-Palazzi L, Neuhauser SN, Orchard S, Perreau V, Roechert B, van Eijk K, Hermjakob H. The IntAct molecular interaction database in 2010. Nucl Acids Res. 2010;38(suppl_1):D525–31.
18. Salgado H, Peralta-Gil M, Gama-Castro S, Santos-Zavaleta A, Muñiz-Rascado L, García-Sotelo JS, Weiss V, Solano-Lira H, Martínez-Flores I, Medina-Rivera A, Salgado-Osorio G, Alquicira-Hernández S, Alquicira-Hernández K, López-Fuentes A, Porrón-Sotelo L, Huerta AM, Bonavides-Martínez C, Balderas-Martínez YI, Pannier L, Olvera M, Labastida A, Jiménez-Jacinto V, Vega-Alvarado L, del Moral-Chávez V, Hernández-Alvarez A, Morett E, Collado-Vides J.

Regulondb v8.0: omics data sets, evolutionary conservation, regulatory phrases, cross-validated gold standards and more. Nucleic Acids Res. 2013;41(D1):203–13. doi:10.1093/nar/gks1201.

19. Wang E. Cancer systems biology. Chapman & Hall/CRC Mathematical and Computational Biology. 2010.

20. Gibson MA, Bruck J. Efficient exact stochastic simulation of chemical systems with many species and many channels. J Phys Chem A. 2000;104(9):1876–89. doi:10.1021/jp993732q.

21. Barabási AL, Albert R. Emergence of scaling in random networks. Science. 1999;206:509–12.

22. de Matos Simoes R, Emmert-Streib F. Bagging statistical network inference from large-scale gene expression data. PLOS ONE. 2012;7(3): 1–11. doi:10.1371/journal.pone.0033624.

23. Altay G, Emmert-Streib F. Inferring the conservative causal core of gene regulatory networks. BMC Syst Biol. 2010;4(1):132. doi:10.1186/1752-0509-4-132.

24. Gama-Castro S, Salgado H, Santos-Zavaleta A, Ledezma-Tejeida D, Muniz-Rascado L, García-Sotelo JS, Alquicira-Hernández K, Martínez-Flores I, Pannier L, Castro-Mondragón JA, Medina-Rivera A, Solano-Lira H, Bonavides-Martínez C, Pérez-Rueda E, Alquicira-Hernández S, Porrón-Sotelo L, López-Fuentes A, Hernández-Koutoucheva A, Moral-Chávez VD, Rinaldi F, Collado-Vides J. Regulondb version 9.0: high-level integration of gene regulation, coexpression, motif clustering and beyond. Nucleic Acids Res. 2016;44(D1):133. doi:10.1093/nar/gkv1156.

25. Mendes P, Sha W, Ye K. Artificial gene networks for objective comparison of analysis algorithms. Bioinformatics. 2003;19:122–9.

26. Hache H, Wierling C, Lehrach H, Herwig R. Genge: systematic generation of gene regulatory networks. Bioinformatics. 2009;25(9):1205–7. doi:10. 1093/bioinformatics/btp115. http://bioinformatics.oxfordjournals.org/content/25/9/1205.full.pdf+html.

27. Haynes BC BM. Benchmarking regulatory network reconstruction with grendel. Bioinformatics. 2009;25(6):801–7.

28. Roy S, Werner-Washburne M, Lane T. A system for generating transcription regulatory networks with combinatorial control of transcription. Bioinformatics. 2008;24(10):1318–20. doi:10.1093/bioinformatics/btn126. http://bioinformatics.oxfordjournals.org/content/24/10/1318.full.pdf+html.

# 14

# Parallel tiled Nussinov RNA folding loop nest generated using both dependence graph transitive closure and loop skewing

Marek Palkowski[*] and Wlodzimierz Bielecki

## Abstract

**Background:** RNA secondary structure prediction is a compute intensive task that lies at the core of several search algorithms in bioinformatics. Fortunately, the RNA folding approaches, such as the Nussinov base pair maximization, involve mathematical operations over affine control loops whose iteration space can be represented by the polyhedral model. Polyhedral compilation techniques have proven to be a powerful tool for optimization of dense array codes. However, classical affine loop nest transformations used with these techniques do not optimize effectively codes of dynamic programming of RNA structure predictions.

**Results:** The purpose of this paper is to present a novel approach allowing for generation of a parallel tiled Nussinov RNA loop nest exposing significantly higher performance than that of known related code. This effect is achieved due to improving code locality and calculation parallelization. In order to improve code locality, we apply our previously published technique of automatic loop nest tiling to all the three loops of the Nussinov loop nest. This approach first forms original rectangular 3D tiles and then corrects them to establish their validity by means of applying the transitive closure of a dependence graph. To produce parallel code, we apply the loop skewing technique to a tiled Nussinov loop nest.

**Conclusions:** The technique is implemented as a part of the publicly available polyhedral source-to-source TRACO compiler. Generated code was run on modern Intel multi-core processors and coprocessors. We present the speed-up factor of generated Nussinov RNA parallel code and demonstrate that it is considerably faster than related codes in which only the two outer loops of the Nussinov loop nest are tiled.

**Keywords:** RNA folding, Parallel biological computing, Loop tiling, Transitive closure, Loop skewing

## Background

RNA secondary structure prediction is an important ongoing problem in bioinformatics. RNA provides a mechanism to copy the genetic information of DNA and can catalyze various biological reactions. RNA folding is the process by which a linear ribonucleic acid molecule acquires secondary structure through intra-molecular interactions.

Algorithms to make predictions of the structure of single RNA molecules use empirical models to estimate the free energies of folded structures. This paper focuses on the base pair maximization algorithm developed by Nussinov [1], which predicts RNA secondary structure in a computationally efficient way. Given an RNA sequence $x_1, x_2, \ldots, x_n$, where $x_i$ is a nucleotide from the alphabet {G (guanine), A (adenine), U (uracil), C (cytosine)}, Nussinov's algorithm solves the problem of RNA non-crossing secondary structure prediction by means of computing the maximum number of base pairs for subsequences $x_i, \ldots, x_j$, starting with subsequences of length 1 and building upwards, storing the result of each subsequence in a dynamic programming array.

---
*Correspondence: mpalkowski@wi.zut.edu.pl
West Pomeranian University of Technology, Faculty of Computer Science,
Zolnierska 49, 71-210 Szczecin, Poland

The following Nussinov recursion $S(i,j)$ is defined over the region $1 \leq i < j \leq N$ as

$$
\begin{aligned}
S(i,j) = max(&S(i+1,j-1) + \delta(i,j), \\
&\max_{i \leq k < j}(S(i,k) + S(k+1,j))),
\end{aligned}
\tag{1}
$$

and zero elsewhere, where $S$ is the $N \times N$ Nussinov matrix, and $\delta(i,j)$ is the function which returns 1 if $(x_i, x_j)$ is an AU, GC or GU pair and $i < j$, or 0 otherwise.

Nussinov's algorithm is within nonserial polyadic dynamic programming (NPDP). The term nonserial polyadic stands for another family of dynamic programming (DP) with nonuniform data dependences, which is more difficult to be optimized [2].

On modern computer architectures, the cost of moving data from main memory is orders of magnitude higher than the cost of computation. Improving data locality and extracting loop nest parallelism of NPDP are still challenging tasks, although a number of authors have developed theoretical approaches to accelerating NPDP codes for RNA folding [3–8].

Fortunately, the Nussinov recursion involves mathematical operations over affine control loops whose iteration space can be represented by the polyhedral model [9]. In this paper, we consider a formulation that is suitable for automatically producing parallel and tiled program loop nests from the dependence structure of the program (as would be used in an automatic optimizing compiler).

Loop tiling, or blocking, is a key transformation used for both coarsening the granularity of parallelism and improving code locality. Smaller blocks of loop nest statement instances in a loop nest iteration space (tiles) can improve cache line utilization and avoid false sharing. On the basis of a valid schedule of tiles, parallel coarse-grained code can be generated.

To our best knowledge, well-known loop nest tiling techniques are based on linear or affine transformations [10–13]. However, only the two outer loops from the three ones of the Nussinov code can be tiled by means of standard tiling algorithms implemented in polyhedral tools [14]. For example, the state-of-the-art compiler, Pluto [10], extracting and applying affine transformations, is able to tile and parallelize the two outer loops of the considered Nussinov code and is not able to tile the innermost loop. The iterations of this loop can be executed only in serial order that prevents enhancing code locality and parallelism degree.

Moreover, classical affine transformations have commonly known limitations [9, 14, 15], which complicate extraction of available parallelism and locality improvement in NPDP codes. Mullapudi and Bondhugula presented dynamic tiling for Zuker's optimal RNA folding[1] in paper [9]. They have explored techniques for tiling codes that lie outside the domain of standard tiling techniques.

3D iterative tiling for dynamic scheduling is calculated by means of reduction chains. Operations along each chain find maximum and can be reordered to eliminate cycles. Their approach involves dynamic scheduling of tiles, rather than the generation of a static schedule. At this time, a precise characterization of the relative domains of this technique is not available.

Wonnacott et al. introduced 3D tiling of "mostly-tileable" loop nests of the Nusinov algorithm in the paper [14]. The "mostly-tileable" term means the iteration space is dominated by non-problematic iterations (iterations of loops 'i' and 'j'). This approach tiles non-problematic iterations with classic tiling strategies while problematic iterations of loop ('k') are peeled off and executed later. Generated code is serial and the authors do not present any parallelization of this code.

Rizk et al. [16] provide an approach to produce efficient GPU code for RNA folding, but they do not consider any loop nest tiling. Tang et al. [17] presented the Pochoir compiler for automatic parallelization and cache performance optimization of stencil computations. Pochoir computes the optimal cost of aligning a pair of DNA or RNA sequences by means of Gotoh's algorithm. It transforms computation to obtain diamond-shaped grid that can be evaluated as a stencil, but it can tile only two of the three loops of original code. Stivala et al. [18] describe a lock-free algorithm for parallel dynamic programming. However, code locality improvement is not considered.

Paper [15] introduces a new technique to generate parallel code applying the power $k$ of a relation representing a dependence graph, but that paper does not consider generation of tiled code and does not concern any RNA folding. Paper [19] considers runtime scheduling of RNA folding for untiled program loops with known bounds.

Motivated by the deficiency of the mentioned techniques, we developed and present in this paper a novel approach for tiling and parallelization of the Nussinov loop nest. To generate valid tiles in all three dimensions, we apply the exact transitive closure of loop nest dependence graphs. It allows for generating target tiles such that there is no cycle in a corresponding inter-tile dependence graph. It is well-known that for such a case, a valid schedule of target tiles exists, i.e., a valid serial or parallel tiled code can be generated [9]. Such a tiling can be applied to bands of original loops not being fully permutable. To parallelize generated serial tiled code, we use the loop skewing transformation and prove its application validity.

## Methods
### Brief introduction

An introduced approach uses the dependence analysis proposed by Pugh and Wonnacott [20] where dependences are represented by relations with constraints defined by means of the Presburger arithmetic using logical and existential operators. A dependence relation is a tuple relation of the form [*input list*]→[*output list*]: *formula*, where *input list* and *output list* are the lists of variables and/or expressions used to describe input and output tuples and *formula* describes the constraints imposed upon *input list* and *output list*. Such a relation is a mathematical representation of a data dependence graph whose vertices correspond to loop statement instances while edges connect dependent instances. The input and output tuples of a relation represent dependence sources and destinations, respectively; the relation constraints specify instances which are dependent.

Standard operations on relations and sets are used, such as intersection (∩), union (∪), difference (−), domain (dom $R$), range (ran $R$), relation application ($S' = R(S)$: $e' \in S'$ iff exists $e$ s.t. $e \to e' \in R$, $e \in S$). In detail, the description of these operations is presented in papers [20, 21].

The positive transitive closure for a given lexicographically forward relation $R$, $R^+$, is defined as follows [21]:

$$R^+ = \{e \to e' : e \to e' \in R \ \lor$$
$$\exists e'' s.t. \ e \to e'' \in R \ \land \ e'' \to e' \in R^+\}.$$

It describes which vertices $e'$ in a dependence graph (represented by relation $R$) are connected directly or transitively with vertex $e$.

Transitive closure, $R^*$, is defined as below:

$$R^* = R^+ \cup I,$$

where $I$ is the identity relation. It describes the same connections in a dependence graph (represented by $R$) that $R^+$ does plus connections of each vertex with itself. Figure 1 presents $R^+$ and $R^*$ in a graphical way.

In the sequential loop nest, the iteration $i$ executes before $j$ if $i$ is *lexicographically less* than $j$, denoted as

$$i \prec j, i.e., i_1 < j_1 \lor \exists k \geq 1 : i_k < j_k \land i_t = j_t, \ for \ t < k. \quad (2)$$

A *schedule* is a function $\sigma : LD \to \mathbb{Z}$ which assigns a discrete time of execution to each loop nest statement instance or tile. A schedule is *valid* if for each pair of dependent statement instances, $s_1(I)$ and $s_2(J)$, satisfying the condition $s_1(I) \prec s_2(J)$, the condition $\sigma(s_1(I)) < \sigma(s_2(J))$ holds true, i.e. the dependences are preserved when statement instances are executed in an increasing order of schedule times.

**The Nussinov loop nest**

The Nussinov recurrence is challenging to accelerate because of its non-local dependency structure shown in Fig. 2. Cell $S(i,j)$ is depended to adjacent cells of the dynamic programming matrix as well as to non-local cells. These non-local dependences are affine, that is, $S(i,j)$ depends on other cells $S(r,s)$ such that the differences $i-r$ or $j-s$ are not constant but rather depend on $i$ and $j$. Therefore, the Nussinov data dependences result in a nonuniform structure [5]. Equation 1 leads directly to the form of the $\mathcal{O}(n^3)$ Nussinov loop nest presented in Listing 1. The loop nest is imperfectly-nested and is comprised of two statements, *s0* and *s1*.

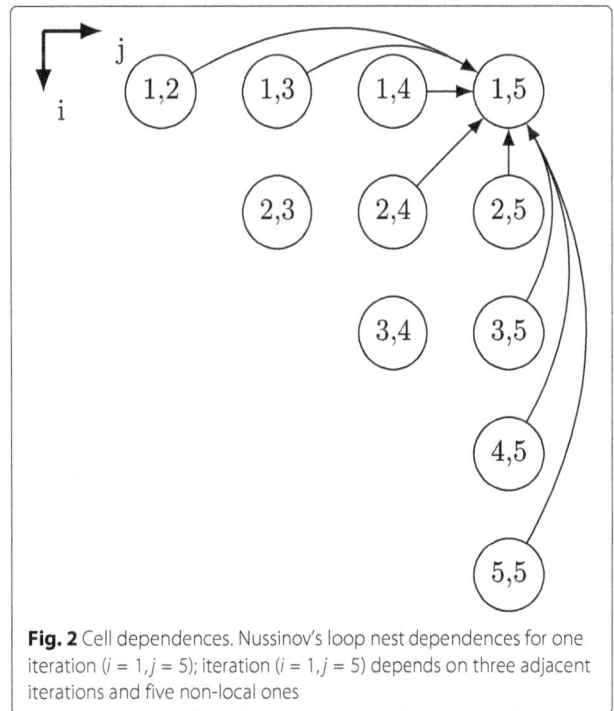

**Fig. 2** Cell dependences. Nussinov's loop nest dependences for one iteration ($i = 1, j = 5$); iteration ($i = 1, j = 5$) depends on three adjacent iterations and five non-local ones

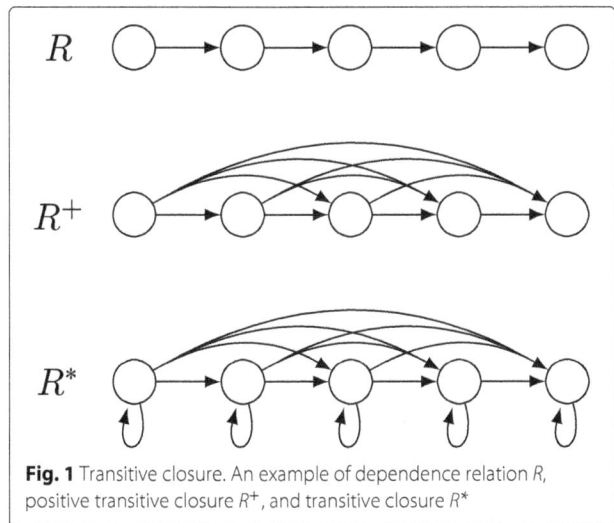

**Fig. 1** Transitive closure. An example of dependence relation $R$, positive transitive closure $R^+$, and transitive closure $R^*$

**Listing 1** Nussinov loop nest

```
for (i = N-1; i >= 0; i--) {
  for (j = i+1; j < N; j++) {
    for (k = 0; k < j-i; k++) {
      S[i][j] = max(S[i][k+i] + S[k+i+1][j], S[i][j]);        // s0
    }
    S[i][j] = max(S[i][j], S[i+1][j-1] + delta(i,j));         // s1
  }
}
```

The following sub-section discusses how to generate serial tile code by means of the transitive closure of dependence graphs.

### Loop nest tiling based on the transitive closure of dependence graphs

To generate valid tiled code, we apply the approach presented in paper [22] based on the transitive closure of dependence graphs. We briefly present the steps of that technique for tiling the Nussinov loop nest. Dependence relations for this loop nest, including non-uniform ones, can be extracted with Petit (the Omega project dependence analyser) [20] and they are presented below.

$$
R = \begin{cases}
s0 \to s0 : \{[i,j,k] \to [i,j',j-i] : j < j' < N \land \\
\quad 0 \le k \land i+k < j \land 0 \le i\} \cup \\
\quad \{[i,j,k] \to [i',j,i-i'-1] : \\
\quad 0 \le i' < i \land j < N \land 0 \le k \land i+k < j\} \cup \\
\quad \{[i,j,k] \to [i,j,k'] : 0 \le k < k' \land j < N \\
\quad \land 0 \le i \land i+k' < j\} \\
s0 \to s1 : \{[i,j,k] \to [i-1,j+1] : j \le N-2 \land \\
\quad 0 \le k \land i+k < j \land 1 \le i\} \cup \\
\quad \{[i,j,k] \to [i,j] : j < N \land 0 \le k \land \\
\quad i+k < j \land 0 \le i\} \\
s1 \to s0 : \{[i,j] \to [i,j',j-i] : 0 \le i < j < j' < N\} \\
\quad \cup \{[i,j] \to [i',j,i-i'-1] : \\
\quad 0 \le i' < i < j < N\} \\
s1 \to s1 : \{[i,j] \to [i-1,j+1] : 1 \le i < j \le N-2\}.
\end{cases}
$$

Next, we calculate the exact transitive closure of the union of all dependence relations, $R^+$, applying the modified Floyd-Warshall algorithm [23]. For brevity, we skip the mathematical representation of $R^+$.

Let vector $I = (i,j,k)^T$ represent indices of the Nussinov loop nest, vector $B = (b_1,b_2,b_3)^T$ define an original tile size, vectors $II = (ii,jj,kk)^T$ and $II' = (iip,jjp,kkp)^T$ specify tile identifiers. Each tile identifier is represented with a non-negative integer, i.e., the constraints $II \ge 0$ and $II' \ge 0$ have to be satisfied.

Below, the mathematical representation of original rectangular tiles for the Nussinov loop nest with the tile size defined with vector $B$ is presented.

$$
TILE = \begin{cases}
i : N-1-b_1 * ii \ge i \ge max(-b_1 * (ii+1), \\
\quad N-1) \land ii \ge 0 \\
j : b_2 * jj + i + 1 \le j \le min(b_2 * (jj+1)+1, \\
\quad N-1) \land jj \ge 0 \\
k : \begin{cases} s0 : b_3 * kk \le k \le min(b_3 * (kk+1)-1, \\
\quad j-i-1) \land kk \ge 0 \\
s1 : k = 0. \end{cases}
\end{cases}
$$

Let us note that for index $i$, the constraints are defined inversely because the value of index $i$ is decremented.

For the tile identifiers, we define constraints, $CONSTR(II,B)$, which have to be satisfied for given values $b1$, $b2$, $b3$, defining a tile size, and parameter $N$ specifying the upper loop index bound.

$$
CONSTR(II,B) = \begin{cases}
ii,b_1 : N-1-b1 * ii >= 0 \\
jj,b_2 : (i+1) + b2 * jj <= N-1 \\
kk,b_3 : b3 * kk + 0 <= j-i-1.
\end{cases}
$$

(3)

In accordance with formula (2), we present below the lexicographical ordering $II < II'$ on vectors $II, II'$ defining tile identifiers as follows.

$$
II' < II = \begin{cases}
s0 : \begin{cases} s0 : ii > iip \lor (ii = iip \land jj > jjp) \lor \\
\quad (ii = iip \land jj = jjp \land kk > kkp)) \\
s1 : ii > iip \lor (ii = iip \land jj > jjp) \end{cases} \\
s1 : \begin{cases} s0 : ii > iip \lor (ii = iip \land jj > jjp) \lor \\
\quad (ii = iip \land jj = jjp)) \\
s1 : ii > iip \lor (ii = iip \land jj > jjp). \end{cases}
\end{cases}
$$

Next, we build sets $TILE\_LT$ and $TILE\_GT$ that are the unions of all the tiles whose identifiers are lexicographically less and greater than that of $TILE(II, B)$, respectively:

$TILE\_LT(GT) = \{[I]| \exists II' : II' < (>)II \land II \ge 0 \land$
$CONSTR(II,B) \land II' \ge 0 \land CONSTR(II',B) \land I \in$
$TILE(II',B)\}$.

Using the exact form of $R^+$, we calculate set, $TILE\_ITR$, as follows.

$$TILE\_ITR = TILE - R^+(TILE\_GT).$$

This set does not include any invalid dependence target, i.e., it does not include any dependence target whose source is within set $TILE\_GT$.

The following set

$$TVLD\_LT = (R^+(TILE\_ITR) \cap TILE\_LT)$$
$$- R^+(TILE\_GT)$$

includes all the iterations that i) belong to the tiles whose identifiers are lexicographically less than that of set $TILE\_ITR$, ii) are the targets of the dependences whose sources are contained in set $TILE\_ITR$, and iii) are not any target of a dependence whose source belong to set $TILE\_GT$.

Target valid tiles are defined by the following set

$$TILE\_VLD = TILE\_ITR \cup TVLD\_LT.$$

To generate serial tiled code, we first form set $TILE\_VLD\_EXT$ by means of inserting i) into the first positions of the tuple of set $TILE\_VLD$ elements of vector $\boldsymbol{II}$ : $ii, jj, kk$; ii) into the constraints of set $TILE\_VLD$ the constraints defining tile identifiers $\boldsymbol{II} \geq 0$ and $CONSTR(\boldsymbol{II}, \boldsymbol{B})$.

The following step is to use the original schedule of the original Nussinov loop nest statement instances, $SCHED\_ORIG$, to form a target set allowing for re-generation of serial valid code. The original schedule can be extracted by means of the Clan tool [24] and is as shown below.

$$SCHED\_ORIG = \begin{cases} s0 : 0, i, 0, j, 0, k \\ s1 : 0, i, 0, j, 1, k. \end{cases}$$

Next we enlarge that schedule with indices $ii, jj, kk$ (responsible for tile identifiers) repeating the same sequence of elements as that for indices $i, j, k$ in the original schedule to get the following schedule.

$$SCHED = \begin{cases} s0 : 0, ii, 0, jj, 0, kk, 0, i, 0, j, 0, k \\ s1 : \begin{cases} s0 : 0, ii, 0, jj, 1, kk, 0, i, 0, j, 0, k \\ s1 : 0, ii, 0, jj, 1, kk, 0, i, 0, j, 1, k. \end{cases} \end{cases}$$

Let us note that tiles, formed for statement $s0$, include only instances of statement $s0$, while those generated for statement $s1$ comprise instances of both statement $s0$ and statement $s1$.

In the next step, we form relation, $Rmap_{s0}$, for the subset of set $TILE\_VLD\_EXT$ representing tiles for statement $s0$, as follows

$$Rmap_{s0} = \left\{ \begin{array}{l} TILE\_s0[ii, jj, kk] \rightarrow \\ \quad [0, ii, 0, jj, 0, kk, 0, i, 0, j, 0, k] \end{array} \right\},$$

and relation, $Rmap_{s1}$, for the sub-set of set $TILE\_VLD\_EXT$ representing tiles for statement $s1$, as follows

$$Rmap_{s1} = \left\{ \begin{array}{l} TILE\_s0[ii, jj, kk] \rightarrow \\ \quad [0, ii, 0, jj, 1, kk, 0, i, 0, j, 0, k]; \\ TILE\_s1[ii, jj, kk] \rightarrow \\ \quad [0, ii, 0, jj, 1, kk, 0, i, 0, j, 1, k] \end{array} \right\},$$

and finally, form target set, $TILE\_VLD\_EXT'$, as bellow

$$TILE\_VLD\_EXT' = Rmap(TILE\_VLD\_EXT),$$

where $Rmap = Rmap_{s0} \cup Rmap_{s1}$.

Sequential tiled code is generated by means of applying the isl AST code generator [25] allowing for scanning elements of set $TILE\_VLD\_EXT'$ in lexicographic order.

### Tiled code parallelization

To parallelize generated serial tiled code, we apply the well-known loop skewing transformation [26]. Loop skewing is a transformation that has been used to remap an iteration space by creating a new loop whose index is a linear combination of two or more loop indices. This results in code whose outermost loop is serial while the other loops can be parallelized.

We use the following skewing transformation: $ii' = ii + jj$, where $ii'$ is the new loop index, $ii, jj$ are the indices of the first two loops in tiled code. Figure 3 illustrates the loop skewing technique applying to the Nussinov loop nest. Iterations lying on each horizontal line can be executed in parallel while time partitions should be enumerated serially.

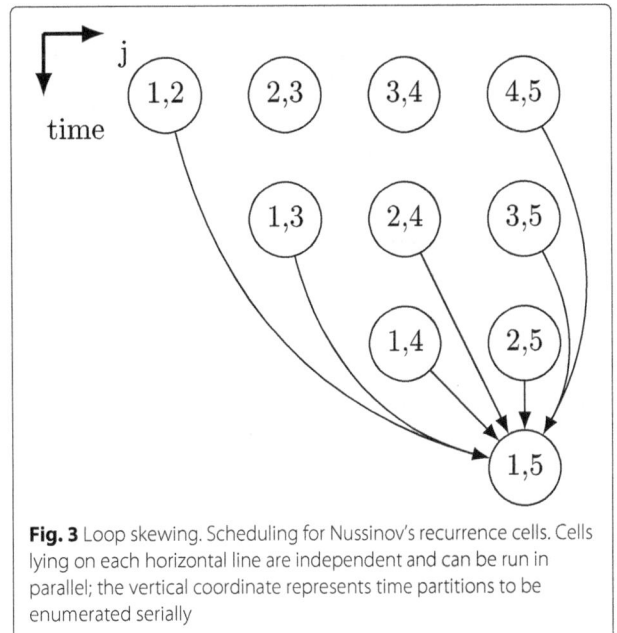

**Fig. 3** Loop skewing. Scheduling for Nussinov's recurrence cells. Cells lying on each horizontal line are independent and can be run in parallel; the vertical coordinate represents time partitions to be enumerated serially

To apply the loop skewing transformation, we create the following relation

$$R\_SCHED = \{[0, ii', 0, jj, \ldots, 0, i, 0, j, \ldots] \rightarrow$$
$$[0, ii + jj, 0, jj, \ldots, 0, -i, 0, j, \ldots] :$$
$$constraints\ of\ set\ TILE\_VLD\_EXT' \},$$

and apply it to set $TILE\_VLD\_EXT'$.

Applying the loop skewing transformation is not always valid. To prove the validity of this transformation applied to generated serial tiled code, we form the following relation, $R\_VALID$, which checks whether all original inter-tile dependences will be respected in parallel code.

$$R\_VALID = \{[II] \rightarrow [JJ] | \; \exists I, J :$$
$$\underbrace{I \in domain\ R \; \wedge \; J = R(I)}_{(*)} \; \wedge$$
$$\underbrace{I \in TILE(II) \wedge J \in TILE(JJ)}_{(**)} \; \wedge$$
$$\underbrace{R\_SCHED(II) \; \geq \; R\_SCHED(JJ)}_{(***)} \},$$

where:

(*) means that $J$ is the destination of the dependence whose source is $I$,

(**) means that $I, J$ belong to the tiles with identifiers $II$ and $JJ$, respectively,

(***) means that the schedule time of tile $II$ is greater or the same as that of tile $JJ$, i.e., the schedule is invalid because the dependence $I \rightarrow J$ is not respected.

This relation returns the empty set when all original inter-tile dependences are respected, otherwise it represents all the pairs of the tile identifiers for which original ones are not respected. Figure 4 presents the case of an invalid schedule, where $I$ and $J$ are vectors representing the source and destination of a dependence, respectively, within the tiles with identifiers $II$ and $JJ$. Relation $R\_VALID$ is empty for the generated serial tiled Nussinov code, this proves the validity of applying the loop skewing transformation.

Target pseudo-code is generated by means of applying the isl AST code generator [25] allowing for scanning elements of set $R\_SCHED(TILE\_VLD\_EXT')$ in lexicographic order. Then we postprocess this code replacing pseudo-statements for the original loop nest statements and insert the work-sharing OpenMP *parallel for* pragmas [27] before the second loop in the generated code to make it parallel. Listing 2 presents the target code for the Nussinov loop nest (Listing 1) tiled with the tiles of the size 16x16x16. The first loop in this code enumerates serially time partitions while the second one scans all the tiles to be executed in parallel for a given time defined with the first loop.

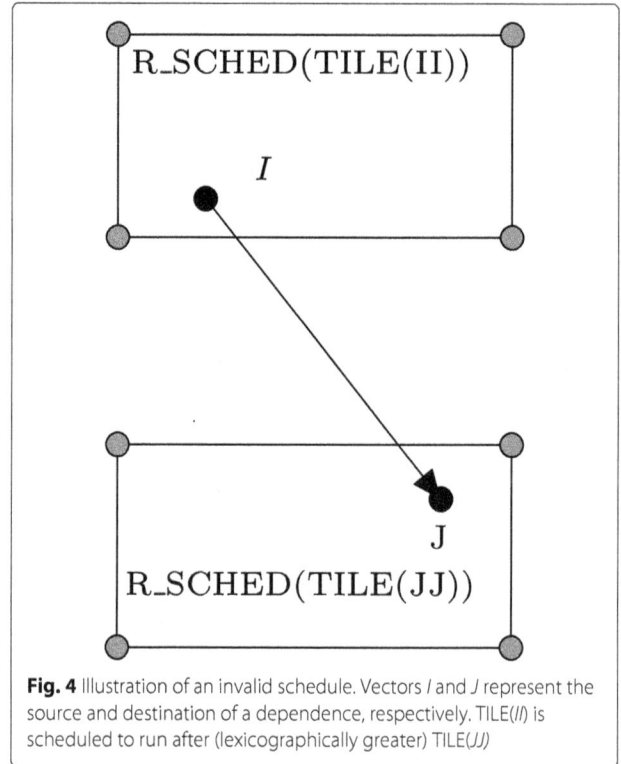

**Fig. 4** Illustration of an invalid schedule. Vectors *I* and *J* represent the source and destination of a dependence, respectively. TILE(*II*) is scheduled to run after (lexicographically greater) TILE(*JJ*)

## Results and discussion

The presented approach has been implemented as a part of the polyhedral TRACO compiler[2]. It takes on input an original loop nest in the C language, a tile size, and affine transformations for each loop nest statement to parallelize serial tiled code. Then TRACO generates serial valid tiled code and checks whether the affine transformations are valid by means of calculating relation $R\_VALID$. If so, parallel tiled code is generated.

All parallel Nussinov tiled codes were generated by means of the Intel C++ Compiler (*icc* 17.0.1) with the -O3 flag of optimization.

This section presents speed-up of generated parallel tiled code. To carry out experiments, we used machines with two processors Intel Xeon E5-2699 v3 (3.6 Ghz, 32 cores, 45MB Cache), four coprocessors Intel Xeon Phi 7120P (1.238 GHz, 61 cores, 30.5 MB Cache), and 128 GB RAM.

Problem sizes 2200 and 5000 were chosen because they are the average and the longest lengths of randomly generated RNA strands (from the {ACGU} alphabet) in human body to illustrate any additional advantages for medium and larger instances, respectively [14]. Furthermore, we used several mRNAs and lncRNAs from the NCBI database[3] for homo sapiens. Analyzing the program code, we expected there should be no difference, performance wise, between actual sequences versus randomly generated sequences. To confirm this fact, we measured

**Listing 2** 3D-tiled and parallel NPDP in the Nussinov algorithm.

```
for( c1 = 0; c1 <= floord(N - 2, 8); c1 += 1) //ii
#pragma omp parallel for shared(c1, S) private(c2,c3,c4,c5,c7,c9,c10,c11) schedule(dynamic,1)
 for( c3 = max(0, c1 - (N + 15) / 16 + 1); c3 <= c1 / 2; c3 += 1) // ii+jj
  for( c4 = 0; c4 <= 1; c4 += 1) { // SCHED for s0 and s1
   if (c4 == 1) {    // SCHED for s1
     for( c7 = max(-N + 16 * c1 - 16 * c3 + 1, -N + 16 * c3 + 2);   c7 <= min(0,
          -N + 16 * c1 - 16 * c3 + 16); c7 += 1)   // i
       for( c9 = 16*c3-c7+1; c9 <= min(N - 1, 16*c3 - c7 + 16); c9++)  // j
         for( c10 = max(0, 16 * c3 - c7 - c9 + 2); c10 <= 1; c10 += 1) {  // 0 for s0, 1 for s1
           if (c10 == 1) {
             S[-c7][c9] = max(S[-c7][c9], S[-c7+1][c9-1] + delta(-c7, c9));  // s1
           } else {
             if (N + 16 * c3 + c7 >= 16 * c1 + 2)
               for( c11 = 0; c11 <= 16 * c3; c11 += 1) // k
                 S[-c7][c9] = max(S[-c7][c11-c7] + S[c11-c7+1][c9], S[-c7][c9]);  // s0
             for( c11 = 16 * c3 + 1; c11 < c7 + c9; c11 += 1)  // k
               S[-c7][c9] = max(S[-c7][c11-c7] + S[c11-c7+1][c9], S[-c7][c9]);  // s0
           }
         }
   } else // SCHED for s0
    for( c5 = 0; c5 <= c3; c5 += 1)  // kk
     for( c7 = max(-N + 16 * c1 - 16 * c3 + 1, -N + 15 * c1 - 14 * c3 + 2); c7 <=
          min(0, -N + 16 * c1 - 16 * c3 + 16); c7++) {  // i
      if (N + 16 * c3 + c7 >= 16 * c1 + 2)
        for( c11 = 16*c5; c11 <= min(15*c3 + c5, 16*c5 + 15); c11++)  // k
          S[-c7][16*c3-c7+1] = max(S[-c7][c11+-c7] + S[c11+-c7+1][16*c3-c7+1],
               S[-c7][16*c3-c7+1]);  // s0
      } else
        for( c9 = N - 16*c1 + 32*c3; c9 <= N - 16*c1 + 32*c3 + 15; c9++)  // j
          for( c11 = 16*c5; c11 <= min(15*c3 + c5, 16 * c5 + 15); c11++)  // k
           S[N-16*c1+16*c3-1][c9] = max(S[N-16*c1+16*c3-1][c11+N-16*c1+16*c3-1] +
                S[c11+N-16*c1+16*c3-1+1][c9], S[N-16*c1+16*c3-1][c9]);  // s0
      }
   }
```

the summary time of calling bonding function $\delta(i,j)$. It takes less than 0.2 percent of the whole tiled code running time regardless of the sequence type, for example, 0.017 seconds for the problem size equal to 5000 (over 12 mln calls) on an Intel Xeon E5-2699 v3 platform. It can be therefore concluded that the studied algorithm performance does not change based on the strings themselves, but it depends on the size of a string.

For generated tiled code, we empirically recognized that the best tile size is 16x16x16 and the most efficient work-sharing is achieved by applying the OpenMP *for* directive [27] with the dynamic scheduling of loop iterations and the chunk size equal to 1.

Table 1 presents the execution times of the serial original and parallel tiled Nussinov loop nest from one to 64 threads for Intel Xeon E5-2699 v3 processors and from one to 244 threads for Intel Xeon Phi 7120P coprocessors. As we can see, for all cases, the execution time of the tiled codes is shorter than that of the original code and it reduces with increasing the number of threads. Speed-up is illustrated in Figs. 5 and 6 in a graphical way for multi-core processors and coprocessors, respectively.

Those figures also present the speed-up of parallel 2D tiled code produced with the state-of-the-art Pluto+ [28] optimizing compiler, which does not enable to tile the third loop in the Nussinov loop nest[4]. From Figs. 5 and 6, we may conclude that the tiled code generated with

**Table 1** Execution times (in seconds) of the tiled Nussinov loop nest

| Platform | Threads | Times | |
|---|---|---|---|
| | | N=2200 | N=5000 |
| Intel Xeon | 1 (original) | 12.28 | 334.32 |
| E5-2699 v3 | 1 | 8.25 | 225.23 |
| | 2 | 4.76 | 147.30 |
| | 4 | 2.37 | 76.79 |
| | 8 | 1.66 | 39.81 |
| | 16 | 0.75 | 21.49 |
| | 32 | 0.44 | 11.90 |
| | 64 | 0.37 | 10.50 |
| Intel Xeon | 1 (original) | 235.38 | 2879.66 |
| Phi 7120P | 1 | 166.92 | 2556.65 |
| | 8 | 29.29 | 339.15 |
| | 16 | 15.09 | 266.34 |
| | 32 | 8.38 | 124.51 |
| | 64 | 4.84 | 72.56 |
| | 128 | 3.78 | 48.81 |
| | 244 | 3.72 | 37.75 |

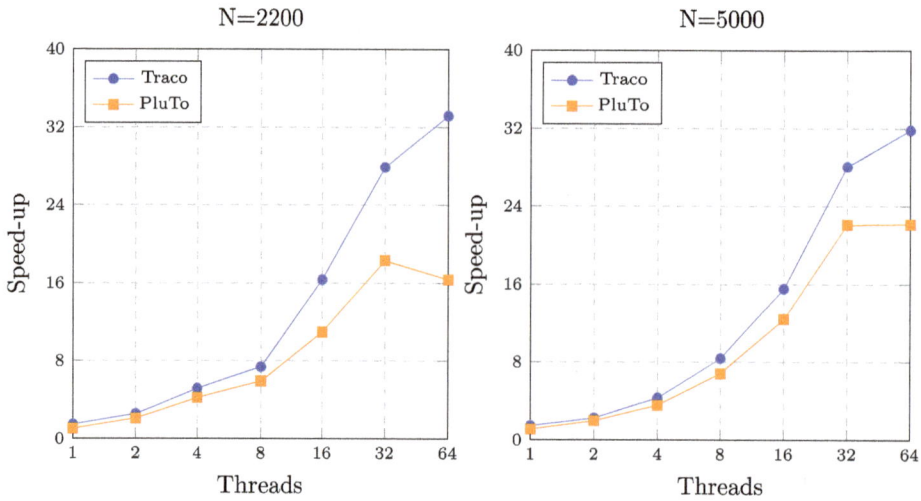

**Fig. 5** Speed-up of parallel codes using two 32-core processors Intel Xeon E5-2699 v3. The horizontal coordinate represents number of threads and the vertical one shows the speedup of codes generated with the TRACO and PluTo compilers for two problem sizes of RNA folding

the proposed approach outperforms that generated with standard affine transformations extracted and applied with Pluto+ for both Intel multi-core processors and coprocessors.

The parallel code presented in the paper is not synchronization free (to our best knowledge, there does not exist any synchronization-free code for Nussinov's loop nest), after each parallel iteration multiple tasks must be synchronized. Synchronization usually involves waiting by at least one task, and can therefore cause a parallel applications wall clock execution time to increase, i.e., it introduces parallel program overhead. Any time one task spends waiting for another is considered synchronization overhead. Synchronization overhead grows with increasing the number of synchronization events and the

number of threads and tends to grow rapidly (in a non-linear manner) as the number of tasks in a parallel job increases, it is the most important factor in obtaining good scaling behavior for the parallel program. Synchronization overhead leads to non-linear character of speed-up when the numbers of threads grows (see Figs. 5 and 6). When the number of threads are less than 16, the code presented in the paper and that generated with PLUTO, have comparable synchronization overhead and locality, but for the number of threads more than 16, our code has less synchronization overhead and better locality that results in higher speed-up.

It is worth noting that the generated tiled serial code has improved locality in comparison with that of the serial original code. This results in about 1.5 and 1.4 higher

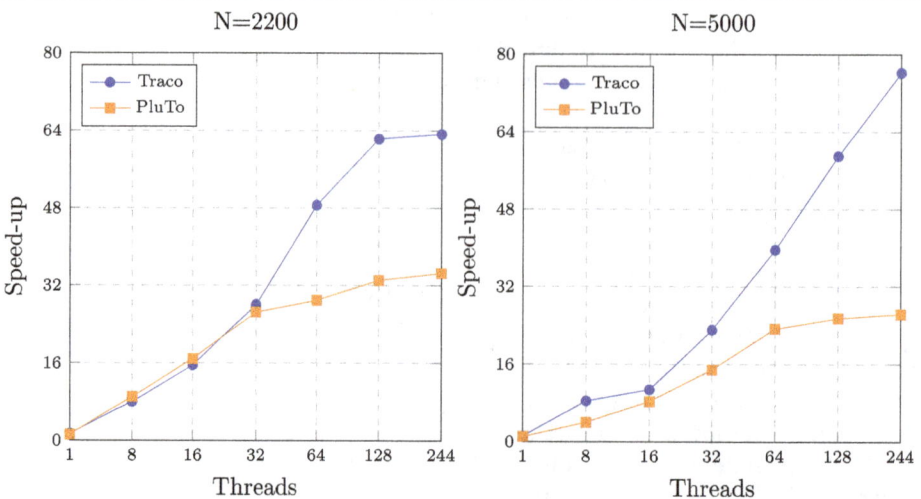

**Fig. 6** Speed-up of parallel codes using four 61-core coprocessors Intel Xeon Phi 7120P. The horizontal coordinate represents number of threads and the vertical one shows the speedup of codes generated with the TRACO and PluTo compilers for two problem sizes of RNA folding

serial tiled code performance for the used Intel muticore processors and co-processors, respectively. Below, we compare the speed-up achieved for the tiled code generated by the presented technique with that of related code.

In paper [7], the authors write: "We have developed GTfold, a parallel and multicore code for predicting RNA secondary structures that achieves 19.8 fold speedups over the current best sequential program". This speed-up is achieved on 32 threads. The code, presented in our paper, outperforms this code (for 32 threads, it yields 28.1 speed-up for the problem size equal to 5000). We also present speed-up for 64 threads for an Intel Xeon E5-2699 v3 platform and from one to 244 threads for Intel Xeon Phi 7120P coprocessors. The higher performance of our code is achieved due to applying loop nest tiling.

Rizk et al. [16] provide an efficient GPU code for RNA folding, but they do not consider any loop nest tiling. The authors give a table which shows that the maximal speedup, using a graphical card GTX280, is 33.1. Applying Intel Xeon Phi 7120P coprocessors for running our code, we reach the maximal speed-up 75.6 for 244 threads (the problem size is equal to 5000). This demonstrates that tiling allows for considerable improving code locality that leads to significant increasing parallel code speed-up.

Pochoir [17] computes the optimal cost of aligning a pair of DNA or RNA sequences by means of diamond-shaped grid that can be evaluated as a stencil, but it can tile only two of the three loops of original code, i.e., tiled code is of maximum 2-d dimension. This results in only 4.5 speedup of the RNA code generated with Pochoir on 12 cores – the maximal number of cores that the authors examined.

Summing up, we conclude that the presented approach allows for generation of a parallel tiled Nussinov loop nest which considerable reduces execution time in comparison with related codes. The code presented in our paper is dedicated to be run on high performance computer systems with the large number of cores. Since the number of cores tends to grow, in our opinion, the presented code is very actual because it has improved scalability and can be run on computer systems with the large number of cores.

## Conclusion

The paper presents automatic tiling and parallelization of the Nussinov program loop nest. The transitive closure of dependence graphs is used to tile this code, whereas for extracting parallelism in the tiled loop nest, the loop skewing transformation is applied, which is within the affine transformation framework. To the best of our knowledge, the presented approach is the first attempt to generate static parallel 3D tiled code for Nussinov's prediction. An experimental study demonstrates significant parallel tiled code speed-up achieved on modern multi-core computer systems.

The presented approach is an important starting point for future research aimed at effective tiling and parallelization of other NPDP codes, in particular the detailed energy models used by Zuker's algorithm.

We are going to examine how the presented approach based on both the transitive closure of dependence graph and affine transformations can be applied to tile and parallelize other important applications of bioinformatics.

## Endnotes

[1] Zuker's algorithm has the same dependence patterns as Nussinov's algorithm [9].

[2] http://traco.sourceforge.net

[3] https://www.ncbi.nlm.nih.gov/

[4] Pluto 0.11.4 BETA and Pluto+ generate the same tiled code for the Nussinov loop nest.

## Abbreviations
AST: Abstract syntax tree; DP: Dynamic programming; GPU: Graphics processing unit; NPDP: Nonserial polyadic dynamic programming

## Acknowledgements
Not applicable.

## Authors' contributions
MP proposed the main concept of the presented technique, implemented it in the TRACO optimizing compiler, and carried out the experimental study. WB checked the correctness of the presented technique, participated in its implementation and the analysis of the results of the experimental study. Both authors read and approved the final manuscript.

## Competing interests
The authors declare that they have no competing interests.

## References
1. Nussinov R, Pieczenik G, Griggs JR, Kleitman DJ. Algorithms for loop matchings. SIAM J Appl Math. 1978;35(1):68–82.
2. Liu L, Wang M, Jiang J, Li R, Yang G. Efficient nonserial polyadic dynamic programming on the cell processor. In: IPDPS Workshops. Anchorage, Alaska: IEEE; 2011. p. 460–71.
3. Almeida F, et al. Optimal tiling for the rna base pairing problem. In: Proceedings of the Fourteenth Annual ACM Symposium on Parallel Algorithms and Architectures. SPAA '02, New York: ACM; 2002. p. 173–82. doi:10.1145/564870.564901.
4. Tan G, Feng S, Sun N. Locality and parallelism optimization for dynamic programming algorithm in bioinformatics. In: SC 2006 Conference, Proceedings of the ACM/IEEE. Tampa: IEEE, Conference Location; 2006. p. 41–1.

5.  Jacob A, Buhler J, Chamberlain RD. Accelerating Nussinov RNA secondary structure prediction with systolic arrays on FPGAs. In: Proceedings of the 2008 International Conference on Application-Specific Systems, Architectures and Processors. ASAP '08, Washington: IEEE Computer Society; 2008. p. 191–6. doi:10.1109/ASAP.2008.4580177.

6.  Markham NR, Zuker M. In: Keith JM, editor. UNAFold. Totowa, NJ: Humana Press; 2008, pp. 3–31.

7.  Mathuriya A, Bader DA, Heitsch CE, Harvey SC. Gtfold: A scalable multicore code for rna secondary structure prediction. In: Proceedings of the 2009 ACM Symposium on Applied Computing. SAC '09, New York: ACM; 2009. p. 981–8.

8.  Jacob AC, Buhler JD, Chamberlain RD. Rapid rna folding: Analysis and acceleration of the zuker recurrence. In: Field-Programmable Custom Computing Machines (FCCM), 2010 18th IEEE Annual Int. Symp. On. Charlotte: IEEE, Conference Location; 2010. p. 87–94.

9.  Mullapudi RT, Bondhugula U. Tiling for dynamic scheduling In: Rajopadhye S, Verdoolaege S, editors. Proceedings of the 4th International Workshop on Polyhedral Compilation Techniques. Vienna, Austria; 2014. http://impact.gforge.inria.fr/impact2014/papers/impact2014-mullapudi.pdf.

10. Bondhugula U, Hartono A, Ramanujam J, Sadayappan P. A practical automatic polyhedral parallelizer and locality optimizer. SIGPLAN Not. 2008;43(6):101–13. doi:10.1145/1379022.1375595.

11. Griebl M. Automatic Parallelization of Loop Programs for Distributed Memory Architectures: University of Passau; 2004. Habilitation thesis.

12. Lim A, Cheong GI, Lam MS. An affine partitioning algorithm to maximize parallelism and minimize communication. In: Proceedings of the 13th ACM SIGARCH Int. Conf. on Supercomputing. Portland: ACM Press; 1999. p. 228–37.

13. Xue J. On tiling as a loop transformation. Parallel Process Lett. 1997;7(4):409-424.

14. Wonnacott D, Jin T, Lake A. Automatic tiling of "mostly-tileable" loop nests. In: IMPACT 2015: 5th International Workshop on Polyhedral Compilation Techniques. Amsterdam; 2015. http://impact.gforge.inria.fr/impact2015/papers/impact2015-wonnacott.pdf.

15. Bielecki W, Palkowski M, Klimek T. Free scheduling for statement instances of parameterized arbitrarily nested affine loops. Parallel Comput. 2012;38(9):518–32.

16. Rizk G, Lavenier D. Gpu accelerated rna folding algorithm In: Allen G, Nabrzyski J, Seidel E, van Albada G, Dongarra J, Sloot PA, editors. Computational Science – ICCS 2009. Lecture Notes in Computer Science, Baton Rouge, LA, USA: Springer; 2009. p. 1004–1013.

17. Tang Y, Chowdhury RA, Kuszmaul BC, Luk CK, Leiserson CE. The pochoir stencil compiler. In: Proceedings of the 23rd ACM Symposium on Parallelism in Algorithms and Architectures. SPAA '11, New York: ACM; 2011. p. 117–28. doi:10.1145/1989493.1989508.

18. Stivala A, Stuckey PJ, Garcia de la Banda M, Hermenegildo M, Wirth A. Lock-free parallel dynamic programming. J Parallel Distrib Comput. 2010;70(8):839–48.

19. Palkowski M. Finding Free Schedules for RNA Secondary Structure Prediction, Springer Int. Publishing, Rutkowski et al., Artificial Intelligence and Soft Computing: ICAISC 2016, Poland, Proceedings, Part II. Zakopane: Springer International Publishing; 2016, pp. 179–88.

20. Pugh W, Wonnacott D. In: Banerjee U, Gelernter D, Nicolau A, Padua D, editors. An exact method for analysis of value-based array data dependences. Berlin, Heidelberg: Springer; 1994, pp. 546–66.

21. Kelly W, Maslov V, Pugh W, Rosser E, Shpeisman T, Wonnacott D. The omega library interface guide. Technical report, College Park, MD, USA 1995.

22. Bielecki W, Palkowski M. Tiling of arbitrarily nested loops by means of the transitive closure of dependence graphs. Int J Appl Math Comput Sci (AMCS). 2016;26(4):919–939.

23. Bielecki W, Kraska K, Klimek T. Using basis dependence distance vectors in the modified floyd–warshall algorithm. J Comb Optim. 2015;30(2):253–75.

24. Bastoul C. Code generation in the polyhedral model is easier than you think. In: PACT'13 IEEE International Conference on Parallel Architecture and Compilation Techniques. Juan-les-Pins: IEEE Computer Society; 2004. p. 7–16.

25. Verdoolaege S. Integer set library - manual, Technical report 2016. http://isl.gforge.inria.fr/manual.pdf. Accessed 27 May 2017.

26. Wolfe M. Loops skewing: The wavefront method revisited. Int J Parallel Programm. 1986;15(4):279–93.

27. OpenMP Architecture Review Board. OpenMP Application Program Interface Version 4.5. 2015. http://www.openmp.org/wp-content/uploads/openmp-4.5.pdf. Accessed 27 May 2017.

28. Bondhugula U, Acharya A, Cohen A. The pluto+ algorithm: A practical approach for parallelization and locality optimization of affine loop nests. ACM Trans Program Lang Syst. 2016;38(3):12–11232. doi:10.1145/2896389.

# 15

# Assessment of genome annotation using gene function similarity within the gene neighborhood

Se-Ran Jun[1]* (iD), Intawat Nookaew[1], Loren Hauser[2] and Andrey Gorin[3]

## Abstract

**Background:** Functional annotation of bacterial genomes is an obligatory and crucially important step of information processing from the genome sequences into cellular mechanisms. However, there is a lack of computational methods to evaluate the quality of functional assignments.

**Results:** We developed a genome-scale model that assigns Bayesian probability to each gene utilizing a known property of functional similarity between neighboring genes in bacteria.

**Conclusions:** Our model clearly distinguished true annotation from random annotation with Bayesian annotation probability >0.95. Our model will provide a useful guide to quantitatively evaluate functional annotation methods and to detect gene sets with reliable annotations.

**Keywords:** Genome functional annotation, Gene function similarity, Gene neighborhood, Bayesian probability

## Background

During recent years, technological advances have enabled the rapid and affordable sequencing of organisms from all kingdoms of life. In 2011 the volume of the NCBI Sequence Read Archive crossed a remarkable size of 100 TB [1], and more than 22,000 complete or nearly complete genomes are available for bacterial organisms with the number increasing by >1000 each month [2, 3]. Functional annotation of bacterial genomes is an obligatory and crucially important step of information processing from the genome sequences toward insights into cellular mechanisms, putative ecological roles, or predictive models of a given organism or microbial community. Numerous software packages, databases, platforms, and score filters involve computational pipelines that assign functions to the genes [4]. However, the sequence information is only as good and useful as the functional annotation when it has functional annotation attached to it. The function of genes is central for all biological insights, including interpretation and design of experiments and comparative genomic analysis, as well as the

input data for metabolic and regulatory models [5, 6]. The manual curation or experimental verification [7] is unlikely to be feasible when >1000 genomes are added each month. Accordingly, there is a greater urgency to have computational tools for genome annotation validation [8].

In the literature, "annotation quality" sometimes refers to the precision of finding an exact start site for the genes in the genome [8, 9]. When the location of a gene is determined incorrectly, it follows that functional annotation will more likely be incorrect as well. Therefore, the gene finding problem is an important part of the process for genome annotation. In this work, we aim to address annotation consistency at the level where genes are found and annotated by standard protein function annotation, Gene Ontology (GO) terms, organized in a hierarchical fashion [10]. The benefits of function annotation by GO are a systematic control vocabulary that enables cross-comparison over different genomes and a higher percentage of genes in the genome that can be annotated because of different levels of information of GO hierarchy.

In an approach described by Skunca et al. [11], the authors measured the annotation quality of individual GO terms using experimental verifications and estimated the annotation quality of the database UniProt-GOA

* Correspondence: sjun@uams.edu
[1]Department of Biomedical Informatics, College of Medicine, University of Arkansas for Medical Sciences, Little Rock, AR 72205, USA
Full list of author information is available at the end of the article

over time. This approach dealt with relatively small datasets composed of model organisms because it was dependent on experimental verifications. Alternatively, the occurrence of annotation terms was used in a recent computational study [12], which indicated that the manually curated annotations have more natural lexical properties than automatically generated ones, but this method was a bulk analysis within the annotation database and it does not describe the annotation quality of any particular genome. In other studies, authors have used multiple tools and performed manual analysis of the problematic annotations [13, 14]. These are reliable approaches, but they are clearly not scalable to dozens of genomes.

Our approach to the validation of gene annotation utilized a well-known and fundamental property of the bacterial genomes: functionally coordinated genes tend to be physically closer on a chromosome than the average gene [15–17]. However, this property was rarely used by others except in a semiquantitative way [18], which used the property to find functional annotations especially for difficult cases of hypothetical proteins. The novel idea of our work (described in Methods in detail)

is illustrated in Fig. 1. In this study, a gene neighborhood is defined as three left and right genes of a given gene along the chromosome. We developed an analytical approach to measure gene function similarity (GFS) for each neighboring pair of genes, applied Bayesian statistics to integrate gene neighborhood information of annotation, and then finally, computed the probability of annotation confidence (PAC) for each gene that has at least one GFS score available within its neighborhood, given that functional assignment with very few and well-controlled empirical assumptions is correct. Our method provides genome annotation assessment through the annotation evaluation of all individual genes in the genome.

## Results
### Probability of annotation confidence
We applied our methodology to *Escherichia coli* and *Clostridium thermocellum* to calculate the PAC for NCBI annotation (assumed to be a well annotation) and compared it with "random" annotation. For each gene with an annotation in *E. coli*, the random annotation was generated by assigning a random annotation selected from 8

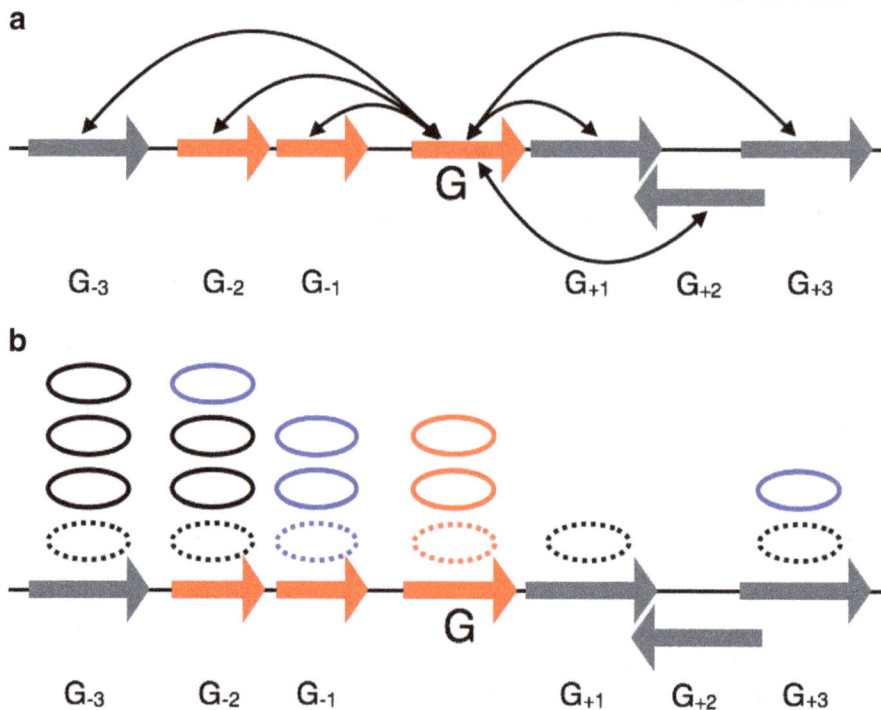

**Fig. 1** Gene neighborhood and gene function similarity. **a** Gene neighborhood. **b** Gene function similarity. **a** In this study, we looked at three genes in the upstream and downstream directions for neighboring genes of a given gene G. For a gene G, the neighboring gene at +2 is from an opposite strand upstream and genes colored in *red* are organized onto the same operon with the gene G. The functional relationship with neighboring genes within the neighborhood of [−3, 3] is integrated into the formula to calculate PAC where strand and operon information can be integrated into the Eq. (4) (described in Methods). **b** For a pair of two GO terms, function similarity (GOsim) measures how much detailed functional information (low-level GO terms on a GO graph) is shared. All *dotted ovals* represent GO terms assigned to genes where the +2 gene does not have a GO term assigned to it, such that GOsim(G+2, G) is not available. All *ovals* over the *dotted ovals* represent predecessor GO terms of assigned GO terms to genes excluding root GO terms on a GO graph. The *ovals* lined in *black* mean that corresponding GO terms do not occur, and the *ovals* lined in *blue* mean that corresponding GO terms occur in a set of predecessor GO terms of a given gene G

million bacterial and archael proteins from UniProtKB/Swiss and UniProtKB/TreEMBL [19] and the NCBI Reference Sequence databases [20]. Note that the random annotation may happen to be correct or partially correct by chance. Figure 2a shows histograms of PAC values (which are Bayesian annotation probabilities described in Methods) for *E. coli* and Fig. 2b for *C. thermocellum* for NCBI annotations and simulated random annotations. For the study in Fig. 2, the simplest model was considered where the independence of function similarities within the gene neighborhood was assumed and information for the operon and strand was not integrated. Note that conditional probabilities derived from each genome were applied to the genome, respectively, for the PAC

calculations in Fig. 2. The total number of genes considered in Fig. 2a was 3117 (of 4147 genes), among which 1021 genes had a probability range from 0.95 to 1.00. The distribution of probabilities of the random annotations showed only 49 genes in the probability bin [0.95, 1]. The NCBI annotations with lower PAC values may come from an insufficient number of detectable function similarities with genes in the neighborhood that were derived from the uncovered knowledge of GO annotation and graph structure. We proposed to use a fraction of genes in the probability bin [0.95, 1] as the annotation quality score (AQS) showing distinct differences between NCBI annotation and random annotation. Hence, the NCBI annotation

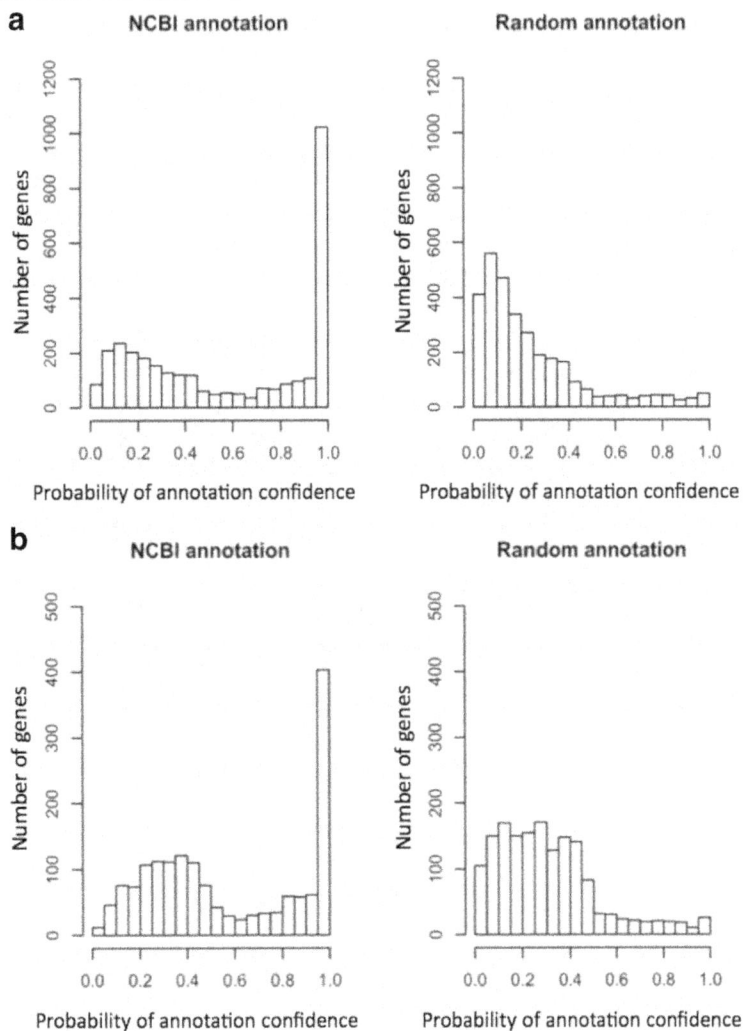

**Fig. 2** Distributions of PAC values of NCBI and random annotations for (**a**) *E. coli* and (**b**) *C. thermocellum*. **a** Using conditional probabilities derived from a given genome and observed gene function similarities, we calculated PAC values for NCBI annotation (assumed to be correct) and random annotation (assumed to be incorrect) for the *E. coli* strain K-12 substrain MG1655. The probability bin [0.95, 1] has 1021 genes for NCBI annotation and 49 genes for random annotation of 3117 genes applicable to PAC calculation. **b** We applied the same methodology to *C. thermocellum*. The probability bin [0.95, 1] contains 403 genes for NCBI annotation and 25 genes for random annotation among 1617 genes applicable to PAC calculation

of *E. coli* has an AQS of 0.33 (= 1021/3117) and the random annotation of *E. coli* has an AQS of 0.016 (= 49/3117). The analogous distributions to *C. thermocellum* were plotted in Fig. 2b, and the AQS for *C. thermocellum* NCBI annotation amounted to 0.24, whereas its random annotation had a similar score to *E. coli*, 0.015. We used *C. thermocellum* as an example of a genome that is evolutionarily distant from *E. coli* and most certainly is more difficult to annotate as comprehensively as *E. coli*. The *C. thermocellum* annotation contained a large number of hypothetical genes (~31% of the genome), as well as genes with annotations not fitted into GO classification (~16%). As a result of those adverse factors, only 1617 genes were applicable to a PAC calculation, such that it is reasonable for the AQS for *C. thermocellum* to be lower than the one for *E. coli*, but the difference is not overwhelmingly huge. Figure 3 provides another important assessment for checking the developed methodology. Figure 3a and b accumulated all collected annotations (correct plus incorrect annotation) for each probability bin. The *x*-axis represents the right-end PAC value (Bayesian annotation probability) for a bin and the *y*-axis represents the fraction of true annotations among annotations collected for the bin. On both plots, our model showed a slight overestimation (points over diagonal) and underestimation (points under diagonal) of the sensitivity. However, the probability bin [0.95, 1] showed sensitivity fairly close to the diagonal. Furthermore, both diagonal plots looked almost identical, suggesting the robust properties of the developed methodology even though the annotation of *C. thermocellum* showed sparse functional annotation compared to *E. coli*.

### Operon structure inclusion into the PAC

So far, we have shown results generated from the simplest model, which used gene function similarities within the gene neighborhood that are assumed to be independent of each other, and clearly distinguished a good quality of annotation from random annotation with the PAC. Yet, a simple integration of the operon structure, which would introduce a separate uncertainty factor in the analysis, could be done by a hybrid system that uses operon-derived conditional probabilities for the genes that are certainly in the same operons and another set of probabilities for the genes that are not. However, in this study, we explored operon structure into PAC by counting only the neighboring genes that are deemed to be on the same operon with a given gene in the formula (4) in Methods. For *E. coli*, inclusion of the operon structure showed rather dramatic changes in the distribution of PAC values in Fig. 4. First, the number of genes with assigned probabilities was reduced significantly because pairs of genes on the same operon were only considered when calculating gene function similarity. The probabilities were assigned only to 1816 genes of 3117 genes in the "no-operon" model. However, there were still 916 genes found in the highly reliable category [0.95, 1] compared to 1021 for the no-operon model (50% of genes for the operon model versus 33% of genes for the no-operon model in the bin [0.95, 1]). The distribution of PAC values in Fig. 4 was much cleaner in a sense that a lower number of genes with PAC values <0.95 were found but still showed a similar shift. However, the distribution for the random annotations had a peak around 0 probability. Summarizing the statement above, Fig. 5 represents the normalized number of genes with PAC

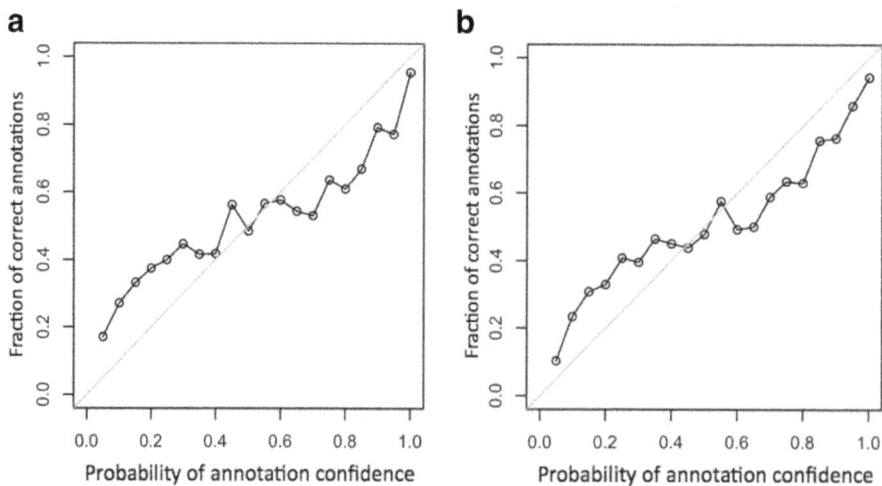

**Fig. 3** Diagonal plots of fractions of correct annotations for (**a**) *E. coli* and (**b**) *C. thermocellum*. The *x*-axis represents the right-end PAC value for a given bin, and the *y*-axis represents a fraction of correct annotations (NCBI annotations) among all annotations (correct and incorrect) collected for the bin. The points over and under the diagonal indicate overestimation and underestimation of fractions of correct annotations, respectively. In general, we observed points fairly close to the diagonal with both plots

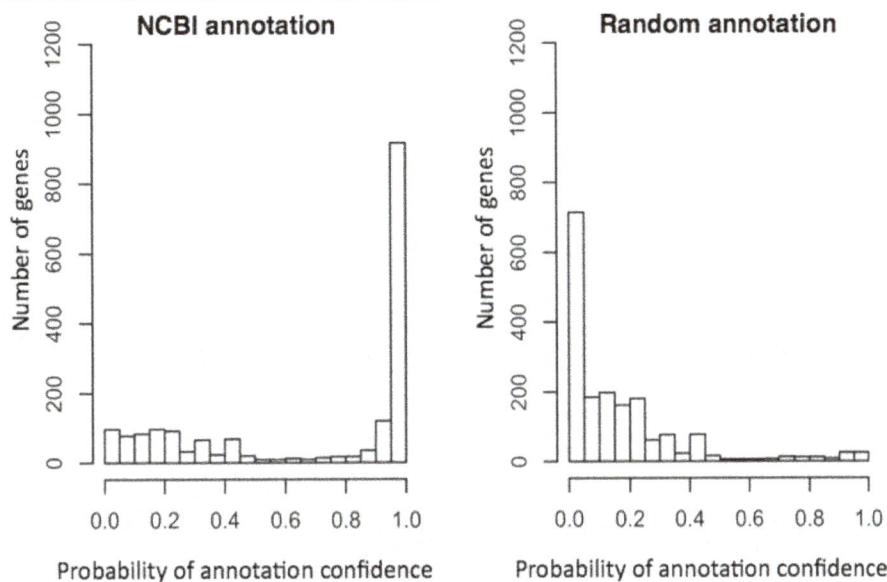

**Fig. 4** Operon structure inclusion into annotation probability with *E. coli*. The predicted operon information of *E. coli* was integrated in PAC values by considering genes on the same operons for NCBI and random annotation

values by the total number of genes applicable to PAC calculation for the no-operon and operon models, respectively. Both plots clearly showed that inclusion of the operon structure into our model contributes to a better distinction between NCBI annotation and random annotation.

### Experiments with gene shuffling

To investigate how our model for annotation validation responds to the increased number of incorrect annotations, we generated annotations with "almost correct functional predictions" through "disturbances by gene shuffling" with NCBI annotation of *E. coli*. In each experiment, we randomly selected *Nr* pairs of genes with

annotations by GO terms and exchanged annotations of the selected pairs where annotations were only used once for shuffling. The shuffling procedure was repeated 100 times for each *Nr*. Figure 6 represents distributions of PAC values of the shuffled annotations where each column shows the average number of genes within a probability bin over 100 repeats and the error bars show 1 standard deviation (SD). Figure 6a was constructed for *Nr* = 100, such that 200 genes likely had the wrong annotations. We did not make any additional check on the shuffling process to determine whether it is possible that the shuffling process would swap close or even identical annotations. The SD was small for all probability bins. For example, the average and SD for the probability bin

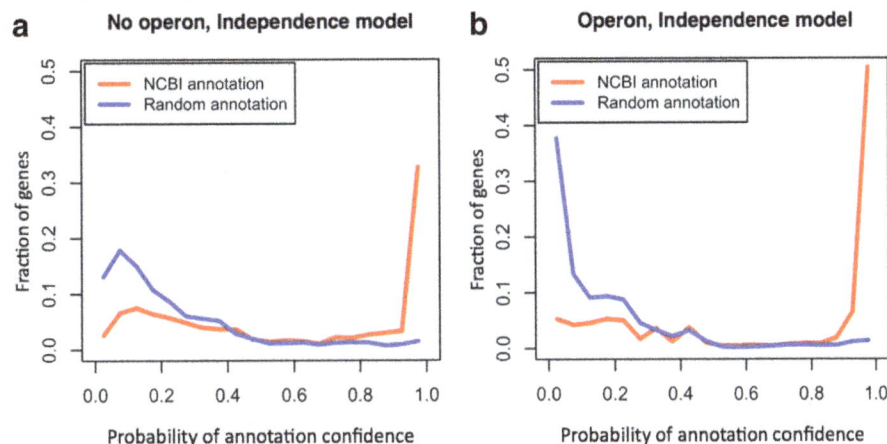

**Fig. 5** Comparison of no-operon and operon models with *E. coli*. The *y*-axis represents the normalized number of genes within a probability bin by the total number of genes applicable for PAC calculation (**a**) without and (**b**) with operon structure inclusion

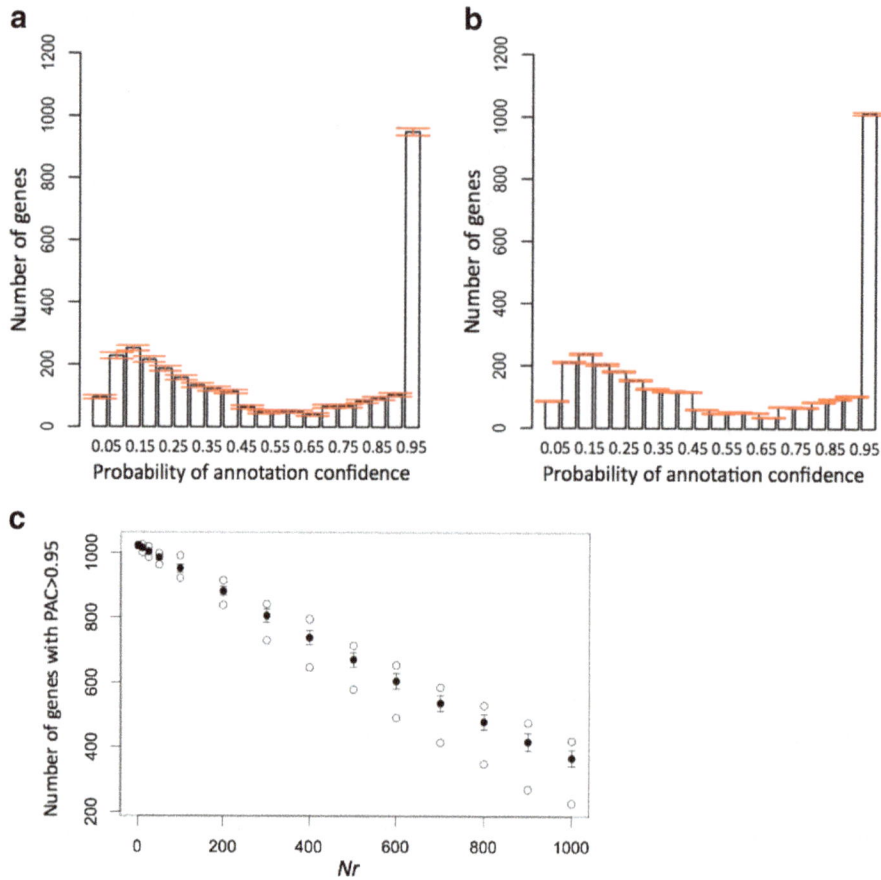

**Fig. 6** Gene shuffling experiments with *E. coli*. **a** Shuffle for *Nr* = 100. **b** Shuffle for *Nr* = 10. **c** Shuffle summary. **a** and **b** The distributions of PAC values were plotted for the shuffled assignments with *E. coli*. In each experiment, the *Nr* pairs of genes with annotations by GO terms were randomly selected and gene annotations in each pair were exchanged. For each *Nr*, the experiment was repeated 100 times, and the plots represent the average number of genes with the SD observed for each probability bin. **c** The average (*black dot*), SD (*vertical lines*), and maximum and minimum (*white dot*) number of genes were presented for the probability bin [0.95, 1.00] for *Nr* = 10, 25, 50, 100, 200, and up to 1000 (shown on the *x*-axis)

[0.95, 1] were 950.5 and 11.6, respectively, which it is about 6 SD away from the value observed for canonical annotation (1021 genes). In Fig. 6b, we observed that our model remains very sensitive to the annotation disturbance of only 20 genes (*Nr* = 10) for the *E. coli* genome composed of >4000 genes. We had 1013.9 on average with 4.4 SD in the bin [0.95, 1], which is still ~2 SD away from the undisturbed annotation (1021 genes). In Fig. 6c, the average (black dot), SD (vertical line), and maximum and minimum (white dot) number of genes for the probability bin [0.95, 1.00] were presented for *Nr* = 10, 25, 50, 100, 200, and up to 1000 (shown on the *x*-axis). Overall, a linear dependency between the number of shuffling, *Nr*, and a decrease in the (average) number of the genes with highly reliable annotations was observed.

## Discussion

Here we discuss possible enhancements and further developments with potential gains in the model performance:

(1) one could explore distance to define neighboring genes as a parameter. For example, one can use basepairs of physical distance along the chromosome as a threshold to define gene neighbors instead of 3 genes upstream and downstream, which is currently used. (2) We treated all genes equally in the current experiments, but in reality the annotations of some genes would be absolutely certain. It would not be difficult to include into our system as another category of genes, "annotation anchors", and then compute a separate set of conditional probabilities of gene function similarities for such genes. (3) We appended another gene neighborhood structure, "strand information", into the Bayesian formula with *E. coli* for which we derived conditional probabilities for a set of genes on the same strand and another set of genes not on the same strand. In the Additional file 1: Figure S1 represents PAC distributions calculated from strand-integrated conditional probabilities for NCBI and random annotations, which showed a slightly better performance than those obtained

from the model without strand information, in a sense that 1042 genes were found in the bin [0.95, 1] for NCBI annotation, whereas 42 genes for random annotation were found in the bin [0.95, 1]. (4) For all results shown, we extracted the conditional probabilities from Eq. (4) in Methods (likelihood in Bayes' rule) derived from a given genome. However, *C. thermocellum* was not annotated by functional terms as much as *E. coli* comprehensively, which led to a much lower number of gene pairs with functional annotations, that might not produce enough data to estimate conditional probabilities (likelihood in Bayes' formula) for probabilistic modeling. To further evaluate robustness toward conditional probabilities, we applied conditional probabilities derived from *E. coli* to calculate the PAC of genes in *C. thermocellum* for NCBI annotation and random annotation. We observed distributions of the PAC values obtained with conditional probabilities derived from *E. coli* similar to those obtained with conditional probabilities derived from *C. thermocellum* in Additional file 1: Figure S2. In the future, we plan to specifically explore this question for a large number of bacterial genomes, yet the result with *C. thermocellum* was very encouraging, even though it is evolutionarily rather distant from *E. coli*. (5) We explored the COG database [21] to annotate genes by functional terms and generated PAC values. Ignoring a poorly characterized functional category, the COG functional terms are organized into three hierarchical levels where the first level consists of three functional classes (Information Storage and Processing, Cellular Processes and Signaling, Metabolism), the finer sub-functional classes (23 functional classes at the second level), and COG terms at the third level. Note that some COG terms belong to more than one functional class. To generate random COG annotation for each protein with an assigned COG term, we assigned a COG term for a protein randomly chosen within the genome to the given protein. The conditional probability of an observation profile given correct and incorrect annotation was calculated for each functional category at the first level where gene COG function similarity takes two values: 0 if two genes share a COG term, and 1 otherwise. In Additional file 1: Figure S3, which represents PAC distributions for NCBI annotation and random annotation with *E. coli*, we obtained an AQS of 0.17 (419/2498 where 2498 proteins were applicable to PAC calculation) for NCBI annotations and an AQS of 0.04 (95/2498) for random annotations that COG annotation showed a less obvious distinction between NCBI and random annotation than GO annotation in the probability bin [0.95, 1]. In the future, we will explore other functional annotation databases including KEGG Orthology [22] and PFAM [23] and compare corresponding PAC distributions for genome annotation validation. (6) So far, we discussed experiments under the "independent" Bayesian model. For example, we approximated the conditional probability of GFSs in the neighborhood as a product of conditional probabilities of individual GFSs within the gene neighborhood. To investigate the influence of the assumption of independence on the AQS, we formulated Bayesian annotation probability under the dependent model, which is described in detail in the Additional files 1 and 2. For the dependent model, we assumed that observations made downstream and upstream depend on only a given gene, and an observation $O_i$ depends on an observation $O_{i+1}$ in the downstream and $O_{i-1}$ in the upstream. The distributions of PAC values under the dependent model for *E. coli* are presented in Additional file 1: Figure S4. Under the dependent model considered in this study, we did not observe any gain in terms of the AQS, which is probably due to the assumption not fitting the biological expectation and not enough data to reliably estimate dependency. The main incentive to use it, in any case, is to avoid overestimation and underestimation of PAC calculation, which was not a problem as shown in Fig. 3.

Currently, we envision three possible application directions for the proposed genome-scale model. First, when the different annotation pipelines annotate the same bacterial genomes, our model should be able to compute a measure of consistency for each annotation pipeline; i.e., AQS, the fraction of the genes with a PAC value >0.95. The workflow with a better score would likely have more correct assignments because our genome-scale probabilistic model sensitively captures the small difference in annotations as shown in the Experiments with gene shuffling section. For example, we compared two *C. thermocellum* genomes annotated at different times where one (called old annotation) was annotated on Feb 14, 2007 at GenBank, and the other genome downloaded from NCBI on May 2013 (called new annotation) was used in this study. The old annotation had 1658 proteins (of 3198 total proteins) annotated with GO terms among which 1582 proteins were applicable to PAC calculation, which resulted in 349 proteins in the bin [0.95, 1] leading to an AQS of 0.22 (= 349/1582). The new annotation had 1671 proteins (of 3173 total proteins) annotated with GO terms applicable to PAC calculation, which resulted in 403 proteins in the bin [0.95, 1] leading to an AQS of 0.24 (= 403/1671). The comparison of *C. thermocellum* genomes annotated at different times may support that our model could be a quantitative tool for genome annotation validation. Second, we plan to measure the annotation consistency for many different bacteria (possibly for 32,000 genomes stored by Land et al. [3]), and such research should provide reasonable estimates of which values are reliable for various branches of the tree of life. Finally, individual PAC values should be valuable for the evaluation of hypothetical protein annotation unless functional inference of hypothetical proteins does not exploit gene neighborhood information as happened in other studies [17, 24].

## Conclusions

Sequencing technologies continue to develop rapidly, and the list of genes with assigned functions is the main product of the sequencing efforts, as it is used to further research. However, there is a lack of methods to evaluate the quality of the obtained functional assignments. We developed a genome-scale probabilistic model that quantitatively measures annotation consistency relying on the well-established property of bacterial genomes; i.e., genes lying in physical adjacency on a chromosome tend to be associated functionally. To our knowledge, this is the first tool that provides both a quality value for the whole set of genes as well as probability of the annotation confidence for individual genes in the set. We have tested our method by simulating large and small "disturbances" of the functional assignments, and the method proved to be sensitive for both cases. The range of potential applications is wide including evaluation and comparison of standard annotation methods for functional assignment. This will lead to more biological insights and more precise cellular models as both use functional assignments as input information.

## Methods
### Data

In this study, the genome-scale probabilistic model was first applied to assess the annotation of two genomes: E. coli str. K-12 substrain MG1655 (NC_000913.faa) and C. thermocellum ATCC 27405 (NC_009012.faa) downloaded from NCBI. The background comparison by random annotation of a genome was performed by randomly picking a protein annotated by functional terms from the protein sequence database. The protein sequence database for random assignments was downloaded from the UniProtKB/Swiss, UniProtKB/TreEMBL [19], and NCBI Reference Sequence [20] databases, which included 8 million bacterial and archeal proteins. The most current version of the same dataset is at least five times as large, but this factor is not important for our particular study.

### GO for functional annotation

To quantitatively assess the annotations, we translated annotations using a controlled vocabulary system, the GO project [10]. The approach to use GO for an evaluation of gene function similarities has been used previously [11, 25], but to our knowledge it has not been used for comprehensive evaluation of genome annotation quality. The GO project describes the ontology of defined GO terms representing gene product properties structured as a directed acyclic graph. The directed graph can be retrieved from "gene_ontology.1_2.obo.txt"

[26] which contains GO terms annotated by both the experimental and computational evidence codes. The directed GO graph covers biological process, molecular function, and cellular component, which are mutually exclusive domains each represented by the root GO terms separately. The directed relationships between GO terms represent either "is-a", "part of", or "regulates" where child terms are more specialized and parent terms are less specialized. Some GO terms may have more than one parent term unlike a hierarchy. In this work, we considered directed edges, which represent only the "is-a" subclass relationship. The UniProt Gene Ontology Annotation (UniProt-GOA) database provides high-quality GO annotations to proteins through the UniProt Knowledgebase. To annotate NCBI annotations by GO terms, we first assigned NCBI GI numbers to the UniprotKB identifier using "idmapping.dat" [26], and then assigned a UniprotKB identifier into GO terms using "gene_association.goa_uniprot" [27]. Note that the mapping between NCBI GI numbers and UniprotKB identifiers is not one-to-one, and some NCBI GI numbers are not mapped into a UniprotKB identifier.

### Gene function similarity

We introduced GO similarity to compare quantitatively functional annotations described by GO terms. To calculate functional similarity between two GO terms ($GO_1$, $GO_2$), we first identified a set of all predecessor GO terms of $GO_1$ ($GO_2$) on the directed GO graph including $GO_1$ ($GO_2$) but excluding the root, denoted by $S_1$ ($S_2$), respectively. Then, the similarity between two GO terms was defined based on overlapping GO terms between sets $S_1$ and $S_2$ as follows:

$$GOsim(GO_1, GO_2) = \frac{|S_1 \cap S_2|}{|S_1 \cup S_2|} \tag{1}$$

where $|S_1 \cap S_2|$ and $|S_1 \cup S_2|$ are the cardinalities of an intersection and the union of $S_1$ and $S_2$, respectively. The normalized GO similarity, which falls in the range of 0 to 1, implicitly measures more than just the detailed functions (low-level GO terms) that are shared. For instance, in Fig. 1b, all dotted ovals represent GO terms assigned to genes where the +2 gene does not have a GO term assigned to it, such that $GOsim(G_{+2}, G)$ is not available. All ovals over the dotted ovals represent predecessor GO terms of assigned GO terms to genes excluding the root GO term on a directed GO graph. The ovals lined in black mean that corresponding GO terms do not occur, and the ovals lined in blue mean that corresponding GO terms occur in a set of predecessor GO terms of a gene G. Therefore, GO similarities between neighboring genes and gene G are as follows: $GOsim(G_{-3}, G) = 0$, $GOsim(G_{-2}, G) = 1/6$, $GOsim(G_{-1}, G) = 1$,

GOsim($G_{+1}$, G) = 0, GOsim($G_{+3}$, G) = 1/4. However, genes can be annotated with more than one GO term because proteins can have multiple functional roles. Let's say that gene $G_1$ is annotated with $A_1$ = {$GO_i$ |$i$ = 1,..,M} and gene $G_2$ with $A_2$ = {$GO_j$ |$j$ = 1,..,N}, the GFS between genes $G_1$ and $G_2$ is defined as the maximum among GO similarities between two GO terms from different genes:

$$GFS(G_1, G_2) = \max_{\substack{1 \le i \le M \\ 1 \le j \le N}} GOsim(GO_i, GO_j)$$

(2)

where $GO_i$ is from gene $G_1$ and $GO_j$ is from gene $G_2$. The maximum of GO similarities takes into account different numbers of GO terms assigned to different proteins. We calculated the GFS associated with each biological process, molecular function, and cellular component separately.

### Gene neighborhood structure

In this study, we explored three different gene neighborhood structures: gene order on a chromosome, operon structure, and strand information. The strand information of genes was retrieved through the NCBI Entrez Programming Utilities. For the predicted operon structure of *E. coli*, we used the Database of Prokaryotic Operons [28]. For each gene G and each functional category (biological process, molecular function, and cellular component) in a given genome, we calculated GFS(G, $G_i$) between G and its neighbor gene $G_i$ at $i$th neighborhood, $i$ = −3, −2, −1, +1, +2, +3, where the minus and plus signs represent upstream and downstream neighborhoods (Fig. 1a).

### Deriving PAC through Bayes' rule

Here we derived the probability that annotation of a gene G is correct in given observations {$O_i$| $i$ = −3,...,+3} with neighbor genes $G_i$, $i$ = −3,...,+3 (called an observation profile), under the assumption that observations are independent of each other within the gene neighborhood. First, we calculated conditional probability (likelihood in Bayes' rule) that an observation $O_i$ is observed at the $i$th neighborhood given the correct annotation, denoted by $Pr(O_i|A_c)$, where $A_c$ represents correct annotation, for which NCBI annotation and corresponding functional annotation by GO terms were all assumed to be correct. Then, we calculated the probability that an observation $O_i$ is observed at the $i$th neighborhood given the incorrect annotation, denoted by $Pr(O_i|A_{inc})$, where $A_{inc}$ represents incorrect annotation, for which we generated an annotation for each protein with assigned GO terms by randomly drawing a protein with assigned GO terms from the database of 8 million proteins, and then assigning the GO terms of the randomly drawn protein

to the given protein. For each protein, we calculated gene function similarity with gene neighbors using the given gene's random annotation, leading to $Pr(O_i|A_{inc})$. If we formulate conditional probabilities using gene function similarity, then a random variable $O_i$ takes $GFS_i$, where $GFS_i$ represents gene function similarity between genes separated by ($i$ - 1) genes on a chromosome. The use of combinatorial information of gene neighborhood structures can be easily integrated into the formula. Based on Bayes' rule along with the assumption of independence of neighbor observations, the probability that an annotation is correct given an observation profile is described as follows:

$$Pr(A_c, |O_i, i = -3, \cdots, +3)$$

$$= \frac{Pr(O_i, i = -3, \cdots, +3, |A_c) \, Pr(A_c)}{Pr(O_i, i = -3, \cdots, +3, |A_c) \, Pr(A_c) + Pr(O_i, i = -3, \cdots, +3, |A_{inc}) \, Pr(A_{inc})}$$

$$= \frac{\prod_{i=-3}^{i=+3} Pr(O_i, |A_c) \, Pr(A_c)}{\prod_{i=-3}^{i=+3} Pr(O_i, |A_c) \, Pr(A_c) + \prod_{i=-3}^{i=+3} Pr(O_i, |A_{inc}) \, Pr(A_{inc})},$$

(3)

where $Pr(A_c)$ and $Pr(A_{inc})$ are prior probabilities of correct and incorrect annotations respectively, which were set to 0.5 in this study. By considering all three functional categories concurrently, the Bayesian annotation probability (called the PAC in this study) is described as follows:

$$Pr(A_c|O_i^{BP}, O_i^{MF}, O_i^{CC}, i = -3, \cdots, +3)$$

$$= \frac{Pr(O_i^{BP}, O_i^{MF}, O_i^{CC}, i = -3, \cdots, +3|A_c) \, Pr(A_c)}{Pr(O_i^{BP}, O_i^{MF}, O_i^{CC}, i = -3, \cdots, +3)}$$

$$= \frac{\prod_{i=-3}^{i=+3} \prod_{j=BP}^{CC} Pr(O_i^j|A_c) \, Pr(A_c)}{\prod_{i=-3}^{i=+3} \prod_{j=BP}^{CC} Pr(O_i^j|A_c) \, Pr(A_c) + \prod_{i=-3}^{i=+3} \prod_{j=BP}^{CC} Pr(O_i^j|A_{inc}) \, Pr(A_{inc})}$$

(4)

where BP indicates biological process; MF, molecular function; and CC, cellular component. For example, if a random variable $O_i$ takes a two-dimensional vector of gene function similarity and strand information for each category, then Bayesian annotation probability in the formula (1) is derived from an 18-dimensional observation vector. In most cases, we do not have all neighbor genes with assigned GO terms for all categories. The nonexistent information elements are silently ignored in the formula (4) under the assumption that non-existent information occurs equally in correct annotation and incorrect annotation.

### Filtering abundant GO terms

The GFS is affected by GO terms with an abundant occurrence due to their general functional description;

for example, GO:0016020, which describes a membrane in a category of the cellular component. Therefore, the GO terms with high frequency can cause random pairs of genes that are not neighbors on a chromosome to share functions, eventually yielding high Bayesian annotation probability. In the Additional file 1; Figure S5 represents the frequency of GO terms in a percentage of proteins with assigned GO terms in the protein sequence database. To avoid false causality with Bayesian annotation probability, we filtered out GO terms whose frequencies were >5%. For 10,000 random protein pairs with assigned GO terms in the protein sequence database, Additional file 1: Figure S6A represents histograms of GFS values before filtering abundant GO terms and Additional file 1: Figure S6B shows GFS values after filtering abundant GO terms with a frequency > 5% in each functional category. In the Additional file 1: Table S1 lists GO terms that were filtered out with a functional description and a 5% of frequency cutoff. All results shown in our study were derived after filtering GO terms with a 5% of frequency cutoff.

## Abbreviations

AQS: Annotation Quality Score; GFS: Gene function similarity; GO: Gene Ontology; PAC: Probability of Annotation Confidence

## Acknowledgements

This manuscript was edited by the Office of Grants and Scientific Publications at the University of Arkansas for Medical Sciences. This work was supported by the Plant–Microbe Interfaces Scientific Focus Area in the Genomic Science Program, United States Department of Energy, Office of Science, Biological and Environmental Research. Oak Ridge National Laboratory is managed by UTBattelle, LLC, for the United States Department of Energy under Contract DEAC05-00OR22725.

## Funding

No funding was obtained for this study.

## Authors' contributions

SJ and AG conceived the project, designed the study, participated in method design, and drafted the manuscript. SJ wrote the program. IN and LH participated in method design, data analysis, and writing. All authors suggested ideas for additional validations that were not included into this publication, and participated in discussions. All authors read and approved the final manuscript.

## Competing interests

The authors declare that they have no competing interests.

## Author details

¹Department of Biomedical Informatics, College of Medicine, University of Arkansas for Medical Sciences, Little Rock, AR 72205, USA. ²Comparative Genomics Group, Biosciences Division, Oak Ridge National Laboratory, Oak Ridge, TN 37831, USA. ³Computer Science and Mathematics Division, Oak Ridge National Laboratory, Oak Ridge, TN 37831, USA.

## References

1. Kodama Y, Shumway M, Leinonen R. INSD. The sequence read archive: explosive growth of sequencing data. Nucleic Acids Res. 2012;40:D54–6.
2. Leggett RM, Ramirez-Gonzalez RH, Clavijo BJ, Waite D, Davey RP. Sequencing quality assessment tools to enable data-driven informatics for high throughput genomics. Front Genet. 2013;4:288.
3. Land ML, Hyatt D, Jun S-R, Kora GH, Hauser LJ, Lukjancenko O, Ussery DW. Quality scores for 32,000 genomes. Stand Genomic. 2014;9:20.
4. Médigue C, Moszer I. Annotation, comparison and databases for hundreds of bacterial genomes. Res Microbiol. 2007;158:724–36.
5. Monk JM, Charusanti P, Aziz RK, Lerman JA, Premyodhin N, Orth JD, Feist AM, Palsson BO. Genome-scale metabolic reconstructions of multiple Escherichia Coli strains highlight strain-specific adaptations to nutritional environments. Proc Natl Acad Sci U S A. 2013;110:20338–43.
6. Caspi R, Altman T, Billington R, Dreher K, Foerster H, Fulcher CA, Holland TA, Keseler IM, Kothari A, Kubo A, et al. The MetaCyc database of metabolic pathways and enzymes and the BioCyc collection of pathway/genome databases. Nucleic Acids Res. 2014;42:D459–71.
7. White O, Kyrpides N. Meeting report. Towards a critical assessment of functional annotation experiment (CAFAE) for bacterial genome annotation. Stand Genomic. 2010;3:240–2.
8. Nelson BK. WRAPS: a system for determining the probability of prokaryotic protein annotation correctness (dissertation, University of Nebraska at Omaha, Department of Computer Science). 2013.
9. Loevenich SN, Brunner E, King NL, Deutsch EW, Stein SE, FlyBase C, Aebersold R, Hafen E, Gelbart W, Bitsoi L, et al. The Drosophila Melanogaster PeptideAtlas facilitates the use of peptide data for improved fly proteomics and genome annotation. BMC Bioinformatics. 2009;10:59.
10. Ashburner M, Ball CA, Blake JA, Botstein D, Butler H, Cherry JM, Davis AP, Dolinski K, Dwight SS, Eppig JT, et al. Gene ontology: tool for the unification of biology. The Gene Ontology Consortium Nat Genet. 2000;25:25–9.
11. Skunca N, Altenhoff A, Dessimoz C. Quality of computationally inferred gene ontology annotations. PLoS Comp Biol. 2012;8:e1002533.
12. Bell MJ, Gillespie CS, Swan D, Lord P. An approach to describing and analysing bulk biological annotation quality: a case study using UniProtKB. Bioinformatics. 2012;28:i562–8.
13. Eilbeck K, Moore B, Holt C, Yandell M. Quantitative measures for the management and comparison of annotated genomes. BMC Bioinformatics. 2009;10:67.
14. Bakke P, Carney N, Deloache W, Gearing M, Ingvorsen K, Lotz M, McNair J, Penumetcha P, Simpson S, Voss L, et al. Evaluation of three automated genome annotations for Halorhabdus utahensis. PLoS One. 2009;4:e6291.
15. Tamames J, Casari G, Ouzounis C, Valencia A. Conserved clusters of functionally related genes in two bacterial genomes. J Mol Evol. 1997;44:66–73.
16. Rogozin IB, Makarova KS, Wolf YI, Koonin EV. Computational approaches for the analysis of gene neighbourhoods in prokaryotic genomes. Brief Bioinform. 2004;5:131–49.
17. Yin Y, Zhang H, Olman V, Xu Y. Genomic arrangement of bacterial operons is constrained by biological pathways encoded in the genome. Proc Natl Acad Sci U S A. 2010;107:6310–5.
18. Yelton AP, Thomas BC, Simmons SL, Wilmes P, Zemla A, Thelen MP, Justice N, Banfield JF. A semi-quantitative, synteny-based method to improve functional predictions for hypothetical and poorly annotated bacterial and archaeal genes. PLoS Comput Biol. 2011;7:e1002230.
19. UniProt C, Apweiler R, Bateman A, Martin MJ, O'Donovan C, Magrane M, Alam-Faruque Y, Alpi E, Antunes R, Arganiska J, et al. Activities at the universal protein resource (UniProt). Nucleic Acids Res. 2014;42:D191–8.
20. Tatusova T, Ciufo S, Fedorov B, O'Neill K, Tolstoy I. RefSeq microbial genomes database: new representation and annotation strategy. Nucleic Acids Res. 2014;42:D553–9.
21. Tatusov RL, Fedorova ND, Jackson JD, Jacobs AR, Kiryutin B, Koonin EV, Krylov DM, Mazumder R, Mekhedov SL, Nikolskaya AN, et al. The COG database: an updated version includes eukaryotes. BMC Bioinformatics. 2003;4:41.
22. Kanehisa M, Goto S, Sato Y, Kawashima M, Furumichi M, Tanabe M. Data, information, knowledge and principle: back to metabolism in KEGG. Nucleic Acids Res. 2014;42:D199–205.
23. Finn RD, Bateman A, Clements J, Coggill P, Eberhardt RY, Eddy SR, Heger A, Hetherington K, Holm L, Mistry J, et al. Pfam: the protein families database. Nucleic Acids Res. 2014;42:D222–30.
24. Zhao S, Kumar R, Sakai A, Vetting MW, Wood BM, Brown S, Bonanno JB,

Hillerich BS, Seidel RD, Babbitt PC, et al. Discovery of new enzymes and metabolic pathways by using structure and genome context. Nature. 2013; 502:698–702.

25. Xu T, Du L, Zhou Y. Evaluation of GO-based functional similarity measures using S. Cerevisiae protein interaction and expression profile data. BMC Bioinformatics. 2008;9:472.

26. The UniProt Consortium. http://www.uniprot.org. Accessed 30 May 2013.

27. The Gene Ontology Consortium. http://www.geneontology.org. Accessed 30 May 2013.

28. Dam P, Olman V, Harris K, Su Z, Xu Y. Operon prediction using both genome-specific and general genomic information. Nucleic Acids Res. 2007;35:288–98.

# 16

# Intervene: a tool for intersection and visualization of multiple gene or genomic region sets

Aziz Khan[1*] and Anthony Mathelier[1,2*]

## Abstract

**Background:** A common task for scientists relies on comparing lists of genes or genomic regions derived from high-throughput sequencing experiments. While several tools exist to intersect and visualize sets of genes, similar tools dedicated to the visualization of genomic region sets are currently limited.

**Results:** To address this gap, we have developed the Intervene tool, which provides an easy and automated interface for the effective intersection and visualization of genomic region or list sets, thus facilitating their analysis and interpretation. Intervene contains three modules: *venn* to generate Venn diagrams of up to six sets, *upset* to generate UpSet plots of multiple sets, and *pairwise* to compute and visualize intersections of multiple sets as clustered heat maps. Intervene, and its interactive web ShinyApp companion, generate publication-quality figures for the interpretation of genomic region and list sets.

**Conclusions:** Intervene and its web application companion provide an easy command line and an interactive web interface to compute intersections of multiple genomic and list sets. They have the capacity to plot intersections using easy-to-interpret visual approaches. Intervene is developed and designed to meet the needs of both computer scientists and biologists. The source code is freely available at https://bitbucket.org/CBGR/intervene, with the web application available at https://asntech.shinyapps.io/intervene.

**Keywords:** Visualization, Venn diagrams, UpSet plots, Heat maps, Genome analysis

## Background

Effective visualization of transcriptomic, genomic, and epigenomic data generated by next-generation sequencing-based high-throughput assays have become an area of great interest. Most of the data sets generated by such assays are lists of genes or variants, and genomic region sets. The genomic region sets represent genomic locations for specific features, such as transcription factor – DNA interactions, transcription start sites, histone modifications, and DNase hypersensitivity sites. A common task in the interpretation of these features is to find similarities, differences, and enrichments between such sets, which come from different samples, experimental conditions, or cell and tissue types.

Classically, the intersection or overlap between different sets, such as gene lists, is represented by Venn diagrams [1] or Edwards-Venn [2]. If the number of sets exceeds four, such diagrams become complex and difficult to interpret. The key challenge is that there are $2^n$ combinations to visually represent when considering $n$ sets. An alternative approach, the UpSet plots, was introduced to depict the intersection of more than three sets [3]. The advantage of UpSet plots is their capacity to rank the intersections and alternatively hide combinations without intersection, which is not possible using a Venn diagram. However, with a large number of sets, UpSet plots become an ineffective way of illustrating set intersections. To visualize a large number of sets, one can represent pairwise intersections using a clustered heat map as suggested in [4].

There are several web applications and R packages available to compute intersection and visualization of up-to six list sets by using Venn diagrams. Although tools exist to perform genomic region set intersections [5–7], there is a

* Correspondence: aziz.khan@ncmm.uio.no; anthony.mathelier@ncmm.uio.no
[1]Centre for Molecular Medicine Norway (NCMM), Nordic EMBL Partnership, University of Oslo, 0318 Oslo, Norway
Full list of author information is available at the end of the article

limited number of tools available to visualize them [5, 6]. To our knowledge no tool exists to generate UpSet plots for genomic region sets. Consequently, there is a great need for integrative tools to compute and visualize intersection of multiple sets of both genomic regions and gene/list sets.

To address this need, we developed Intervene, an easy-to-use command line tool to compute and visualize intersections of genomic regions with Venn diagrams, UpSet plots, or clustered heat maps. Moreover, we provide an interactive web application companion to upload list sets or the output of Intervene to further customize plots.

## Implementation

Intervene comes as a command line tool, along with an interactive Shiny web application to customize the visual representation of intersections. The command line tool is implemented in Python (version 2.7) and R programming language (version 3.3.2). The build also works with Python versions 3.4, 3.5, and 3.6. The accompanying web interface is developed using Shiny (version 1.0.0), a web application framework for R. Intervene uses pybedtools [6] to perform genomic region set intersections and Seaborn (https://seaborn.pydata.org/), Matplotlib [7], UpSetR [8], and Corrplot [9] to generate figures. The web application uses the R package Venerable [10] for different types of Venn diagrams, UpSetR for UpSet plots, and heatmap.2 and Corrplot for pairwise intersection clustered heat maps. The UpSet module of the web ShinyApp was derived from the UpSetR [8] ShinyApp, which was extended by adding more options and features to customize the UpSet plots.

Intervene can be installed by using *pip install intervene* or using the source code available on bitbucket https://bitbucket.org/CBGR/intervene. The tool has been tested on Linux and MAC systems. The Shiny web application is hosted with shinyapps.io by RStudio, and is compatible with all modern web browsers. A detailed documentation including installation instructions and how to use the tool is provided in Additional file 1 and is available at http://intervene.readthedocs.io.

## Results

### An integrated tool for effective visualization of multiple set intersections

As visualization of sets and their intersections is becoming more and more challenging due to the increasing number of generated data sets, there is a strong need to have an integrated tool to compute and visualize intersections effectively. To address this challenge, we have developed Intervene, which is composed of three different modules, accessible through the subcommands *venn*, *upset*, and *pairwise*. Intervene accepts two types of input files: genomic regions in BED, GFF, or VCF format and gene/name lists in plain text format. A detailed sketch of Intervene's

command line interface and web application utility with types of inputs is provided in Fig. 1.

Intervene provides flexibility to the user to choose figure colors, label text, size, resolution, and type to make them publication-standard quality. To read the help about any module, the user can type *intervene < subcommand > --help* on the command line. Furthermore, Intervene produces results as text files, which can be easily imported to the web application for interactive visualization and customization of plots (see "An interactive web application" section).

### Venn diagrams module

Venn diagrams are the classical approach to show intersections between sets. There are several web-based applications and R packages available to visualize intersections of up-to six list sets in classical Venn, Euler, or Edward's diagrams [11–16]. However, a very limited number of tools are available to visualize genomic region intersections using classical Venn diagrams [5, 6].

Intervene provides up-to six-way classical Venn diagrams for gene lists or genomic region sets. The associated web interface can also be used to compute the intersection of multiple gene sets, and visualize it using different flavors of weighted and unweighted Venn and Euler diagrams. These different types include: classical Venn diagrams (up-to five sets), Chow-Ruskey (up-to five sets), Edwards' diagrams (up-to five sets), and Battle (up-to nine sets).

As an example, one might be interested to calculate the number of overlapping ChIP-seq (chromatin immunoprecipitation followed by sequencing) peaks between different types of histone modification marks (H3K27ac, H3K4me3, and H3K27me3) in human embryonic stem cells (hESC) [17] (Fig. 2a, can be generated with the command *intervene venn --test*).

### UpSet plots module

When the number of sets exceeds four, Venn diagrams become difficult to read and interpret. An alternative and more effective approach is to use UpSet plots to visualize the intersections. An R package with a ShinyApp (https://gehlenborglab.shinyapps.io/upsetr/) and an interactive web-based tool are available at http://vcg.github.io/upset to visualize multiple list sets. However, to our knowledge, there is no tool available to draw the UpSet plots for genomic region set intersections. Intervene's *upset* subcommand can be used to visualize the intersection of multiple genomic region sets using UpSet plots.

As an example, we show the intersections of ChIP-seq peaks for histone modifications (H3K27ac, H3K4me3, H3K27me3, and H3K4me2) in hESC using an UpSet plot, where interactions were ranked by frequency (Fig. 2b, can be generated with the command *intervene upset --test).

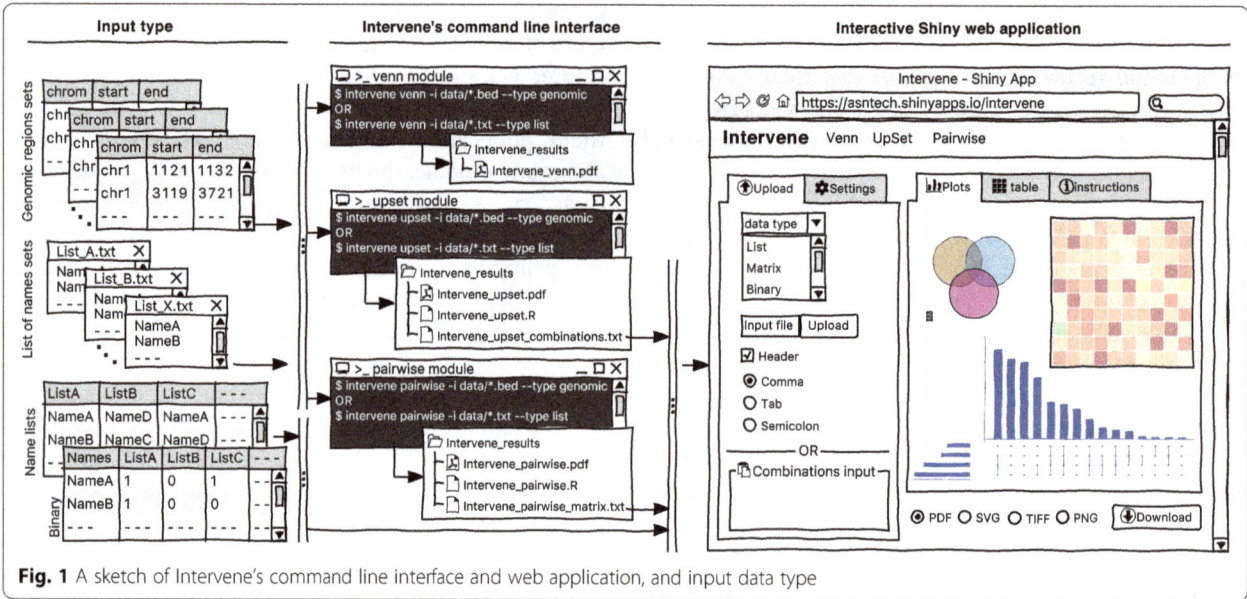

**Fig. 1** A sketch of Intervene's command line interface and web application, and input data type

This plot is easier to understand than the four-way Venn diagram (Additional file 1).

## Pairwise intersection heat maps module

With an increasing number of data sets, visualizing all possible intersections becomes unfeasible by using Venn diagrams or UpSet plots. One possibility is to compute pairwise intersections and plot-associated metrics as a clustered heat map. Intervene's *pairwise* module provides several metrics to assess intersections, including number of overlaps, fraction of overlap, Jaccard statistics, Fisher's exact test, and distribution of relative distances. Moreover, the user can choose from different styles of heat maps and clustering approaches.

As an example, we obtained the genomic regions of super enhancers in 24 mouse cell type and tissues from dbSUPER [18] and computed the pairwise intersections in terms of Jaccard statistics (Fig. 2c). The triangular heat map shows

the pairwise Jaccard index, which is between 0 and 1, where 0 means no overlap and 1 means full overlap. The bar plot shows the number of regions in each cell-type or tissue. This plot can be generated using the command *intervene pairwise –test*).

## An interactive web application

Intervene comes with a web application companion to further explore and filter the results in an interactive way. Indeed, intersections between large data sets can be computed locally using Intervene's command line interface, then the output files can be uploaded to the ShinyApp for further exploration and customization of the figures (Fig. 1).

The ShinyApp web interface takes four types of inputs: (i) a text/csv file where each column represents a set, (ii) a binary representation of intersections, (iii) a pairwise matrix of intersections, and (iv) a matrix of overlap counts. The web application provides several easy and

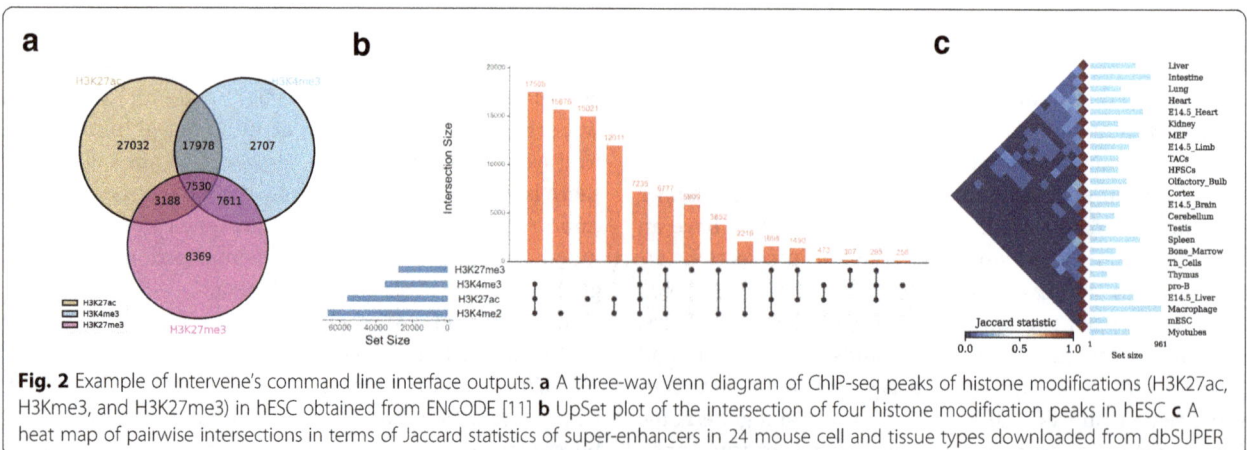

**Fig. 2** Example of Intervene's command line interface outputs. **a** A three-way Venn diagram of ChIP-seq peaks of histone modifications (H3K27ac, H3Kme3, and H3K27me3) in hESC obtained from ENCODE [11] **b** UpSet plot of the intersection of four histone modification peaks in hESC **c** A heat map of pairwise intersections in terms of Jaccard statistics of super-enhancers in 24 mouse cell and tissue types downloaded from dbSUPER

intuitive customization options for responsive adjustments of the figures (Figs. 1 and 3). Users can change colors, fonts and plot sizes, change labels, and select and deselect specific sets. These customized and publication-ready figures can be downloaded in PDF, SVG, TIFF, and PNG formats. The pairwise modules also provides three types of correlation coefficients and hierarchical clustering with eight clustering methods and four distance measurement methods. It further provides interactive features to explore data values; this is done by hovering the mouse cursor over each heat map cell, or by using a searchable and sortable data table. The data table can be downloaded as a CSV file and interactive heat maps can be downloaded as HTML. The Shiny-based web application is freely available at https://asntech.shinyapps.io/intervene.

## Case study: highlighting co-binding factors in the MCF-7 cell line

Transcription factors (TFs) are key proteins regulating transcription through their cooperative binding to the DNA [19, 20]. To highlight Intervene's capabilities, we used the command-line tool and its ShinyApp companion to predict and visualize cooperative interactions between TFs at cis-regulatory regions in the MCF-7 breast cancer cell line. Specifically, we considered (i) TF binding regions derived from uniformly processed TF ChIP-seq experiments compiled in the ReMap database [21] and (ii) promoter and enhancer regions predicted by chromHMM [22] from histone modifications and regulatory factors ChIP-seq [23]. The pairwise module of Intervene was used to compute the fraction of overlap between all pairs of ChIP-seq data sets and regulatory regions. The output

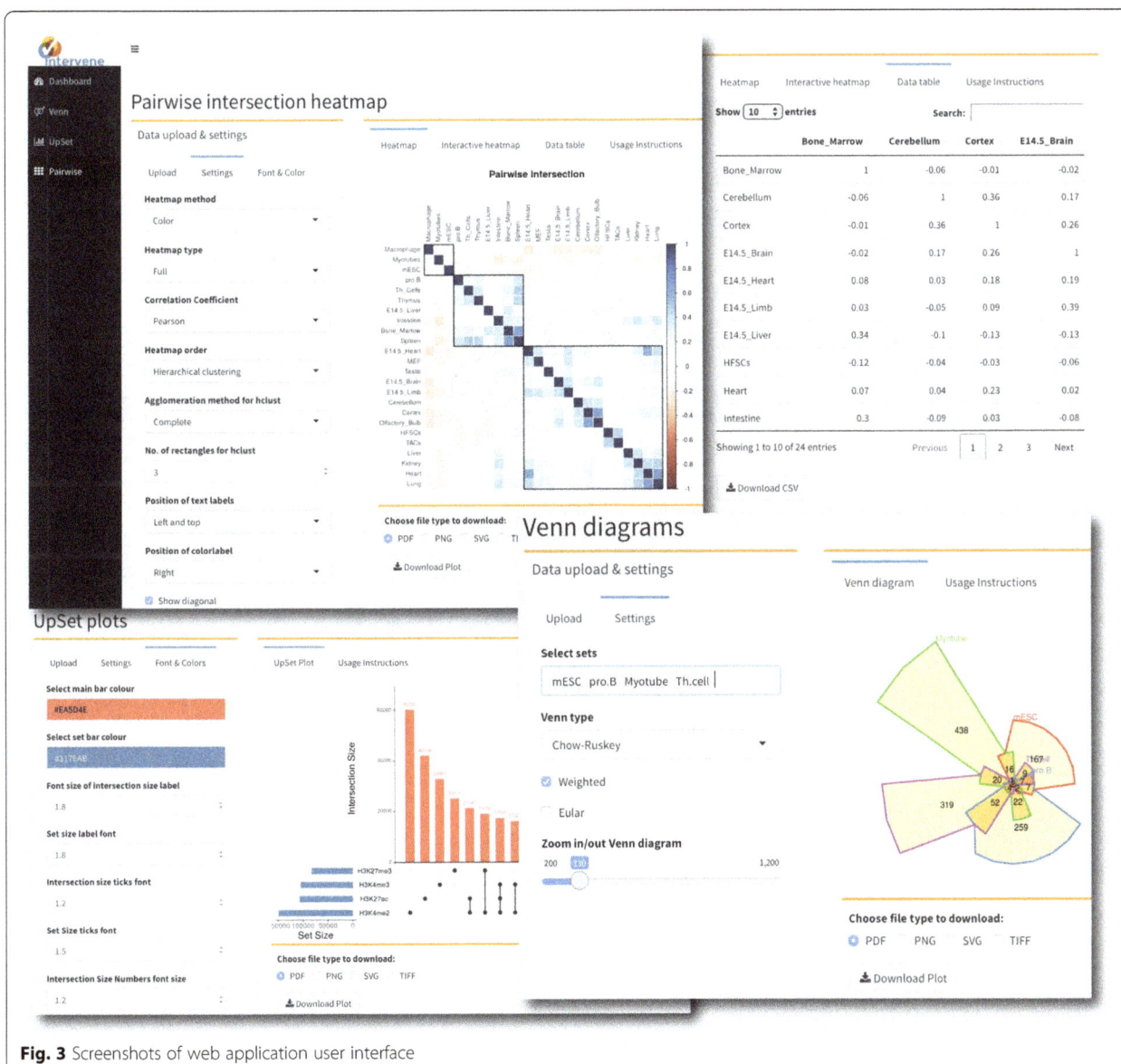

**Fig. 3** Screenshots of web application user interface

matrix was provided to the ShinyApp to compute Spearman correlations of the computed values and to generate the corresponding clustering heat map (default parameters; Fig. 4). The largest cluster (green cluster) was composed of the three key cooperative TFs involved in oestrogen-positive breast cancers: ESR1, FOXA1, and GATA3. They were clustered with enhancer regions where they have been shown to interact [24]. The cluster highlights potential TF cooperators: ARNT, AHR, GREB1, and TLE3. Promoter regions were found in the second largest cluster (red cluster), along with CTCF, STAG1, and RAD21, which are known to orchestrate chromatin architecture in human cells [25]. The last cluster was principally composed by TFAP2C data sets. Taken together, Intervene visually highlighted the cooperation of different sets TFs at MCF-7 promoters and enhancers, in agreement with the literature.

## Discussion

A comparative analysis of different tools to compute and visualize intersections as Venn diagrams, UpSet plots, and pairwise heat maps is provided in Table 1. Most of

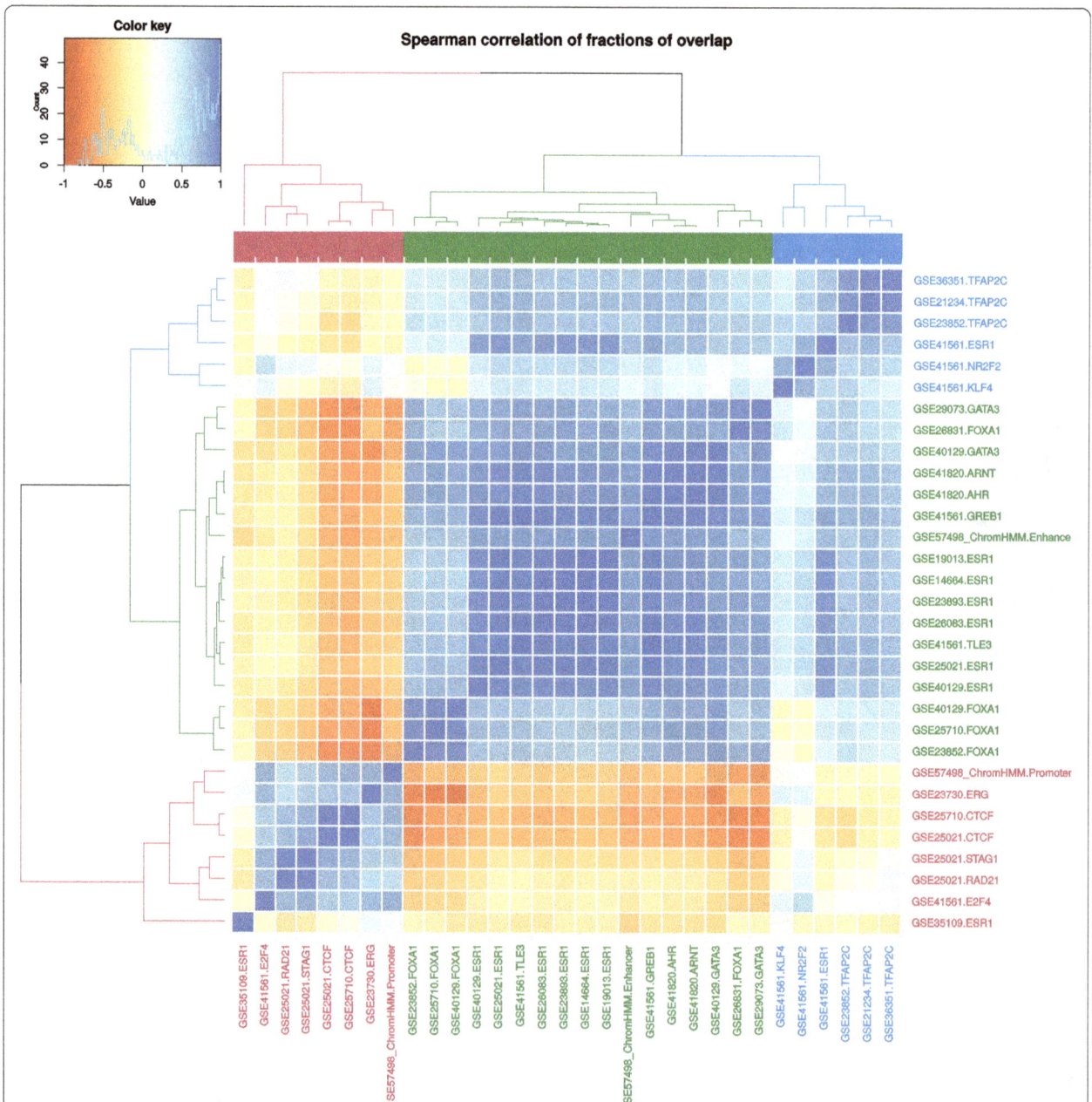

**Fig. 4** MCF-7 cluster heat map. Cluster heat map of the Spearman correlations of fractions of overlap between TF ChIP-seq data sets and regulatory regions in MCF-7. Three clusters (*red*, *green*, and *blue*) are highlighted

**Table 1** Comparison of Intervene with currently available tools to draw Venn diagrams, UpSet plots and pairwise heatmaps

| Application | Venn plot types | Upset plot | Pairwise heat map | Weighted venn | Application type | Input type | No. Of inputs | Output type |
|---|---|---|---|---|---|---|---|---|
| VennDiagramWeb [12] | Classical Venn, Euler | ✗ | ✗ | ✗ | web app | Lists | 5 | TIFF, SVG, PNG, R objects |
| VennPainter [14] | Classical Venn, Edwards, Nested Venn | ✗ | ✗ | ✗ | Stand-alone | Lists | 8 | SVG, text |
| Vennture [15] | Edwards | ✗ | ✗ | ✗ | Stand-alone | Lists | 6 | PowerPoint, Excel |
| BioVenn [11] | Classical Venn | ✗ | ✗ | ✓ | web app | Lists | 3 | SVG, PNG |
| jVenn [13] | Classical Venn, Edwards | ✗ | ✗ | ✗ | web app | Lists | 6 | PNG and SVG |
| InteractiVenn [16] | Edwards | ✗ | ✗ | ✗ | web app | Lists | 6 | SVG, PNG, text |
| UpSetR [3, 8] | ✗ | ✓ | ✗ | ✗ | web app, R package | Lists, binary, counts | Multiple | PDF, PNG |
| ChippeakAnno [5] | Classical Venn | ✗ | ✗ | ✗ | R package | Genomic regions | 5 | PDF, SVG, PNG |
| pybedtools [6] | Classical Venn | ✗ | Matrix only | ✓ | command line | Genomic regions | 3 for Venn, multiple for pairwise | PDF, SVG, PNG |
| Intervene | Classical Venn, Euler, Edwards, Chow-Ruskey, Square, Battle | ✓ | ✓ | ✓ | command line, web app | Genomic regions, lists, binary, counts | 6 for Venn, multiple for upset and pairwise | PDF, SVG, PNG, TIFF, R objects, text |

the tools available currently can only draw Venn diagrams for up-to six list sets. Intervene provides Venn diagrams, UpSet plots, and pairwise heat maps for both list sets and genomic region sets. To the best of our knowledge, it is the only tool available to draw UpSet plots for the intersections of genomic region sets. Intervene is the first of its kind to allow for the computation and visualization of intersections between multiple genomic region and list sets with three different approaches.

In the near future, Intervene will be integrated to the Galaxy Tool Shed to be easily installed to any Galaxy instance with one click. We plan to develop a dedicated web application allowing users to upload genomic region sets for intersections and visualization.

## Conclusion

We described Intervene as an integrated tool that provides an easy and automated interface for intersection, and effective visualization of genomic region and list sets. To our knowledge, Intervene is the first tool to provide three types of visualization approaches for multiple sets of gene or genomic intervals. The three modules are developed to overcome the situations where the number of sets is large. Intervene and its web application companion are developed and designed to fit the needs of a wide range of scientists.

**Abbreviations**
ChIP-seq: Chromatin immunoprecipitation followed by sequencing; ENCODE: The Encyclopedia of DNA Elements; hESCs: Human embryonic stem cells; SEs: Super-enhancers; TFs: Transcription factors

**Acknowledgements**
We thank the developers of the tools we have used to build Intervene and Intervene ShinyApp for sharing their code in open-source software. We thank Marius Gheorghe and Dimitris Polychronopoulos for their useful suggestions and testing the tool, and Annabel Darby for providing suggestions on the manuscript text.

**Funding**
This work has been supported by the Norwegian Research Council, Helse Sør-Øst, and the University of Oslo through the Centre for Molecular Medicine Norway (NCMM), which is part of the Nordic European Molecular Biology Laboratory Partnership for Molecular Medicine.

**Author's contributions**
AK conceived the project. AK and AM designed the tool. AM supervised the project. AK implemented both Intervene and the Shiny web application. AK wrote the manuscript draft and AM revised it. All authors read and approved the manuscript.

**Competing interests**
The authors declare that they have no competing interests.

**Author details**
¹Centre for Molecular Medicine Norway (NCMM), Nordic EMBL Partnership, University of Oslo, 0318 Oslo, Norway. ²Department of Cancer Genetics, Institute for Cancer Research, Oslo University Hospital Radiumhospitalet, 0310 Oslo, Norway.

**References**
1. Venn J. On the diagrammatic and mechanical representation of propositions and reasonings. Philos Mag J Sci. 1880;10:1–18.
2. Edwards AWF. Cogwheels of the mind: the story of venn diagrams. Baltimore: JHU Press; 2004.
3. Lex A, Gehlenborg N, Strobelt H, Vuillemot R, Pfister H. UpSet: visualization of intersecting sets. IEEE Trans Vis Comput Graph. 2014;20:1983–92.
4. Lex A, Gehlenborg N. Points of view: sets and intersections. Nat Meth. 2014;11:779.
5. Zhu LJ, Gazin C, Lawson ND, Pagès H, Lin SM, Lapointe DS, et al. ChIPpeakAnno: a bioconductor package to annotate ChIP-seq and ChIP-chip data. BMC Bioinformatics. 2010;11:237.
6. Dale RK, Pedersen BS, Quinlan AR. Pybedtools: a flexible python library for manipulating genomic datasets and annotations. Bioinformatics. 2011;27:3423–4.
7. Hunter JD. Matplotlib: a 2D graphics environment. Comput Sci Eng. 2007;9:99–104.
8. Conway JR, Lex A, Gehlenborg N: UpSetR: An R package for the visualization of intersecting sets and their properties. *bioRxiv*. 2017. doi: https://doi.org/10.1101/120600.
9. Wei T, Simko V: Corrplot: visualization of a correlation matrix. Volume R package. 2016.
10. Swinton J: Venn diagrams in R with the vennerable package. 2011.
11. Hulsen T, de Vlieg J, Alkema W. BioVenn – a web application for the comparison and visualization of biological lists using area-proportional Venn diagrams. BMC Genomics. 2008;9:488.
12. Lam F, Lalansingh CM, Babaran HE, Wang Z, Prokopec SD, Fox NS, et al. VennDiagramWeb: a web application for the generation of highly customizable Venn and Euler diagrams. BMC Bioinformatics. 2016;17:401.
13. Bardou P, Mariette J, Escudié F, Djemiel C, Klopp C. jvenn: an interactive Venn diagram viewer. BMC Bioinformatics. 2014;15:293.
14. Lin G, Chai J, Yuan S, Mai C, Cai L, Murphy RW, et al. VennPainter: a tool for the comparison and identification of candidate genes based on venn diagrams. PLoS One. 2016;11:e0154315.
15. Martin B, Chadwick W, Yi T, Park S-S, Lu D, Ni B, et al. VENNTURE–A novel venn diagram investigational tool for multiple pharmacological dataset analysis. PLoS One. 2012;7:e36911.
16. Heberle H, Meirelles GV, da Silva FR, Telles GP, Minghim R. InteractiVenn: a web-based tool for the analysis of sets through Venn diagrams. BMC Bioinformatics. 2015;16:169.
17. Dunham I, Kundaje A, Aldred SF, Collins PJ, Davis CA, Doyle F, et al. An integrated encyclopedia of DNA elements in the human genome. Nature. 2012;489:57–74.
18. Khan A, Zhang X. dbSUPER: a database of super-enhancers in mouse and human genome. Nucleic Acids Res. 2016;44(Database issue):D164–71.
19. Papp B, Sabri S, Ernst J, Plath K. Cooperative binding of transcription factors orchestrates reprogramming. Cell. 2017:1–18.
20. Spitz F, Furlong EEM. Transcription factors: from enhancer binding to developmental control. Nat Rev Genet. 2012;13:613–26.
21. Griffon A, Barbier Q, Dalino J, Van Helden J, Spicuglia S, Ballester B. Integrative analysis of public ChIP-seq experiments reveals a complex multi-cell regulatory landscape. Nucleic Acids Res. 2015;43:1–14.
22. Ernst J, Kellis M. ChromHMM: automating chromatin-state discovery and characterization. Nat Methods. 2012;9:215–6.
23. Taberlay PC, Statham AL, Kelly TK, Clark SJ, Jones PA. Reconfiguration of nucleosome-depleted regions at distal regulatory elements accompanies DNA methylation of enhancers and insulators in cancer. Genome Res. 2014;24:1421–32.
24. Theodorou V, Stark R, Menon S, Carroll JS. GATA3 acts upstream of FOXA1 in mediating ESR1 binding by shaping enhancer accessibility. Genome Res. 2013;23:12–22.
25. Zuin J, Dixon JR, van der Reijden MIJA, Ye Z, Kolovos P, Brouwer RWW, et al. Cohesin and CTCF differentially affect chromatin architecture and gene expression in human cells. Proc Natl Acad Sci U S A. 2014;111:996–1001.

# 17

# Robust gene selection methods using weighting schemes for microarray data analysis

Suyeon Kang and Jongwoo Song[*] (ID)

## Abstract

**Background:** A common task in microarray data analysis is to identify informative genes that are differentially expressed between two different states. Owing to the high-dimensional nature of microarray data, identification of significant genes has been essential in analyzing the data. However, the performances of many gene selection techniques are highly dependent on the experimental conditions, such as the presence of measurement error or a limited number of sample replicates.

**Results:** We have proposed new filter-based gene selection techniques, by applying a simple modification to significance analysis of microarrays (SAM). To prove the effectiveness of the proposed method, we considered a series of synthetic datasets with different noise levels and sample sizes along with two real datasets. The following findings were made. First, our proposed methods outperform conventional methods for all simulation set-ups. In particular, our methods are much better when the given data are noisy and sample size is small. They showed relatively robust performance regardless of noise level and sample size, whereas the performance of SAM became significantly worse as the noise level became high or sample size decreased. When sufficient sample replicates were available, SAM and our methods showed similar performance. Finally, our proposed methods are competitive with traditional methods in classification tasks for microarrays.

**Conclusions:** The results of simulation study and real data analysis have demonstrated that our proposed methods are effective for detecting significant genes and classification tasks, especially when the given data are noisy or have few sample replicates. By employing weighting schemes, we can obtain robust and reliable results for microarray data analysis.

**Keywords:** Microarray data, Gene selection method, Significance analysis of microarrays, Noisy data, Robustness, False discovery rate

## Background

Microarray technologies allow us to measure the expression levels of thousands of genes simultaneously. Analysis on such high-throughput data is not new, but it is still useful for statistical testing, which is a crucial part of transcriptomic research. A common task in microarray data analysis is to detect genes that are differentially expressed between experimental conditions or biological phenotype. For example, this can involve a comparison of gene expression between treated and untreated samples, or normal and cancer tissue samples. Despite the rapid change of technology and the affordable cost for conducting whole-genome expression experiments, many past and recent studies still have relatively few sample replicates in each group, which makes it difficult to use typical statistical testing methods. These two problems, high dimensionality and small sample size problems, have triggered developments of feature selection in transcriptome data analysis [1–9]. These feature selection methods can be mainly classified into four categories depending on how they are combined with learning algorithms in classification tasks: filter, wrapper, embedded, and hybrid methods. For details and the corresponding examples of these methods, we refer the reader to several review papers [10–18]. As many researchers commented, filter

* Correspondence: josong@ewha.ac.kr
Department of Statistics, Ewha Womans University, Seoul, South Korea

methods have been dominant over the past decades due to its strong advantages, although they are the earliest in the literature [11–13, 15, 16]. They are preferred by biology and molecular domain experts as the results generated by feature ranking techniques are intuitive and easy to understand. Moreover, they are very efficient because they require short computation time. As they are independent of learning algorithms, they can give general solutions for any classifier [15]. They also have a better generalization property as the bias in the feature selection and that of the classifier are uncorrelated [19]. Inspired by its advantages, we focus on the filter method in this study.

One of the most widely used filter-based test methods is significance analysis of microarrays (SAM) [1]. It identifies genes with a statistically significant difference in expression between different groups by implementing gene-specific modified $t$-tests. In microarray experiments, some genes have small variance so their test statistics become large, even though the difference between the expression levels of two groups is small. SAM prevents those genes from being identified as statistically significant by adding a small positive constant to the denominator of the test statistic. This is a simple but powerful modification for detecting differentially expressed genes, considering the characteristics of microarray data. Since its establishment, the SAM program has been repeatedly updated. The latest version is 5.0 [20].

We also aim to develop methods for detecting significant genes based on a deeper understanding of microarray data. Even when researchers monitor an experimental process and control other factors that might have an influence on the experiment, biological or technical error can still arise in high-throughput experiments. For example, when one sample among a number of replicated samples gives an outlying result owing to a technical problem, variance of the gene expression becomes larger than expected and its test statistic becomes small. This is a major issue because it can lead to biologically informative genes failing to be identified as having a significant effect. Therefore, we here attempt to reduce this increase in variance for such cases by modifying the variance structure of SAM statistics, using two weighting schemes. It is also important to adjust the significance level of tests. Since we generally need to test thousands of genes simultaneously, the multiple testing problem arises. To resolve this problem, several methods have been suggested as replacements for the simple $p$-value; for example, we can use the family-wise error rate (FWER), false discovery rate (FDR) [1, 21], and positive false discovery rate (pFDR) [22]. Among them, FDR, which is the expected proportion of false positives among all significant tests, is a popular method to adjust the significance level. It can be computed by permutation of the original dataset. The test procedures we propose in this paper also use FDR, the same as SAM.

Once a list of significant genes is established by a gene selection method, researchers may carry out further experiments such as real-time polymerase chain reaction to determine whether these reference genes are biologically meaningful. However, many genes may not be tested owing to limitations of time and resources. For example, even if hundreds of genes are included in a list of reference genes for a user-defined significance cutoff, researchers may just select a few top-ranked genes among them for further analyses. Therefore, it is very important that the genes are properly ranked in terms of their significance, especially for top-ranked genes [23, 24]. As such, in this paper, we focus on improving test statistics for each gene and assessing how well each test method identifies significant genes.

For microarray data analysis, a comparison of the performance of gene selection methods is difficult because we generally do not know the "gold standard" reference genes in actual experiments. In other words, we do not know which genes are truly significant. This is a common problem encountered in transcriptome data analysis, so most studies have focused on comparing classification performances, which are determined by the combination of the feature selection and learning algorithm. As these results are clearly dependent on the performance of learning method, we cannot compare the effectiveness of feature selection techniques definitively [16]. Therefore, in this paper, we generate spike-in synthetic data that allow us to determine which genes are truly differentially expressed between two groups. For this, we suggest a data generation method based on the procedure proposed by Dembélé [25]. By performing such simulations, we can see how the performance changes depending on the characteristics of the dataset, such as sample size, the proportion of differentially expressed genes, and noise level. In this study, we focus on comparing performance according to noise level as our goal is to efficiently detect significant genes in a noisy dataset. To verify that our proposed methods can also compete with previous methods for actual microarray data, we use two sets of actual data that have a list of gold standard genes based on previous findings. All of these real datasets are publicly available and can be downloaded from a website [26] and R package [27]. In order to compare different gene selection methods, we also define two performance metrics that can be used when true differentially expressed genes are known.

This paper is organized as follows. In the next section, we review the algorithm of SAM and propose statistical tests for microarray data that are modified versions of SAM, named MSAM1 and MSAM2. In addition, we explain our synthetic data generation method and suggest two performance metrics. In the results section, we describe our simulation studies and real data analysis. We compare SAM, MSAM1, and MSAM2 using 14 types of simulated dataset, which have different noise levels and sample sizes, and two sets of real microarray data. We next discuss the difference between the three methods in detail, focusing on FDR estimation. Additionally, we give the results of classification analysis using some top-ranked genes selected by each method. In the last section, we summarize and conclude this paper.

## Methods

In this section, we briefly review the SAM algorithm [1] and propose new modified versions of SAM, focusing on calculating the test statistic.

### SAM

Let $x_{ij}$ and $y_{ij}$ be the expression levels of gene $i$ in the $j$th replicate sample in states 1 and 2, respectively. For such a two-class case, the states of samples indicate different experimental conditions, such as control and treatment groups. Let $n_1$ and $n_2$ be the numbers of samples in these two groups, respectively. The SAM statistic proposed in [1] is defined as follows:

$$d_i = \frac{\bar{x}_i - \bar{y}_i}{s_i + s_0}$$

where $\bar{x}_i$ and $\bar{y}_i$ are the mean expression of the $i$th gene for each group, $\bar{x}_i = \sum_{j=1}^{n_1} x_{ij}/n_1$ and $\bar{y}_i = \sum_{j=1}^{n_2} y_{ij}/n_2$. The gene-specific scatter $s_i$ is defined as:

$$s_i = \sqrt{a\left\{\sum_{j=1}^{n_1}\left(x_{ij}-\bar{x}_i\right)^2 + \sum_{j=1}^{n_2}\left(y_{ij}-\bar{y}_i\right)^2\right\}}$$

where $a = (1/n_1 + 1/n_2)/(n_1 + n_2 - 2)$ and $s_0$ is a small positive constant called the fudge factor, which is chosen to minimize the coefficient of variation of $d_i$. The computation of $s_0$ is explained in detail in [3].

Now let us consider the overall algorithm. The SAM algorithm proposed in [1] can be stated as follows:

1. Calculate test statistic $d_i$ using the original dataset.
2. Make a permuted dataset by fixing the gene expression data and shuffling the group labels under the $H_0$ where $H_0$: $\bar{x}_i - \bar{y}_i = 0$ for all $i$.
3. Compute test statistics $d_i^*$ using the permuted data and order them according to their magnitudes as $d_{(1)}^* \le d_{(2)}^* \le \cdots \le d_{(n)}^*$, where $n$ is the number of genes.
4. Repeat steps 2 and 3 $B$ times and obtain $d_{(1)}^*(b) \le d_{(2)}^*(b) \le \cdots \le d_{(n)}^*(b)$ for $b = 1, 2, \ldots, B$, where $B$ denotes the total number of permutations.
5. Calculate the expected score $d_{(i)}^E = \sum_{b=1}^{B} d_{(i)}^*(b)/B$.

6. Sort the original statistic from step 1, $d_{(1)} \le d_{(2)} \le \cdots \le d_{(n)}$.
7. For user-specific cutoff $\Delta$, genes that satisfy $|d_{(i)} - d_{(i)}^E| > \Delta$ are declared significant. A gene is defined as being significantly induced if $d_{(i)} - d_{(i)}^E > \Delta$ and significantly suppressed if $d_{(i)} - d_{(i)}^E < -\Delta$.
8. Define $d_{(\text{up})}$ as the smallest $d_{(i)}$ among significantly induced genes and $d_{(\text{down})}$ as the largest $d_{(i)}$ among significantly suppressed genes.
9. The false discovery rate (FDR) is defined as the proportion of falsely significant genes among genes considered to be significant and can be estimated as follows:

$$\widehat{\text{FDR}} = \frac{\sum_{b=1}^{B} \#\left\{i : d_{(i)}(b) \ge d_{(\text{up})} \vee d_{(i)}(b) \le d_{(\text{down})}\right\}/B}{\#\left\{i : d_{(i)} \ge d_{(\text{up})} \vee d_{(i)} \le d_{(\text{down})}\right\}}$$

The algorithm consists of two parts: computation of the test statistic and determination of the cutoff for a given $\Delta$. We will focus on the first of these parts and apply a simple modification to the computation of gene-specific scatter $s_i$ to find a more robust test statistic. The numerator of the modified statistic and that of the original SAM statistic are the same. All of the procedures can be implemented using the *samr* package for Bioconductor in R. [20] described how to use the package and provided technical details of the SAM procedure.

### Modified SAM

From one experiment [28], we observed several cases in which most of the results of gene expression are very close to each other, apart from one substantial outlier. As a result, the ranks of these genes from SAM are lower than expected. This prompted us to propose a new test method that has a different variance structure, leading to robustness on identifying informative genes in the presence of outliers. Throughout the paper, we use the term "outliers" to indicate "unusual observations".

Let us consider two cases with the following data: case 1: (5,5,5,5,8.54) and case 2: (3,4,5,6,7). For these two cases, the variance is the same, inferring that they have the same spread. However, even though the levels of variance are equal, in fact, we cannot say that the data points are similarly distributed. We believe that case 1 is more reliable than case 2. Our goal, therefore, is to propose a test statistic that has a more significant result for case 1 than for case 2. To minimize the effects of outliers among samples, we use the median instead of the mean and employ a weight function $w$ when computing the test statistic, resulting in a less weight on an outlier sample that is far from other samples. A modified $s_i$, $\tilde{s}_i$, is defined as follows:

$$\tilde{s}_i = \sqrt{\sum_{j=1}^{n_1} w(x_{ij})\left(x_{ij} - median_j(x_{ij})\right)^2 + \sum_{j=1}^{n_2} w\left(y_{ij}\right)\left(y_{ij} - median_j\left(y_{ij}\right)\right)^2}$$

Accordingly, our test statistic $\tilde{d}_i$ is defined as follows:

$$\tilde{d}_i = \frac{\overline{x}_i - \overline{y}_i}{\tilde{s}_i + s_0}$$

Methods modified by this approach might be particularly useful when detecting differentially expressed genes from noisy microarray data. The key idea is to reduce the impact of outliers when calculating the test statistic. We propose two different weight functions in this paper. The values of $\tilde{s}_i$ and $\tilde{d}_i$ would differ quite markedly depending on the used weight function.

### Modified SAM1 (Gaussian weighted SAM)

The weight function used in Modified SAM1 (MSAM1) is based on the Gaussian kernel, which is a widely used weight that decreases smoothly to 0 with increasing distance from the center. It is defined as follows:

$$w(x_{ij}; \mu_i, \sigma) = \frac{1}{\sigma}\phi\left(\frac{x_{ij} - \mu_i}{\sigma}\right)$$

where $\phi$ is the probability density function of a standard normal distribution, $\phi(x) = e^{-x^2/2}/\sqrt{2\pi}$. The mean $\mu_i$ is a gene-specific parameter such that $\mu_i = median_j(x_{ij})$ and standard deviation $\sigma$ is a data-dependent constant determined by the following procedure: first, $m$ is

defined as follows. $m = \max(|x_{ij} - median_j(x_{ij})|, |y_{ij} - median_j(y_{ij})|)$. It is calculated from given data. Second, $p$ is a user-defined value between 0 and 1. Finally, given $m$ and $p$, we can find the value of $\sigma$ that satisfies the following equation:

$$m = F^{-1}(1-p; 0, \sigma)$$

where $F$ is the cumulative distribution function of a normal distribution. Therefore, $m$ would approximately be the $100(1 - p)$th percentile point of a normal distribution with mean 0 and standard deviation $\sigma$. As can be seen from Fig. 1, smaller $p$ yields smaller $\sigma$. Therefore, smaller $p$ makes the weight applied to outlier samples smaller. On the other hand, as $p$ increases, the results of original and modified SAMs become similar because the weight on the outlier is very similar to the weight on the non-outliers. In this research, we set $p = 0.001$ since we found that this value is sufficiently small to reduce the effect of outliers.

For a better understanding of MSAM1, we here illustrate the weight function of MSAM1 and its application in detail. Let us consider Leukemia data [29]; for details of this data, see real data analysis section. The data consist of 38 samples (27 from ALL patients and 11 from AML patients) and 7129 genes. For simplicity and clarity, we randomly selected five samples for each sample type and applied SAM, MSAM1 with $p = 0.01$ and MSAM1 with $p = 0.001$. In order to compare weights given by each method, let us take one gene, M96326_rna1_at (Azurocidin). This gene would be a good example to clarify the difference between SAM and MSAM1 because it has an outlier sample. From Fig. 2, we can see that gene expressions in group 1 are

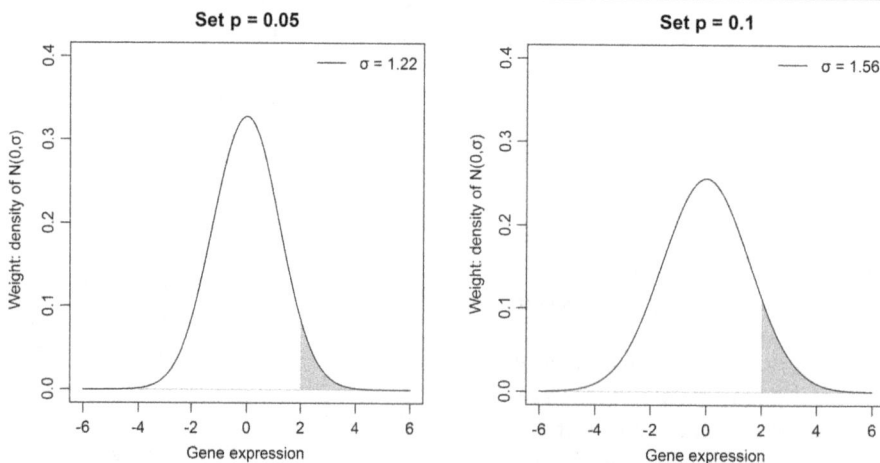

**Fig. 1** Two examples of the weight function for MSAM1 when $m$ is 2. When setting $p = 0.05$, $\sigma$ is determined to be 1.22 (left panel), and when setting $p = 0.1$, it is determined to be 1.56 (right panel). Since $m$ is the $100(1 - p)$th percentile point of $N(0, \sigma)$, the grey-shaded area in each panel is 0.05 and 0.1, respectively

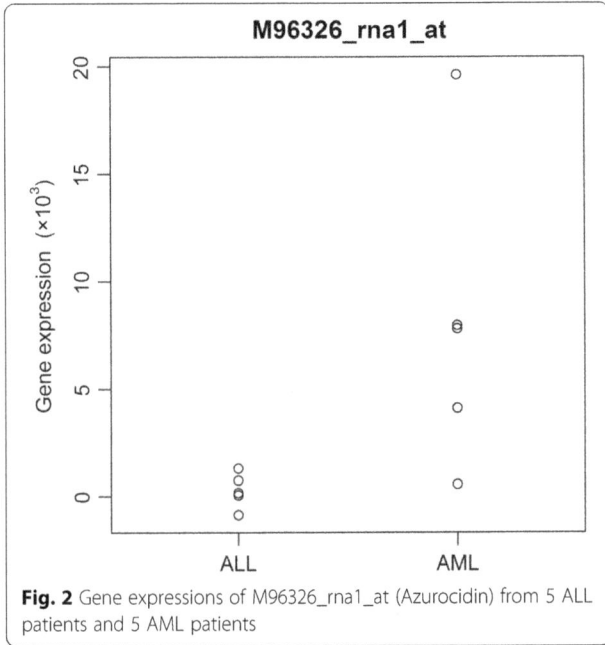

**M96326_rna1_at**

**Fig. 2** Gene expressions of M96326_rna1_at (Azurocidin) from 5 ALL patients and 5 AML patients

replicates (observations) in a group of the first gene ($i = 1$) and their gene expressions are $x_{11}, x_{12}, x_{13}$ and $x_{14}$. Let $w_j$ be the weight on $j$th observation for $j=1, 2, 3$ and 4. In this case, the weights on these observations are as follows.

$$w_1 = \left( \sum_{k=1}^{4} d_E(x_{11}, x_{1k}) \right)^{-1}, \quad w_2 = \left( \sum_{k=1}^{4} d_E(x_{13}, x_{1k}) \right)^{-1},$$

$$w_3 = \left( \sum_{k=1}^{4} d_E(x_{13}, x_{1k}) \right)^{-1}, \quad w_4 = \left( \sum_{k=1}^{4} d_E(x_{14}, x_{1k}) \right)^{-1}$$

If $x_{11}, x_{12}$ and $x_{13}$ are close to each other and $x_{14}$ is far from these 3 values, $w_4$ is much smaller than $w_1, w_2$ and $w_3$. Therefore, by using this weight function, we can give a smaller weight to an outlier. The further away an observation is from the others, the smaller weight is given.

**Synthetic data generation**
To run experiments, we need to generate synthetic gene expression data. These datasets should have characteristics similar to those of real microarray data to ensure that the results are reliable and valid. Two important characteristics of gene expression data, which are reported elsewhere [25, 30, 31] and also considered in this study, are as follows:

1. Under similar biological conditions, the level of gene expression varies around an average value. In rare cases, technical problems would result in values far away from this average.
2. Genes at low levels of expression have a low signal-to-noise ratio.

The 'technical problems' mentioned in the first of these points are one possible explanation for outliers observed in microarray data. Since our goal is to develop methods that detect differentially expressed genes well in a noisy dataset containing outliers, we consider not only a dataset with little noise, but also a noisy dataset with outliers. We ensure that outliers are present at higher probability in several of the datasets to provide a wider range of comparisons among the different test methods. Basically, we follow the microarray data generation model by Dembélé [25], which uses a beta

similar. On the other hand, one of five samples in group 2 is clearly far from others. Table 1 and Fig. 3 show its gene expressions and weights computed by SAM and MSAM1. In Fig. 3, the lengths of 5 red dashed lines indicate the weights on the 5 observations. As we stated above, we can also see that smaller $p$ makes the difference between weights applied to outlier and non-outlier samples greater.

**Modified SAM2 (inverse distance weighted SAM)**
This method uses Euclidean distance among the observations. The weight function used in Modified SAM2 (MSAM2) is defined as follows:

$$w(x_{ij}) = \frac{1}{\sum_k d_E(x_{ij}, x_{ik})}$$

where $d_E(x_{ij}, x_{ik})$ is the Euclidean distance between the $j$th and $k$th samples of gene $i$. The reason that we use this weight function can be explained by the following example. Let us assume that there are 10,000 genes ($i = 1, 2, \ldots, 10000$). Also, suppose there are 4 sample

**Table 1** Comparison of SAM and MSAM1 weights: an informative gene from leukemia data, M96326_rna1_at (Azurocidin)

| | ALL | | | | | AML | | | | |
|---|---|---|---|---|---|---|---|---|---|---|
| Gene expressions ($\times 10^3$) | −0.86 | 0.05 | 0.16 | 0.74 | 1.30 | 0.55 | 4.11 | 7.79 | 7.96 | 19.60 |
| SAM weights | 1.00 | 1.00 | 1.00 | 1.00 | 1.00 | 1.00 | 1.00 | 1.00 | 1.00 | 1.00 |
| MSAM1 weights ($\times 10^{-4}$) for $p = 0.01$ | 0.37 | 0.37 | 0.37 | 0.37 | 0.37 | 0.30 | 0.35 | 0.37 | 0.37 | 0.20 |
| MSAM1 weights ($\times 10^{-4}$) for $p = 0.001$ | 0.49 | 0.50 | 0.50 | 0.50 | 0.49 | 0.33 | 0.45 | 0.50 | 0.50 | 0.17 |

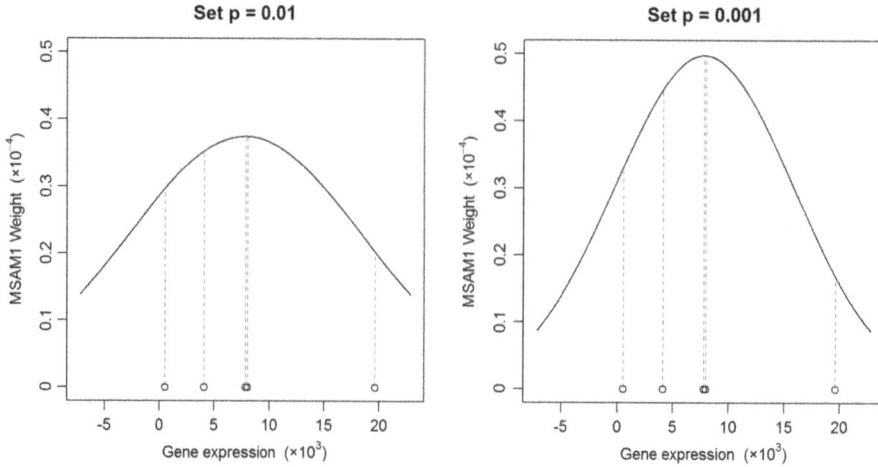

**Fig. 3** The left panel illustrates the weights of MSAM1 when $p$ is 0.01. The right panel is the case when $p$ is 0.001. In each panel, 5 black circle points are gene expressions of M96326_rna1_at (Azurocidin) from 5 AML patients. The lengths of 5 red dashed lines indicate the weights on the 5 observations

distribution. In this article, we employ a beta and a normal distribution to generate data points, assuming that the levels of gene expression essentially follow such distributions. To allow outliers in generated data, we add a technical error term in our model; this term is mentioned in [25], but not used in their model. According to the noise level and distribution type, we consider four different simulation set-ups as follows: Scenario 1, non-contaminated beta; 2, contaminated beta; 3, non-contaminated normal; 4, contaminated normal. Therefore, data used in scenarios 1 and 3 have low noise level, and data used in scenarios 2 and 4 have high noise level. The step-by-step procedure for our data generation method is summarized as follows.

Step 1. Let $n$ be the number of genes and $n_1$ and $n_2$ be control and treatment sample sizes, respectively.

Step 2. Generate $z_i$ from a beta (normal) distribution for $i = 1, 2, \ldots, n$ and transform the values, $\bar{z}_i = lb + ub \times z_i$.

Step 3. For each $\bar{z}_i$, generate $(n_1 + n_2)$ values as follows: $z_{ij} \sim \text{unif}((1-\alpha_i)\bar{z}_i, (1+\alpha_i)\bar{z}_i)$, where $\alpha_i = \lambda_1 e^{-\lambda_1 \bar{z}_i}$.

Step 4. The final model is given by

$$d_{ij} = z_{ij} + s_{ij} + n_{ij} + t_{ij}$$

where the term $s_{ij}$ allows us to define differentially expressed genes. Their values are zero for the control group, $s_{ij} \sim N(\mu_{de}, \sigma_{de}^2)$ for genes with induced expression, and $s_{ij} \sim N(-\mu_{de}, \sigma_{de}^2)$ for genes with suppressed expression, where $\mu_{de} = \mu_{de}^{min} + \text{Exp}(\lambda_2)$. $n_{ij}$ is an additive noise term, $n_{ij} \sim N(0, \sigma_n^2)$. The final term $t_{ij}$ is used to define outlying samples by allowing non-zero values for some genes. The undefined parameters for each step can be set by the users. The values we use in this paper are as follows: $\lambda_1 = 0.13$, $\lambda_2 = 2$, $\mu_{de}^{min} = 0.5$, $\sigma_{de} = 0.5$, $\sigma_n = 0.4$.

For these parameters, the influence of different parameter settings on the generated data is well explained elsewhere [25].

### Scenario 1: Beta with low noise level

In this case, we generate data points from Beta($shape_1$, $shape_2$). $shape_1$ and $shape_2$ are two shape parameters of the beta distribution and we here set $shape_1 = 2$ and $shape_2 = 4$. We also set $lb = 4$, $ub = 14$. The values of $t_{ij}$ are zero for this case.

### Scenario 2: Beta with high noise level

Here, we generate a noisier data than above data. The generation procedure is basically the same as the above case, except for allowing some non-zero $t_{ij}$. To make outlying samples, we contaminate the data by adding gaussian noise to some treatment samples: For genes with induced or suppressed expression,

$$t_{ij} \sim N(0, \sigma_{deo}^2) \text{ for } j = (n_1 + n_2 - n_{deo} + 1), \ldots, (n_1 + n_2.)$$

where $\sigma_{deo}$ is a non-zero constant and $n_{deo}$ is the number of outlying samples. We here set $\sigma_{deo} = 1$ and $n_{deo} = [0.2 \times n_2]$ where $[x] = m$ if $m \le x < m + 1$ for all integer $m$. For example, if there are five sample replicates in a treatment group, there can be one possible candidate as an outlier. Therefore, $\sigma_{deo}$ and $n_{deo}$ control the distribution and noise level of outlying samples. We believe that this set-up is reasonable because it does not destroy the original data structure while controlling the noise level of the data.

### Scenario 3: Normal with low noise level

This scenario assumes that the levels of gene expression essentially follow a normal distribution, instead of a beta distribution. In this research, we use the normal distribution with mean 10 and standard deviation 1.5 for generated data points to be distributed between realistic bounds; the gene expression levels on a log2 scale after robust multichip analysis normalization usually vary between 0 and 20. We set $lb = 0$, $ub = 1$ in Step 2, which means that no transformation is applied.

### Scenario 4: Normal with high noise level
To generate a noisier normal data, we use the same data generation procedure of Scenario 3, except for allowing some non-zero $t_{ij}$ in Step 4. The structure of $t_{ij}$ is the same as in Scenario 2.

### Performance metrics
To compare the performance of several methods, we need several evaluation measures. Since we know which genes are differentially expressed in our simulated datasets, we can define two performance metrics as follows, measuring how well each method identifies these TRUE genes. Prior to define metrics, let $G_{up}=\{i:$ gene $i$ the expression of which is truly significantly induced$\}$ and $G_{down}=\{i:$ gene $i$ the expression of which is truly significantly suppressed$\}$.

### Rank sum (RS)
We define the rank sum (RS) of TRUE genes as follows:

$$\mathrm{RS} = \sum_{i \in G_{up} \cup G_{down}} \sum_{j:d_id_j>0} \mathrm{I}\left(|d_i| \le |d_j|\right)$$

where $\mathrm{I}(\cdot)$ is an indicator function. The reason for determining the ranks of genes with high and low expression is that the SAM procedure uses such a method when detecting genes of the two groups. We use the absolute value of test statistics because test statistics of genes with suppressed expression have negative values. For RS, lower values indicate better performance.

### Top-ranked frequency (TRF)
The top-ranked frequency (TRF) of TRUE genes is computed by

$$\mathrm{TRF}(r) = \#\left\{i \in G_{up} \cup G_{down} : \sum_{j:d_id_j>0} \mathrm{I}\left(|d_i| \le |d_j|\right) \le r\right\}.$$

Here, $r$ denotes the rank cutoff and is set to be smaller than the number of observations in $G_{up}$ and $G_{down}$. For a given cutoff $r$, TRF computes the number of TRUE genes ranked within $r$. For TRF, higher values indicate better performance.

To understand the performance metrics better, let us consider the following case. We have 100 genes and 10 TRUE genes among them. Assume that we obtain a top-ranked gene list as shown in Table 2 by a gene selection

**Table 2** An example list of top-ranked genes

| Gene rank | Rank of true genes | True or false |
|---|---|---|
| 1 | 1 | T |
| 2 | 2 | T |
| 3 | - | F |
| 4 | 4 | T |
| 5 | 5 | T |
| 6 | 6 | T |
| 7 | - | F |
| 8 | - | F |
| 9 | 9 | T |
| 10 | 10 | T |
| 11 | 11 | T |
| 12 | - | F |
| 13 | 13 | T |
| 14 | - | F |
| 15 | 15 | T |
| Rank sum | 76 | |

method. Among the 15 genes in the table, five are false genes (3rd, 7th, 8th, 12th, and 14th genes in the table). In this case, RS = 76, TRF(5) = 4, and TRF(10) = 7.

### Results
#### Simulation studies
In this section, we compare gene selection methods using synthetic datasets. We consider four scenarios described above. For each scenario, we consider 7 different combinations of $n_1$ and $n_2$ in order to take into account the affects of sample size and class imbalance on gene selection performance as follows: $(n_1, n_2) = (5, 5), (5, 10), (10, 5), (10, 10), (10, 15), (15, 10)$ and $(15, 15)$. For all scenarios, we assume that there are 2% target genes (1% up-regulated and 1% down-regulated genes) among the total of 10,000 genes. For simplicity, let us assume that the first 100 genes are downregulated and last 100 genes are upregulated. Then, we can describe the structure of our simulation data as shown in Fig. 4. This example illustrates the structure of noisy data containing outliers. In this case, the last two samples are outlying samples among 10 treatment samples of 200 target genes. There are five different distributions of data points: A, B, C, D, and E. For 9800 nontarget genes, the distributions of the control and treatment samples are the same (A). The first 100 downregulated genes are generated from two distributions (B and C) and the last 100 upregulated genes are also generated from two distributions (D and E). Groups C and E indicate outlier samples. If there are no outliers in the dataset, B is equivalent to C and D is equivalent to E. The empirical density plot of each group is shown in Fig. 5. For

| Gene | Control samples | | | | | | | | | | Treatment samples | | | | | | | | | |
|------|---|---|---|---|---|---|---|---|---|----|---|---|---|---|---|---|---|---|---|----|
|      | 1 | 2 | 3 | 4 | 5 | 6 | 7 | 8 | 9 | 10 | 1 | 2 | 3 | 4 | 5 | 6 | 7 | 8 | 9 | 10 |
| 1%   | | | | | | | | | | | | | | B | | | | | C | |
| 98%  | | | | | A | | | | | | | | | | A | | | | | |
| 1%   | | | | | | | | | | | | | | D | | | | | E | |

**Fig. 4** An example of simulated data structure. Each row and each column of this data frame correspond to a gene and a replicate sample, respectively, so we have a 10,000 × 20 data matrix in this study. We assume that there are 2% target genes (1% up-regulated and 1% down-regulated genes) among the total of 10,000 genes, and ten replicates in each group. There are five different distributions of data points: A, B, C, D, and E; groups C and E indicate outlier samples

visualization, we use 5000 data points to ensure equivalent density of the points for each group (A, B, and C), that is, with a 1:1:1 ratio, not using the original ratio among the three groups.

We conduct simulation studies using synthetic data and compare the results using three metrics; two of them are RS and TRF, which were defined above, and the third is AUC. AUC is the area under a receiver operating characteristic (ROC) curve. Therefore, this value falls between 0 and 1, and higher values indicate better performance. We consider five gene selection methods, named SAM, SAM-wilcoxon, SAM-tbor, MSAM1 and MSAM2. SAM-wilcoxon is the Wilcoxon version of SAM [20, 32]. SAM-tbor is basically the same with SAM, except for applying a simple trim-based outlier removing algorithm to data prior to running SAM. In this study, we remove the largest and smallest observations from each sample type. Figs. 6 and 7 display the average

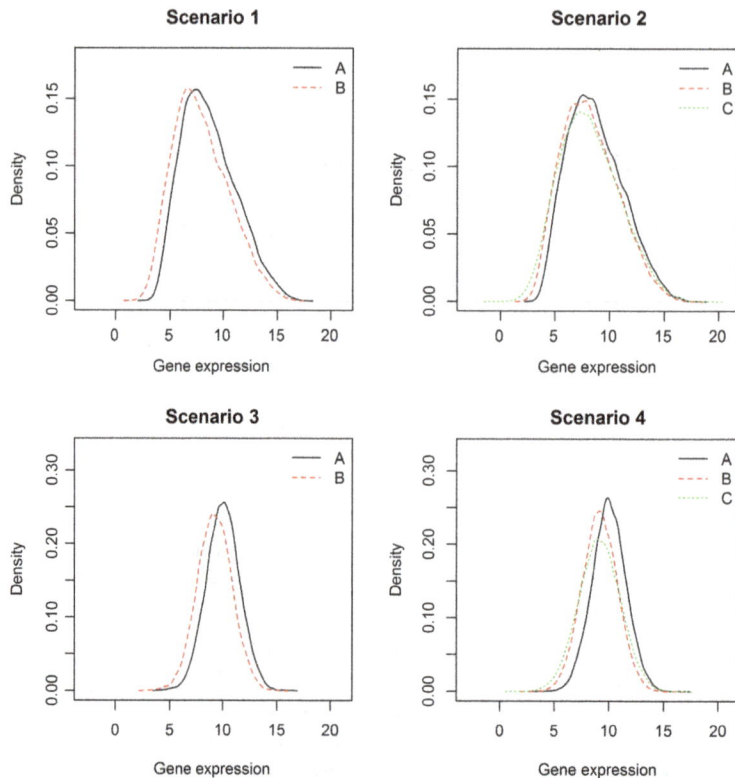

**Fig. 5** Empirical density of data points for scenarios 1, 2, 3, and 4. The solid line (**a**) for each plot is the density of control samples for target genes (**a**). The red dashed line (**b**) and green dotted line (**c**) are the densities of treatment samples for target genes. There are no green dotted lines (**c**) in the top-left and top-right plots because there are no outliers in scenarios 1 and 3

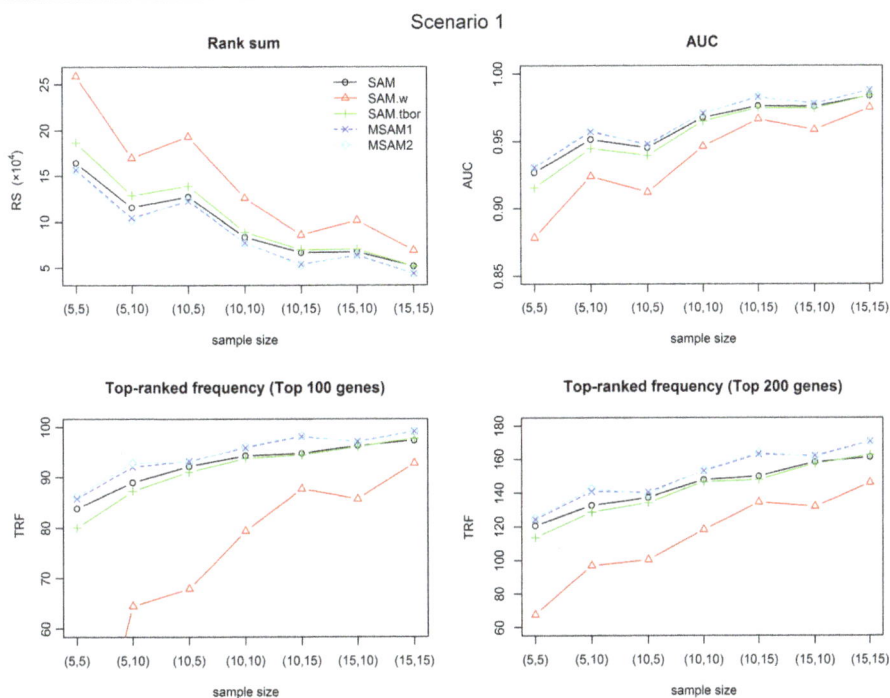

**Fig. 6** Simulation results for Scenario 1. Three solid lines (black, red, and green) indicate the results of three versions of SAM. Two dashed lines (blue and cyan) indicate the results of two versions of modified SAM. For RS, lower values are better. For AUC and TRF, higher values are better

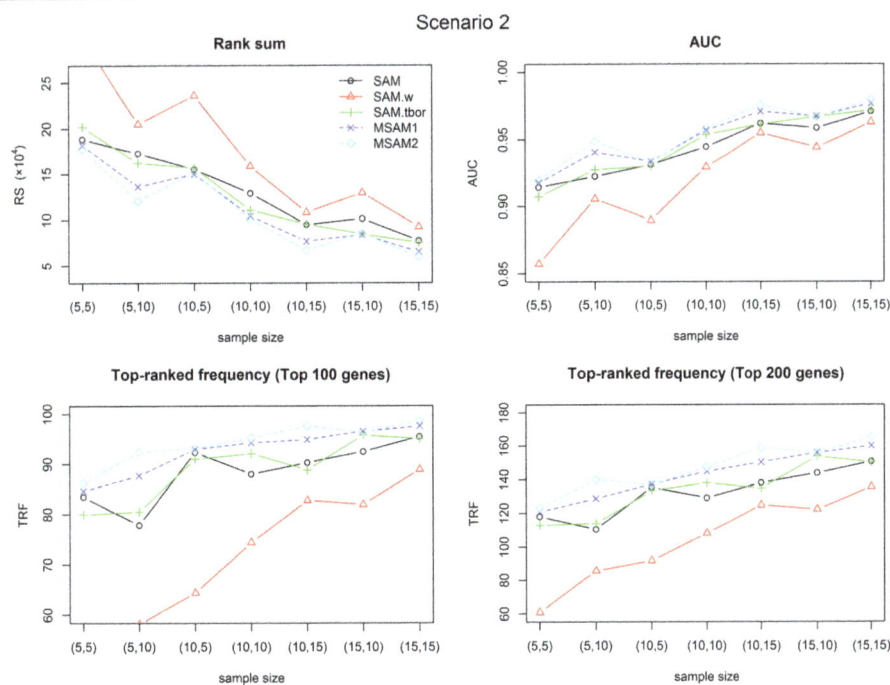

**Fig. 7** Simulation results for Scenario 2

performance of 100 simulations for each method on the three metrics. Table 3 shows numerical results of 4 cases. The best performance on each metric is shown in boldface. In scenario 1, the original SAM always outperform SAM-wilcoxon and SAM-tbor. Although SAM-tbor show better performance than SAM in some cases of scenario 2, its performance is worse than those of MSAMs. As can be seen from the figures and table, our proposed methods show better performance than three versions of SAM in all cases. In particular, modified SAMs are much better when given data is noisy (scenario 2, compared to scenario 1) and is a little better for less noisy cases. We can also see that our methods show more robust performance in all cases. When there is two outliers among ten samples, the number of target genes found by original SAM is reduced by 2–17%, whereas that found by MSAMs is reduced by 1–8%. In particular, when $n_1 = 5$, $n_1 = 10$ in scenario 2, SAM fails to detect 90 genes among the 200 TRUE genes, whereas MSAM2 fails to detect only 60 genes on average. Simulation results of scenarios 3 and 4 are in Additional file 1. These results are very similar with those of scenarios 1 and 2; MSAMs always perform better than three versions of SAM.

**Real data analysis 1: *Fusarium***

The *Fusarium* dataset contains 17,772 genes and nine samples: three each from control, dtri6, and dtri10 groups [28]. Robust multichip analysis algorithm is used for condensing the data for the following [33]: extraction of the intensity measure from the probe level data, background adjustments, and normalization. The post-processed dataset used in [28] are stored at PLEXdb (http://www.plexdb.org) (accession number: FG11) [26]. As this data was from gene mutation experiments, researchers provided a list of genes that are differentially expressed between control and treatment (dtri6, dtri10) groups. These genes are as follows: fgd159-500_at (conserved hypothetical protein), fgd159-520_at (trichothecene 15-O-acetyltransferase), fgd159-540_at (Tri6 trichothecene biosynthesis positive transcription factor), fgd159-550_at (TRI5_GIBZE – trichodiene synthase), fgd159-560_at, fgd159-600_at (putative trichothecene biosynthesis), fgd321-60_at (trichothecene 3-O-acetyltransferase), fgd4-170_at (cytochrome P450 monooxygenase), fgd457-670_at (TRI15 – putative transcription factor), fg03534_s_at (trichothecene 15-O-acetyltransferase), fg03539_at (TRI9 – putative trichothecene biosynthesis gene), and fg03540_s_at (TRI11 – isotrichodermin C-15 hydroxylase).

In real data analysis sections, we only consider SAM, MSAM1, and MSAM2, all of which show good performance in simulation studies; we found that SAM-wilcoxon and SAM-tbor are worse than the original SAM in the previous section. Moreover, we cannot apply SAM-tbor to this data because this data has only three sample replicates in each group. Like this case, we can see that such a trim-based method is limited in its applications.

Tables 4 and 5 show the rank of 11 reference genes that are differentially expressed between the control group and the treatment groups (dtri6 and dtri10, respectively). The last row in each table indicates the rank sum of these 11 genes. As we can see, MSAM2 shows the best performance because the rank sum of this method is the smallest among those of the three gene

**Table 3** Simulation results for 4 cases

| | Scenario 1, $n_1 = 5$, $n_2 = 10$ | | | | Scenario 1, $n_1 = 10$, $n_2 = 10$ | | | |
|---|---|---|---|---|---|---|---|---|
| | RS | AUC | TRF | | RS | AUC | TRF | |
| Rank cutoff | | | 100 | 200 | | | 100 | 200 |
| SAM | 115,542 | 0.95 | 88.93 | 132.67 | 83,544 | **0.97** | 94.28 | 147.74 |
| SAM-w | 169,865 | 0.92 | 64.43 | 96.84 | 125,513 | 0.95 | 79.37 | 118.18 |
| SAM-tbor | 128,588 | 0.94 | 87.21 | 128.67 | 88,765 | 0.96 | 93.72 | 146.42 |
| MSAM1 | 104,236 | **0.96** | 92.02 | 140.70 | 77,317 | **0.97** | 95.84 | 153.08 |
| MSAM2 | **101,705** | **0.96** | **92.81** | **142.47** | **77,109** | **0.97** | **96.05** | **153.58** |
| | Scenario 2, $n_1 = 5$, $n_2 = 10$ | | | | Scenario 2, $n_1 = 10$, $n_2 = 10$ | | | |
| | RS | AUC | TRF | | RS | AUC | TRF | |
| Rank cutoff | | | 100 | 200 | | | 100 | 200 |
| SAM | 172,669 | 0.92 | 77.88 | 110.38 | 128,966 | 0.94 | 87.97 | 129.06 |
| SAM-w | 205,161 | 0.91 | 58.14 | 85.55 | 158,618 | 0.93 | 74.44 | 108.06 |
| SAM-tbor | 162,252 | 0.93 | 80.54 | 113.94 | 110,655 | 0.95 | 92.05 | 137.97 |
| MSAM1 | 136,442 | 0.94 | 87.69 | 128.76 | 104,286 | **0.96** | 94.23 | 144.82 |
| MSAM2 | **120,594** | **0.95** | **92.45** | **139.89** | **100,887** | **0.96** | **95.15** | **147.48** |

Note: the best performance in each case is shown in **bold type**

**Table 4** Rank of genes of interest: control versus dtri6

| Gene | $\bar{x}_i-\bar{y}_i$ | SAM $\tilde{d}_i$ | rank | MSAM1 $\tilde{d}_i$ | rank | MSAM2 $\tilde{d}_i$ | rank |
|---|---|---|---|---|---|---|---|
| fgd457-670_at | −4.82 | −25.45 | 1 | −19.78 | 1 | −8.62 | 3 |
| fgd159-550_at | −5.13 | −24.58 | 2 | −19.20 | 2 | −9.27 | 2 |
| fgd159-600_at | −5.38 | −18.24 | 6 | −14.10 | 6 | −8.12 | 4 |
| fg03534_s_at | −4.48 | −14.70 | 7 | −10.29 | 7 | −5.64 | 7 |
| fg03540_s_at | −3.34 | −13.56 | 8 | −9.97 | 9 | −5.02 | 14 |
| fgd321-60_at | −3.71 | −13.26 | 9 | −9.50 | 10 | −5.28 | 10 |
| fg03539_at | −3.80 | −13.21 | 10 | −10.07 | 8 | −5.19 | 12 |
| fgd159-500_at | −3.66 | −12.49 | 11 | −9.01 | 11 | −5.39 | 9 |
| fgd159-520_at | −5.08 | −11.60 | 12 | −8.70 | 12 | −5.62 | 8 |
| fgd159-540_at | −4.06 | −10.73 | 18 | −8.59 | 13 | −5.03 | 13 |
| fgd4-170_at | −4.98 | −9.22 | 26 | −7.58 | 21 | −5.21 | 11 |
| Rank sum | | | 110 | | 100 | | **93** |

Note: the best performance in terms of rank sum is shown in **bold type**

selection methods. In particular, MSAMs improve the rank of the genes named fgd4-170_at and fgd159-500_at. For each of these genes, the result for one of their treatment samples is far from those for the other two samples. From the analysis, it can be asserted that our proposed methods efficiently identify the genes whose replicate samples contain an outlier, such as fgd4-170_at and fgd159-500_at.

### Real data analysis 2: Leukemia
Leukemia is a cancer of the bone marrow, where blood cells are made. In leukemia, abnormal blood cells are produced in the bone marrow and crowd out other normal blood cells. Depending on the type of abnormal blood cells that are multiplying, leukemia can be classified as acute lymphocytic leukemia (ALL) or acute

**Table 5** Rank of interest genes: control versus dtri10

| Gene | $\bar{x}_i-\bar{y}_i$ | SAM $\tilde{d}_i$ | rank | MSAM1 $\tilde{d}_i$ | rank | MSAM2 $\tilde{d}_i$ | rank |
|---|---|---|---|---|---|---|---|
| fg03539_at | −6.66 | −22.18 | 1 | −17.27 | 1 | −10.46 | 2 |
| fg03534_s_at | −4.00 | −11.24 | 4 | −8.42 | 4 | −5.68 | 4 |
| fgd159-560_at | −2.76 | −9.74 | 5 | −8.01 | 5 | −5.22 | 6 |
| fgd159-600_at | −3.28 | −8.77 | 8 | −6.81 | 8 | −4.92 | 7 |
| fgd159-520_at | −4.17 | −8.12 | 9 | −6.31 | 10 | −4.65 | 8 |
| fgd457-670_at | −3.35 | −6.44 | 16 | −5.05 | 18 | −3.83 | 12 |
| fgd4-170_at | −3.53 | −5.73 | 21 | −4.76 | 19 | −3.68 | 13 |
| fgd159-550_at | −3.22 | −4.89 | 26 | −4.23 | 22 | −3.42 | 16 |
| fg03540_s_at | −2.35 | −4.84 | 27 | −3.88 | 27 | −2.90 | 23 |
| fgd321-60_at | −1.62 | −3.07 | 60 | −2.44 | 64 | −1.88 | 38 |
| fgd159-500_at | −1.73 | −2.88 | 78 | −2.37 | 73 | −1.90 | 37 |
| Rank sum | | | 255 | | 251 | | **166** |

Note: the best performance in terms of rank sum is shown in **bold type**

myeloid leukemia (AML). Identifying the type of leukemia is very important because patients should receive different treatments according to the disease type. [29] studied a generic approach to cancer classification based on gene expression and provided a list of 50 significant genes for classifying ALL and AML. After this study, this dataset has been widely used in transcriptomic analysis, e.g., [34, 35]. This data are available in the *golubEsets* library in Bioconductor [27]. The original data consist of 38 samples (27 from ALL patients and 11 from AML patients) and 7129 genes. We randomly selected five, seven, and ten samples for each sample type and repeated this experiment 100 times for averaging because biological experiments usually have a small number of samples owing to limitations of time and resources. It is thus important that a method shows good performance even if the sample size is small.

The simulation results are shown in Table 6. In this table, RS and TRF values of three gene selection methods, which were computed by using 50 genes that are considered informative in [29] over 100 trials. For each case, the best performance is shown in boldface in the table. As we can see, MSAM1 or MSAM2 performs better than SAM in terms of RS and TRF, regardless of rank cutoff values. The overall performance of SAM and MSAM1 are very similar, but MSAM1 always performs slightly better than SAM. In the point of view of sample size, MSAM2 outperform SAM and MSAM1 when the sample size is very small, e.g., 5, and MSAM1 performs better than SAM and MSAM2 when the sample size is moderate, e.g., 7 and 10. As the sample size increases, all of the three methods identify informative genes better.

### FDR comparison
In this section, we discuss the FDR estimation procedures of SAM, MSAM1, and MSAM2. FDR is used in SAM procedure in order to deal with a multiple testing problem. The SAM interface in R, *samr* package [20], provides a significant gene list based on the FDR value that is estimated by its internal function. We also construct our own interface for MSAMs in R, based on the *samr* package, in order to allow for users to apply our proposed methods to their transcriptome research; see Additional file 2. Users start the procedure by setting their desired FDR value (for example, 0.2). We will call this value 'estimated FDR'. Based on the estimated FDR, our procedure calculates the value of corresponding $\Delta$ and identifies potentially significant genes. In real applications, we do not know TRUE FDR, so the estimated FDR is used as a substitute for TRUE FDR. If the estimated value is different from the true value, the number of genes that are detected using the estimated FDR is larger or smaller than the true number. Therefore, users may be interested in how well SAM and MSAMs

**Table 6** Rank sum and top-ranked frequency of informative genes in Leukemia data

| | # picked samples: 5 | | | # picked samples: 7 | | | # picked samples: 10 | | |
|---|---|---|---|---|---|---|---|---|---|
| | | | | Rank sum | | | | | |
| | SAM | MSAM1 | MSAM2 | SAM | MSAM1 | MSAM2 | SAM | MSAM1 | MSAM2 |
| | 22,287 | 21,585 | **15,815** | 11,924 | **11,790** | 13,286 | 5566 | **5534** | 11,256 |
| | | | | Top-ranked frequency | | | | | |
| $r$ | SAM | MSAM1 | MSAM2 | SAM | MSAM1 | MSAM2 | SAM | MSAM1 | MSAM2 |
| 20 | 5.05 | 5.29 | **7.46** | 7.56 | 7.86 | **9.05** | 11.93 | **12.15** | 10.34 |
| 40 | 7.74 | 8.24 | **12.72** | 12.55 | 12.95 | **14.48** | 19.11 | **20.05** | 16.70 |
| 60 | 10.30 | 11.19 | **16.46** | 16.33 | 17.09 | **18.02** | 25.10 | **25.89** | 20.23 |
| 80 | 12.76 | 13.54 | **19.02** | 19.78 | 20.52 | **20.63** | 29.30 | **30.27** | 22.71 |
| 100 | 14.67 | 15.74 | **21.36** | 22.59 | **23.52** | 23.07 | 32.72 | **33.54** | 24.75 |
| 120 | 16.72 | 17.75 | **23.43** | 24.76 | **25.74** | 25.45 | 35.65 | **36.22** | 26.94 |
| 140 | 18.38 | 19.48 | **25.25** | 26.90 | **27.67** | 27.29 | 37.77 | **38.16** | 28.91 |
| 160 | 20.03 | 20.94 | **26.79** | 28.67 | **29.74** | 28.92 | 39.60 | **39.92** | 30.48 |

Note: the best performance for each rank cutoff is shown in **bold type**

procedures estimate TRUE FDR value. To this end, in this section, we evaluate SAM, MSAM1, and MSAM2, focusing on their FDR estimation performances.

Since we know the number of TRUE significant genes in our simulated datasets, we can compare the estimated FDR and TRUE FDR in simulation study. After 100 simulations, we draw a scatter plot of the TRUE FDR versus the estimated FDR by calculating the average values of the TRUE FDR for each estimated FDR. We next draw a smooth curve close to the scatter plot for scenarios 1 and 2 to find the estimation accuracy at various levels of FDR. In particular, the estimation accuracy at low FDR is important since researchers generally set FDR at a small value so as to avoid having a large proportion of falsely significant genes among the detected genes. For this reason, we only show the results when the estimated FDR is lower than 0.5. Figure 8 displays the results; see the top two plots. As we can see, SAM estimates the TRUE FDR very accurately and two modified SAMs slightly overestimate the TRUE FDR. In other words, our methods have conservative property in their FDR estimation. However, the conservative estimation of FDR may not cause serious problems for the analysis when we use FDR as an upper bound of a tolerable error [36].

For such an analysis, the more important thing is how many non-significant genes are included in the detected genes. Because the truths are known in the simulated data, we can calculate the number of falsely detected genes among the identified genes. With the same number of total positives, the method with the smallest number of false positives is the best [36]. Using the plotting method described above, a smooth curve of the number of false positive genes versus the total number of identified genes are drawn. Figure 8 shows the results From the figure, we can see that MSAM1 and MSAM2 gives

smaller number of false positive genes than SAM across all noise level and the total number of identified genes. From the results, we can say that MSAMs are better than SAM because they includes the less number of false genes in the selected gene subset.

When we estimate FDR, we calculate both median FDR and mean FDR to determine which estimate more closely approximates the true value. Since the original *samr* interface provides the median FDR and 90th percentile FDR only, we modified its estimation function and obtained the median and mean values of FDR. As a result, we found that the median FDR was closer than the mean FDR to the TRUE FDR for all methods. This coincides with results published elsewhere [37], in which the median FDR was recommended as a criterion for gene selection methods when the estimated proportion of differentially expressed genes is greater than 1%, regardless of the sample size. Based on these results, we use the median value instead of the mean value when estimating FDR.

**Classification analysis**

Once important genes are identified from thousands of genes, they can be used to predict two different experimental states or responses (for example, cancer and normal). Therefore, we also examine how well a few top genes selected by each method identify the true classes. We attach these results in Additional file 3. In this file, we introduce 4 datasets we used and explain the construction of classifiers, 6 gene selection methods, 3 performance metrics to be considered in this study. Our comments on the results are also included. As can be seen in the file, our proposed methods, MSAMs, show quite good performances in all cases. In this additional

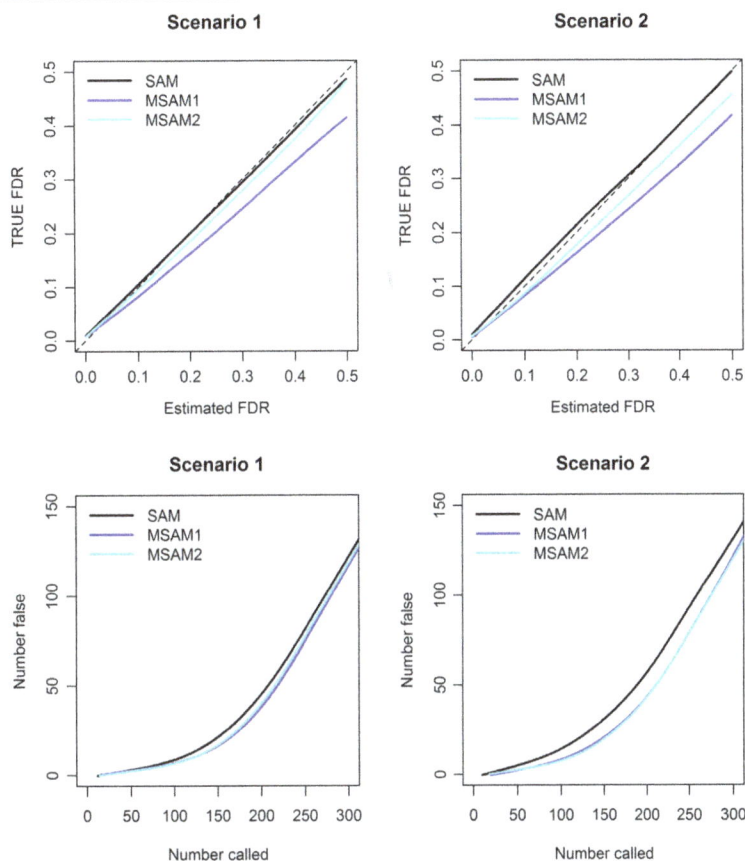

**Fig. 8** The top two plots show TRUE FDR vs. estimated FDR and the bottom two plots show the number of falsely detected genes relative to the total number of detected genes for scenario 1 and 2. In each top plot, the solid lines indicate estimation curves of each method and the dashed line represents $Y = X$

section, we prove their competitiveness in classification tasks, not only in gene selection tasks.

## Discussion

In transcriptome data analysis, most studies have been devoted to developing filter-based methods that are the simplest and fastest, and most computationally efficient. Hybrid methods, which are generally the combination of filter and wrapper methods, have recently gained popularity in the literature [13]. These methods consist of two steps: First, relevant features are selected by a filter method and the remaining features are eliminated. Second, a wrapper method verify these features and determine the final feature set that gives high classification accuracy [16]. In this point of view, filter methods have a lot of flexibility as they can be combined with not only any learning algorithm, but also any gene selection method, such as a wrapper method, resulting in a hybrid method. The performance of a hybrid method relies totally on the combination of filter and wrapper methods as well as the classifier [18]. We believe that accurate gene selection by filter methods clearly allow better

classification accuracy. Therefore, our new filter-based methods will be useful not only in gene selection, but also constructing a good classifier in microarray applications.

Our experiments showed the efficiency of our methods; it was demonstrated that when the same number of genes were selected, our methods included the less number of false genes than the conventional method. Our results also strongly suggest that these newly proposed methods outperform the conventional method and show quite consistent performance, even with a high noise level and a small sample size. Given that noisy data and a small sample size are commonly encountered in microarray studies [30, 38–40], we believe that our methods will prove useful.

This research was based on the existing interface of SAM that was modified to apply our proposed methods. This modified version of the *samr* package is available in Additional file 2. We attempted to find a balance between flexibility and control in the usage of our methods by allowing users to set particular parameters and by minimizing the number of modifications to the original interface. Additional file 2 includes a detailed explanation of what we changed, but users can easily apply our methods to their own datasets without reading the manuscript in the first file, since we provide some simple and useful examples of detecting differentially expressed genes using our methods in Additional file 4. We also provide two real datasets and one simulated dataset used in this study (see Additional files 5, 6 and 7). All of the additional files are also available at author's homepage (http://home.ewha.ac.kr/~josong/MSAM/index.html).

## Conclusions

We have proposed new test methods for identifying genes that are differentially expressed between two groups in microarray data and evaluated their performance using a series of simulated data and two real datasets. The results have demonstrated that our proposed methods identified target genes better than the original method, SAM, for both simulation studies and real data analysis. Using our weighting schemes, significant genes can be selected in a more robust manner by avoiding the overestimation of variance. In particular, these procedures are very effective when the given data are noisy or the sample size is limited. Therefore, they prevent technical or biological problems that can occur in biological experiments and data pre-processing from impeding accurate gene selection. We believe that our proposed methods can be applied to various datasets in other fields if they have characteristics similar to microarray data.

## Additional files

**Additional file 1:** Additional simulation results for scenario 3 and 4

**Additional file 2:** . R code for the modified *samr* package.

**Additional file 3:** Classification analysis section

**Additional file 4:** R code for some examples of our method for detecting genes that are differentially expressed.

**Additional file 5:** Fusarium data

**Additional file 6:** Leukemia data

**Additional file 7:** Simulated data (scenario 2)

## Abbreviations

ALL: acute lymphocytic leukemia; AML: acute myeloid leukemia; AUC: area under the curve; FDR: false discovery rate; MSAM1: modified SAM1; MSAM2: modified SAM2; ROC: receiver operating characteristic; RS: rank sum of true genes; SAM: significance analysis of microarrays; TRF: top-ranked frequency of true genes

## Acknowledgments

The authors would like to thank the editor and three anonymous reviewers for their insightful comments that significantly improve this article.

## Funding

This work was supported by the Ministry of Education of the Republic of Korea and the National Research Foundation of Korea (NRF-2017R1D1A1B03036078).

None of funding bodies played any role in the design or conclusions of this study.

## Authors' contributions

All authors developed the approach, designed the study, wrote the computer code, analyzed the data, conducted the simulation studies and wrote the manuscript. All authors read and approved the final manuscript.

## Competing interests

The authors declare that they have no competing interests.

## References

1. Tusher VG, Tibshirani R, Chu G. Significance analysis of microarrays applied to the ionizing radiation response. Proc Natl Acad Sci U S A. 2001;98(9): 5116–21.
2. Pavlidis P, Weston J, Cai J, Grundy WN. Gene functional classification from heterogeneous data. Proceedings of the fifth annual international conference on Computational biology. 2001:249–55.
3. Mak MW. Kung SY. A solution to the curse of dimensionality problem in pairwise scoring techniques. In neural information processing. Springer Berlin/Heidelberg. 2006:314–23.
4. Efron B. Microarrays, empirical Bayes and the two-groups model. Stat Sci. 2008;23(1):1–22.
5. Sharma A, Imoto S, Miyano S, Sharma V. Null space based feature selection method for gene expression data. Int J Mach Learn Cybern. 2012;3(4):269–76.
6. Sharma A, Imoto S, Miyano S. A between-class overlapping filter-based method for transcriptome data analysis. J Bioinforma Comput Biol. 2012;10(5):1–20.
7. Sharma A, Imoto S, Miyano SA. Top-r feature selection algorithm for microarray gene expression data. IEEE/ACM Trans Comput Biol Bioinform. 2012;9(3):754–64.
8. Ghalwash MF, Cao XH, Stojkovic I, Obradovic Z. Structured feature selection using coordinate descent optimization. BMC bioinformatics. 2016;17(1):158.
9. Sharbaf FV, Mosafer S, Moattar MHA. Hybrid gene selection approach for microarray data classification using cellular learning automata and ant colony optimization. Genomics. 2016;107(6):231–8.
10. Saeys Y, Inza I, Larranaga PA. Review of feature selection techniques in bioinformatics. Bioinformatics. 2007;23(19):2507–17.
11. Ahmad FK, Norwawi NM, Deris S. Othman NH. A review of feature selection techniques via gene expression profiles. In 2008 International Symposium on Information Technology
12. George G, Raj VC. Review on feature selection techniques and the impact of SVM for cancer classification using gene expression profile. arXiv preprint arXiv. 2011:1109–062.
13. Bolon-Canedo V, Sanchez-Marono N, Alonso-Betanzos A, Benitez JM, Herrera FA. Review of microarray datasets and applied feature selection methods. Inf Sci. 2014;282:111–35.
14. Tang J, Alelyani S, Liu H. Feature selection for classification: a review. Data Classification: Algorithms and Applications. 2014;37
15. Ang JC, Mirzal A, Haron H, Hamed HNA. Supervised, unsupervised, and semi-supervised feature selection: a review on gene selection. IEEE/ACM Trans Comput Biol Bioinform. 2016;13(5):971–89.
16. Bolón-Canedo V, Sánchez-Maroño N, Alonso-Betanzos A. Feature selection for high-dimensional data. Prog. Artif. Intell. 2016;5:65–75.
17. Mahajan S, Singh S. Review on feature selection approaches using gene expression data. Imp. J. Interdiscip. Res. 2016;2(3).
18. Aziz R, Verma CK, Srivastava N. Dimension reduction methods for microarray data: a review. AIMS. Bioengineering. 2017;4(1):179–97.
19. Ding C, Peng H. *minimum* Redundancy feature selection from microarray gene expression data. J Bioinforma Comput Biol. 2005;3(2):185–205.
20. Chu G, Narasimhan B. Tibshirani R, and Tusher VG. SAM users guide and technical document: Stanford University Labs; 2005.

21. Benjamini Y, Hochberg Y. Controlling the false discovery rate: a practical and powerful approach to multiple testing. J R Stat Soc Ser B. 1995;57:289–300.
22. Storey JDA. Direct approach to false discovery rates. J R Stat Soc Ser B. 2002;64(3):474–98.
23. Mukherjee SN, Roberts SJ, Sykacek P, Gurr SJ. Gene ranking using bootstrapped p-values. SIGKDD Explor. 2003;5(2):16–22.
24. Boulesteix AL, Slawski M. Stability and aggregation of ranked gene lists. Brief Bioinform. 2009;10(5):556–68.
25. Dembélé DA. flexible microarray data simulation model. Microarrays. 2013;2(2):115–30.
26. Wise RP, Caldo RA, Hong L, Shen L, Cannon EK, Dickerson JA. BarleyBase/PLEXdb: Plant Bioinformatics: Methods and Protocols. 2007:347?63.
27. http://www.bioconductor.org.
28. Seong KY, Pasquali M, Zhou X, Song J, Hilburn K, McCormick S, Dong Y, JR X, Kistler HC. Global gene regulation by fusarium transcription factors Tri6 and Tri10 reveals adaptations for toxin biosynthesis. Mol Microbiol. 2009;72(2):354–67.
29. Golub TR, Slonim DK, Tamayo P, Huard C, Gaasenbeek M, Mesirov JP, Coller H, Loh M, Downing JR, Caligiuri MA, Bloomfield CD, Lander ES. Molecular classification of cancer: class discovery and class prediction by gene expression monitoring. Science. 1999;286(5439):531?7.
30. Kooperberg CF, Aragaki AD, Strand A, Olson JM. Significance testing for small microarray experiments. Stat Med. 2005;24(15):2281–98.
31. Nykter M, Aho T, Ahdesmaki M, Ruusuvuori P, Lehmussola A, Yli-Harja O. Simulation of microarray data with realistic characteristics. BMC Bioinformatics. 2006;7(1):1.
32. Li J, Tibshirani R. Finding consistent patterns: a nonparametric approach for identifying differential expression in RNA-Seq data. Stat Methods Med Res. 2013;22(5):519–36.
33. Irizarry RA, Bolstad BM, Collin F, Cope LM, Hobbs B, Speed TP. Summaries of Affymetrix gene-Chip probe level data. Nucleic Acids Res. 2003;31(4):e15.
34. Pan W. A comparative review of statistical methods for discovering differentially expressed genes in replicated microarray experiments. Bioinformatics. 2002;18(4):546?54.
35. Zhang SA. Comprehensive evaluation of SAM, the SAM R-package and a simple modification to improve its performance. BMC Bioinformatics. 2007;8(1):230.
36. Xie Y, Pan W, Khodursky ABA. Note on using permutation-based false discovery rate estimates to compare different analysis methods for microarray data. Bioinformatics. 2005;21(23):4280–8.
37. Hirakawa A, Sato Y, Hamada D, Yoshimura IA. New test statistic based on shrunken sample variance for identifying differentially expressed genes in small microarray experiments. Bioinform Biol Insights. 2008;2:145–56.
38. Dougherty ER. Small sample issues for microarray?Based classification. Comp Funct Genomics. 2001;2(1):28–34.
39. Marshall E. Getting the noise out of gene arrays. Science. 2004;306(5696):630–1.
40. Cobb K. Microarrays: the search for meaning in a vast sea of data. Biomed. Comput Rev. 2006;2(4):16–23.

# 18

# Methods for discovering genomic loci exhibiting complex patterns of differential methylation

Thomas J. Hardcastle ⓘ

## Abstract

**Background:** Cytosine methylation is widespread in most eukaryotic genomes and is known to play a substantial role in various regulatory pathways. Unmethylated cytosines may be converted to uracil through the addition of sodium bisulphite, allowing genome-wide quantification of cytosine methylation via high-throughput sequencing. The data thus acquired allows the discovery of methylation 'loci'; contiguous regions of methylation consistently methylated across biological replicates. The mapping of these loci allows for associations with other genomic factors to be identified, and for analyses of differential methylation to take place.

**Results:** The `segmentSeq` **R** package is extended to identify methylation loci from high-throughput sequencing data from multiple experimental conditions. A statistical model is then developed that accounts for biological replication and variable rates of non-conversion of cytosines in each sample to compute posterior likelihoods of methylation at each locus within an empirical Bayesian framework. The same model is used as a basis for analysis of differential methylation between multiple experimental conditions with the `baySeq` **R** package. We demonstrate the capability of this method to analyse complex data sets in an analysis of data derived from multiple Dicer-like mutants in *Arabidopsis*. This reveals several novel behaviours at distinct sets of loci in response to loss of one or more of the Dicer-like proteins that indicate an antagonistic relationship between the Dicer-like proteins at at least some methylation loci. Finally, we show in simulation studies that this approach can be significantly more powerful in the detection of differential methylation than many existing methods in data derived from both mammalian and plant systems.

**Conclusions:** The methods developed here make it possible to analyse high-throughput sequencing of the methylome of any given organism under a diverse set of experimental conditions. The methods are able to identify methylation loci and evaluate the likelihood that a region is truly methylated under any given experimental condition, allowing for downstream analyses that characterise differences between methylated and non-methylated regions of the genome. Futhermore, diverse patterns of differential methylation may also be characterised from these data.

**Keywords:** Methylation, DMRs, High-throughput sequencing, Epigenomics, Dicer

## Background

Cytosine methylation, found in most eukaryotes and playing a key role in gene regulation and epigenetic effects [1–3], can be investigated at a genome wide level through high-throughput sequencing of bisulphite treated DNA [4]. Treatment of denatured DNA with sodium bisulphite deaminates unmethylated cytosines into uracil; sequencing this treated DNA thus allows, in principle, not only the identification of every methylated cytosine but an assessment of the proportion of cells in which the cytosine is methylated. Moreover, by comparing these quantitative methylation data across experimental conditions, genomic regions displaying differential methylation can be detected.

The data available for methylation locus finding from bisulphite treated DNA are generated from a set of sequencing libraries. Each library consists of a set of sequenced reads which can be aligned and summarised [5] to report at each cytosine the number of sequenced reads in which the cytosine is methylated, and the number

Correspondence: tjh48@cam.ac.uk
Department of Plant Sciences, University of Cambridge, Downing Street, CB2 3EA Cambridge, UK

in which the cytosine is unmethylated [6]. Several methods have been proposed to detect differential methylation at the cytosine level, and to identify contiguous differentially methylated cytosines defined as differentially methylated regions (DMRs) [7]. However, these approaches, in not identifying non-differentially methylated regions and unmethylated regions, preclude many strategies for downstream identification of the biological significance of differential methylation.

We propose here a new method for methylation analysis based on the notion of methylation 'loci'; genomic regions defined by the presence of contiguous cytosines whose methylation is correlated across experimental conditions (Fig. 1). Cytosines within a given locus may thus be assumed to share biogenesis and functional properties [8]. Furthermore, the identification of methylation 'loci' and the quantification of methylation within a locus increases statistical power to detect differential methylation. We show that by taking a novel approach in which methylation loci are identified and subsequently analysed for patterns of differential behaviour, we are able to achieve high levels of accuracy in identifying differential

methylation. The empirical Bayesian methods employed also allow analysis of multiple patterns of differential methylation to be identified within complex data sets, allowing for detailed downstream analysis of biological mechanisms. We achieve this by adapting our previously described methods for defining small RNA (sRNA) loci from high-throughput sequencing of sRNAs [9], and for a generalised analysis of high-throughput sequencing data [10]. These methods allow for an analysis of differential behaviour in the methylome that accounts both for biological variation between replicates and systemic differences between samples caused by variations in the conversion rates of bisulphite treatment.

We demonstrate that this approach, in addition to allowing greater flexibility in analysis, offers better performance than a number of existing methods for the analysis of cytosine methylation data. We achieve this by simulating data using WGBSSuite [11], a recently developed stochastic method for generating simulated single base resolution DNA methylation data, with simulations based on parameters derived from both plant and mammalian systems. We demonstrate high performance under a variety

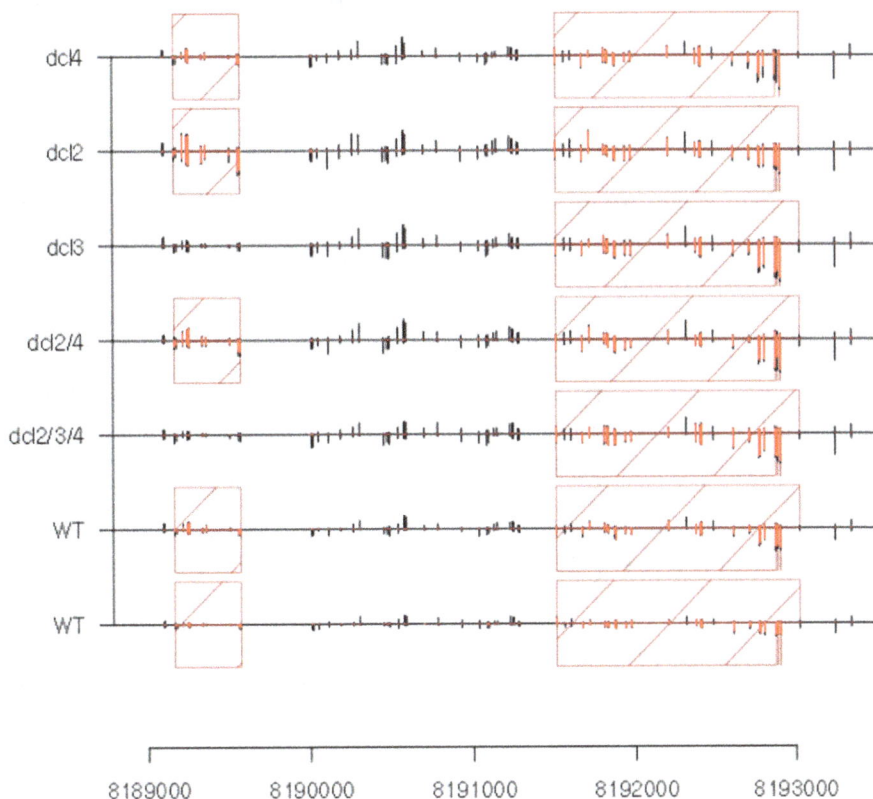

**Fig. 1** Examples of methylation loci in a set of *Arabidopsis thaliana* mutants in the Dicer-like (dcl) proteins, and wild-type samples [12]. Each row represents a single biological sample; the height of the bars represents the number of sequenced reads in which unmethylated cytosines (black) and methylated cytosines (red) appear. Values above the horizontal lines represent cytosines on the positive strand while values below the horizontal lines represent cytosines on the negative strand. The dark red hatched boxes indicate the identified loci. On the left, a methylation locus is identified that is methylated in all samples except the *dcl3* and *dcl2/3/4* mutants, on the right, a locus that is methylated in all samples

of simulation conditions, with substantial improvements over existing methods in the analysis of small changes in methylation between experimental conditions and for low numbers of biological replicates.

We further demonstrate these methods in an analysis of methylation in all contexts in mutants of the Dicer-like (DCL) proteins in *Arabidopsis* [12]. In higher plants, Dicer or Dicer-like proteins form a small gene family of sometimes overlapping function in the biogenesis of small RNAs [13]. In *Arabidopsis*, four different DCL proteins exist, acting in a partially redundant manner [14]. Predominantly, DCL1 is involved in the production of 21-nt miRNAs. DCL2 and DCL4 act redundantly and perhaps hierarchically to produce 22 and 21-nt sRNAs. DCL3 produces 24-nt sRNAs, previously identified as the key component of RNA-directed DNA methylation (RdDM). Recent work [15–17] has however emphasised the importance of 21 and 22-nt sRNAs, and hence of the DCL2 and DCL4 proteins, in regulating methylation at at least some loci. By applying the methods developed here we are able to identify multiple patterns of differential behaviour between the Dicer-like mutants and the loci which correspond to these.

## Methods
### Candidate loci and nulls
We analyse these data by adapting our previous methods for the identification of small RNA loci [9]. We begin by defining a set of *candidate loci* which may plausibly represent some methylation loci. A candidate locus begins and ends at some cytosine with a minimal proportion $p_{min}$ of methylation in at least one sequencing library. Considering all such loci is computationally infeasible and so filters are required to exclude implausible candidates and reduce the computational effort required. If two cytosines with a proportion of methylation above $p_{min}$ are within some minimal distance $d_{min}$ they are assumed to lie within the same locus. We further restrict the set by removing from consideration any candidate locus containing a region greater than $\lambda_{max}$ that contains no cytosine with a proportion of methylation above $p_{min}$. Candidate loci may be defined with respect to a single strand (by default), or combine the observed data from both strands. In the analyses presented here, we use $p_{min} = 0.05$, $d_{min} = 2$, $\lambda_{max} = 1000$. These parameters have been found by experience to well characterise the methylation loci under most circumstances; however, the results identified will in most cases be robust to relatively large changes to these parameters.

We define the set of *candidate nulls*, regions which may represent a region without significant methylation, by considering the gaps between candidate loci. We refer to the regions separating each candidate locus from its nearest neighbour (in either direction) as 'empty'. Candidate nulls consist of the union of the set of 'empty' regions, the set of candidate loci extended into the empty region to their left, the set of candidate loci extended into the empty region to their right, and the set of candidate loci extended into the empty regions to both the left and right.

### Classification of candidate loci
The data pertaining to the candidates defined above are the number of methylated and un-methylated cytosines sequenced and aligning to these regions for each sample. Biological replication is defined in terms of *replicate groups*, non-intersecting sets of biological replicates. Thus the samples may be thought of as the set $\{A_1, \cdots, A_m\}$ with a replicate structure defined by $\mathcal{R} = \{R_1, \cdots, R_n\}$ where $j \in R_q$ if and only if sample $A_j$ is a member of replicate group $q$. We then identify those candidates which represent at least part of a true locus of methylation, given the observed data for each replicate group.

For a replicate group $R_q$ and locus $i$ we consider the total number of methylated and unmethylated cytosines $u_{iq} = \sum_{j \in R_q} u_{ij}$ and $u'_{iq} = \sum_{j \in R_q} u'_{ij}$ respectively. For the purposes of identifying the methylation loci, we assume that these data are described by a binomial distribution with parameter $p_{iq}$ which has a beta prior distribution with parameters $(\alpha, \beta)$; we use an uninformative Jeffreys prior of $\alpha = \beta = \frac{1}{2}$. This assumption implicitly neglects biological variability between samples, which could be better modelled by a beta-binomial distribution. However, the computational cost involved in using a beta-binomial distribution is considerable. We therefore make this simplifying assumption in order to identify the loci but apply a model based on the beta-binomial distribution downstream to evaluate the likelihood that an identified locus represents a true methylation locus within a set of biological replicates.

The posterior distribution of the parameter $p_{iq}$ is then a beta distribution with parameters $\left(\alpha + u_{iq}, \beta + u'_{iq}\right)$. A segment is identified as a methylation 'locus' if the posterior likelihood that $p_{iq} > t$ exceeds some critical value. Similarly, we can classify candidate nulls as true representatives of a null region by identifying those candidates with a posterior likelihood that $p_{iq} < t$ exceeding some critical value. By default, we use a required likelihood of 90%. The parameter $t$ is intended to provide a threshold distinguishing regions of 'true' or biologically relevant methylation from low level background methylation attributable to biological or technical noise. The appropriate value for this parameter is contingent on many factors, including organism, methylation context, heterogeneity of sample and the assignment of biological meaning; we use a value of 20% here across all analyses with the caveat that this may be more or less appropriate to any individual experiment.

The above analysis neglects the effect of non-conversion rates on the observed values for $u_{iq}$ and $u'_{iq}$. The data observed for a given sample may be defined as the number of methylated ($C_{ij}$) and unmethylated ($T_{ij}$) sequenced cytosines at the $i$th locus and $j$th sample. However, if non-conversion of cytosines occurs, we might expect that the true number of methylated cytosines is somewhat lower, and the true number of unmethylated cytosines somewhat higher than the observed values.

We can find no closed form expression for the posterior if the effects of non-conversion rates on the distribution of the data are accounted for. However, we can adjust the observed data by the expected non-conversion rates by setting $u_{ij} = C_{ij} - \frac{Q_j}{1-Q_j}T_{ij}$ and $u'_{ij} = T_{ij} + \frac{Q_j}{1-Q_j}T_{ij}$, where $Q_j$ is the non-conversion rate for sample $j$.

## Consensus loci

Given a classification on the set of candidate loci and nulls, we identify a set of consensus loci given the classifications on sets of overlapping candidates in a similar manner to that described for sRNA loci [18]. We begin by assuming that a true locus of methylation should not contain a null region within a replicate group in which the locus is methylated. Thus, if some candidate locus $l_i$ is classified as a locus in a set of replicate groups $\Psi$, and there exists some candidate null $n_j$ that lies completely within $l_i$ and is classified as a null in one or more of the replicate groups in the set $\Psi$, we discard the locus $l_i$. Of the remaining candidate loci, we then rank those that remain by the number of replicate groups in which they are classified as a locus, settling ties by giving higher rank to the longest candidate loci. The consensus loci are then formed by choosing all those candidate loci that do not overlap with a higher ranked candidate, giving a non-overlapping set of loci on each strand. 'Null' loci are defined as the contiguous regions of the genome containing no identified locus.

## Likelihood of data

We can compute posterior likelihoods of methylation and differential methylation on the identified loci through application of the empirical Bayesian methods described in Hardcastle (2016) [10]. Since the set of identified loci is considerably smaller than the set of all possible loci, we are able to incorporate biological variability into our models at this stage without the computational cost becoming excessive.

We achieve this by defining a distribution on the data accounting for biological variation between replicates. Ignoring issues of non-conversion, we would assume that the data in equivalently methylated samples are beta-binomially distributed as in a straightforward analysis of paired data [18].

$$\mathbb{P}(D_{ij}|p_q,\phi) = \binom{C_{ij}+T_{ij}}{C_{ij}} \frac{B(\alpha+C_{ij},\beta+T_{ij})}{B(\alpha,\beta)} \quad (1)$$

Equation 1 defines the density function of the observed data $D_{ij}$, given a proportion of methylation $p_q$ for each equivalence class $E_q$ and a dispersion parameter $\phi$ capturing the level of variation between biological replicates. Then $\alpha = p_q \frac{1-\phi}{\phi}$, $\beta = (1-p_q)\frac{1-\phi}{\phi}$. Following our previous work [10], a joint distribution on $\{p_q,\phi\}$ may be empirically estimated by repeatedly sampling individual loci (without replacement) and estimating for each replicate group $q$ the values $\{p_q,\phi\}$ by maximum likelihood estimation based on the data observed at that locus. The dispersion parameter $\phi$ is assumed to be preserved across replicate groups and $p_q$ is not.

If non-conversion rates are estimable, we can first normalise the observed data as above and proceed assuming a beta-binomial distribution. However, this does not fully account for the stocasticity of non-conversion events at each cytosine. A full analysis incorporating non-conversion events requires that the data within each sample $j$ are assumed to be the sum of a binomial distribution with success parameter $Q_j$ (the rate of non-conversion) and a beta-binomial distribution with parameters $p_q$ (the expected proportion of methylated cytosines) and dispersion parameter $\phi$. Then the likelihood of the observed data $D_{ij}$ at a single locus $i$ for a sample $j$ is given by

$$\mathbb{P}(D_{ij}|Q_j,p,\phi) = \sum_{m=0}^{C_{ij}} \binom{T_{ij}+m}{m} Q_j^m (1-Q_j)^{T_{ij}} \times \binom{C_{ij}+T_{ij}}{C_{ij}-m} \frac{B(\alpha+C_{ij}-m,\beta+T_{ij}+m)}{B(\alpha,\beta)} \quad (2)$$

where $m$ is the number of unconverted unmethylated cytosines and the remaining parameters are as in Eq. 1. As before, we can then estimate an empirical distribution on the parameters $\{p_q,\phi\}$ may be empirically estimated by repeatedly sampling individual loci (without replacement) and fitting values for a sampled locus by maximum likelihood methods.

## Posterior likelihoods of methylation

Given the empirically estimated joint distributions on the parameters of our distribution, we can estimate posterior likelihoods of methylation for each replicate group and locus using the methods described in Hardcastle (2016) [10]. We derive two empirical distributions on the parameters, one by sampling regions identified as methylation loci, and one by sampling regions identified as null regions.

Given these two distributions, we are able to calculate posterior likelihoods of methylation for each locus and replicate group by taking the product of the probability of the observed data at a locus under each distribution and an empirically determined prior on the likelihood of an arbitrary region being methylated or not in a given replicate group [10]. Regions exhibiting various patterns of differential methylation can be similarly identified using the density function defined in Eq. 2 or Eq. 1 (neglecting non-conversion rates) in the baySeq **R** package.

Since the definition of differential methylation is primarily concerned with a shift in ratios between the number of methylated cytosines and the number of unmethylated cytosines, it is possible for long regions of low methylation to exhibit patterns of differential methylation. We thus find improved performance by combining the likelihood of differential methylation within a locus with the likelihood of that locus being methylated in at least one replicate group. The final statistic used to identify DMRs with this method is thus:

$$\mathbb{P}(M|D_{ij}) \left( 1 - \prod_q \left( 1 - \mathbb{P}\left(L_q|D_{E_q}\right) \right) \right) \tag{3}$$

where $\mathbb{P}(M|D_{ij})$ is the likelihood of a model $M$ of differential methylation given the observed data in each sample at the $i$th defined region, and $\mathbb{P}(L_{iq}|D_{E_q})$ is the likelihood that the $i$th region defines an expressed locus within replicate group $q$.

## Results

### Analysis of the methylome in *dcl* mutants of *Arabidopsis*

We demonstrate the value of this approach in a reanalysis of the methylome in the Dicer-like mutants from the Stroud et al. (2013) [12] dataset. We identify in a single analysis methylation loci in the *dcl2*, *dcl3*, *dcl4*, *dcl2/4* and *dcl2/3/4* mutants, together with wild-type samples and discover complex patterns of differential methylation that exist between these mutants and the wild-type samples.

We begin with a standard pipeline for read alignment and summarisation [5]. Reads were aligned and summarised for each methylation context using the Bismark caller [19] with default settings. Since cytosine methylation should be absent in the chloroplast and mitochondrial genomes, we can estimate non-conversion rates as the ratio of sequenced cytosines to thymines in the reads aligning to these genomes. This was done for each sample and incorporated into the analysis at the distributional level.

We separate the data into the three major contexts of methylation; CpG, CHG, and CHH. For each context of methylation, we identify a set of loci and estimate posterior likelihoods that any given locus is truly methylated in each of the experimental conditions. Figure 2 summarises the input data and expected numbers of loci in each mutant, based on the posterior likelihoods. The genomewide trends in methylation remain relatively constant in all of the *dcl* mutants relative to the wild-type samples,

with some minor loss of methylation (relative to wild-type) at this scale in the CHH context in the *dcl2/3/4* and *dcl3* mutants, and some gain of CHH methylation in *dcl4* mutant at the centromeric regions. The total number of methylation loci in each condition may be estimated by summing the posterior likelihoods of loci (Fig. 2d). Relative to wild-type, expected numbers of loci do not alter substantially for *dcl2/4* loci in any condition, or for CpG methylation in *dcl2/3/4*, while all the single mutants show lower numbers of methylation in all contexts. The numbers of methylation loci discovered in the CHG context are substantially lower than for other contexts; however, the loci discovered are generally longer, as shown by the estimated portion of the genome covered by loci in each context (Fig. 2e), which shows roughly equivalent coverage for CpG and CHG with a minor reduction in CHH context.

We next consider patterns of differential methylation at the level of the identified loci. For each region of the genome, posterior likelihoods of difference are identified, and adjusted by the likelihood that the region is a methylation locus in at least one condition. From these posterior likelihoods, we can estimate the expected number of loci belonging to each model of equivalence and difference between the conditions as the sum of the posterior likelihoods for this model over all loci. We can also select specific loci by controlling the false discovery rate (FDR) estimated from the posterior likelihoods. Ten patterns of differential methylation (Fig. 3) are identified with an estimated number of loci greater than one thousand and a number of loci with an FDR < 0.05 greater than two hundred in at least one methylation context. A fuller list of potential models and the numbers of loci corresponding to these is available in Additional file 1: Table S1.

The ten models selected for further consideration can be roughly partitioned into five classes based on their definitions and the contexts in which they are most commonly found. Model A represents those methylation loci which show no differential methylation. Unsurprisingly, these are common in all contexts of methylation, as the DCL-dependent methylation makes up a small proportion of the total methylation on the *Arabidopsis* genome. The next class is of the models B, C, D & E, describing loci that show some loss of methylation in one or more of the single *dcl* mutants, but not in either the double *dcl2/4* or triple *dcl2/3/4* mutants. These loci are predominantly found in the CpG context. Model F is also predominantly found in CpG context methylation, and describes loci which show a gain in methylation in all *dcl* mutants relative to the wild-type samples. Similarly, Model G represents a gain in methylation in the majority of the *dcl* mutants over wild-type and the *dcl2* mutant, and is found with confidence only in the CHG methylation context. Models H, I & J represent the canonical changes in sRNA-linked methylation

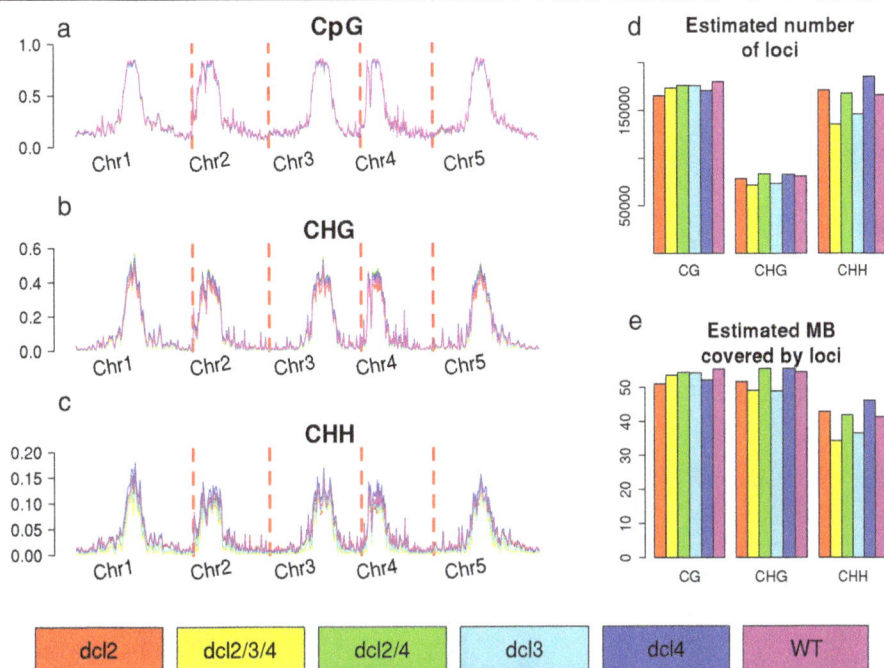

**Fig. 2** Genome wide profiles of methylation for the various Dicer-like mutants, and wild-type, in CpG (**a**) CHG (**b**) and CHH (**c**) contexts, adjusted for non-conversion rates. The estimated number of loci identified in each condition are shown in (**d**), while the estimated length of the genome covered by loci are shown in (**e**)

[20], in which there is loss of methylation in either *dcl3* and *dcl2/3/4* relative to wild-type, with Model I somewhat exceptional in that it does not represent a loss of methylation in the *dcl3* mutant but only in the *dcl2/3/4* triple mutant. These loci are predominantly found in CHG and CHH contexts, conforming to the expectation that DCL3 is particularly relevant to the CHG and CHH methylation pathways.

The level of change in methylation varies considerably between models and contexts (Fig. 4). For example, the average loss of methylation specific to the *dcl2* mutant (model B) in the CpG mutant is substantial, whereas that specific to the *dcl4* mutant (model C) is much lower (though still detectable at large numbers of loci). Gain in methylation in some or all of the *dcl* mutants can also be substantial (model F; all contexts) or marginal (model G). For CpG context methylation, several of the more significant changes in methylation occur in loci with a short average width (Additional file 1: Figure S1), notably those in models B, E & F, though this does not necessarily negate their biological significance [21].

Some evidence for the biological relevance of the identified classes can be acquired by examining the average methylation profiles for these loci across a range of additional mutants from the Stroud et al. (2013) [12] dataset (Additional file 1: Figure S2). In the CpG context, loci representing models B, C & D, in which methylation is lost in the *dcl2*, *dcl4* or *dcl2/4* mutants also show

a substantial average loss of methylation in the *met1* heterozygous mutant, while this effect is much reduced in the non-differential loci (model A) and those loci showing a loss of methylation in the *dcl3* mutant (model E). This effect appears even stronger in loci showing a gain in methylation in all *dcl* mutants over wildtype. Conversely, loci representing models C & D show a reduced loss of CpG methylation in the *ddm1* mutant, perhaps implying a partial independence of these loci from the chromatin remodelling methylation pathway [22].

In the CHG context, it is notable that the loci showing small gains in average methylation observed in all *dcl* mutants except *dcl2* over wild-type (model G) show a similar gain in the *ago4* and *nrpd1* mutants, supporting the role of the sRNA pathways in repressing methylation at these loci. Also of note is the relative independence of methylation from CMT3 in the loci representing model J, coupled with an increased dependendence on DRM1/2. This suggests a refinement of the model for redundant maintainence of CHG methylation by DRM and CMT3 proposed in Cao et al. (2003) [23] as it indicates that for some RdDM loci it is DRM that is primarily required, and that this correlates with specific patterns of differential behaviour between *dcl3* and *dcl2/3/4*. In the CHH context, perhaps the most notable feature is the partial maintainence of methylation in the *met1/cmt3* mutant in the RdDM loci (models H, I & J). Notably, the average methylation across these loci in the *met1/cmt3* mutant is

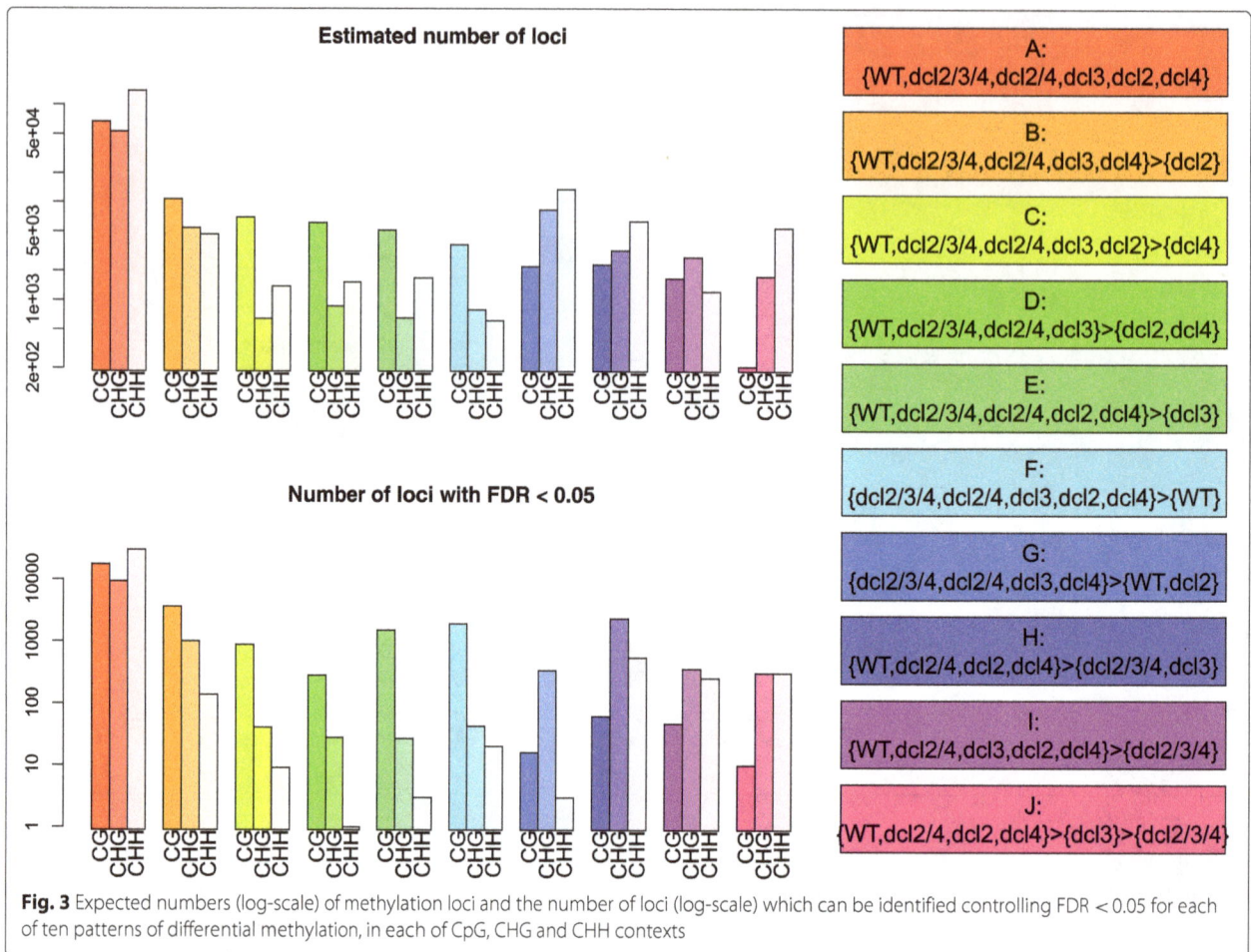

**Fig. 3** Expected numbers (log-scale) of methylation loci and the number of loci (log-scale) which can be identified controlling FDR < 0.05 for each of ten patterns of differential methylation, in each of CpG, CHG and CHH contexts

greatest in those loci affected only in *dcl2/3/4* and not *dcl3* (model I), perhaps indicating that methylation at these loci is more strongly regulated at the establishment phase by 21/22-nt sRNAs [20].

Variation is also marked in the genomic localisation of these models (Additional file 1: Figure S3). Loci

representing models B, C & D, in which methylation is reduced in either or both of the *dcl2* and *dcl4* mutants, but neither of the double (*dcl2/4*) or triple (*dcl2/3/4*) mutants are found ubiquitously across the genome in the CpG context but are heavily centromeric in CHG and CHH contexts. Conversely, those loci in which methylation is

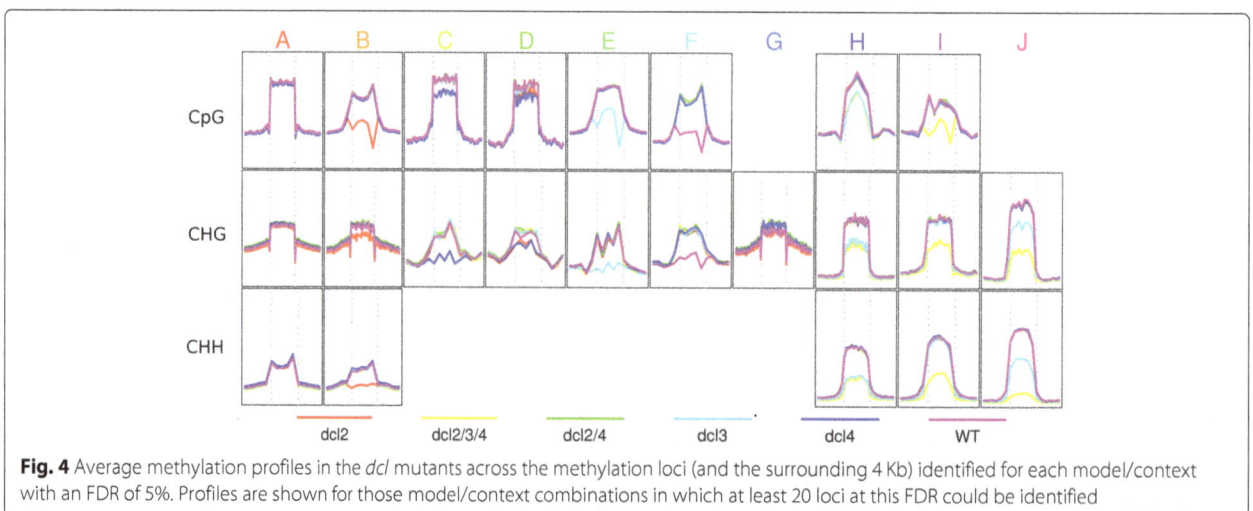

**Fig. 4** Average methylation profiles in the *dcl* mutants across the methylation loci (and the surrounding 4 Kb) identified for each model/context with an FDR of 5%. Profiles are shown for those model/context combinations in which at least 20 loci at this FDR could be identified

reduced only in the *dcl3* mutant is centromeric in the CpG context. Gains in methylation in some or all of the *dcl* mutants appear evenly distributed across the genome in the CpG context but are strongly centromeric in the CHG context.

## Simulated data

We next compare the performance of the approach developed here to several existing methods for detection of differential methylation using simulation data generated by WGBSSuite [11]. This tool simulates differentially methylated regions based on a complex parameter set allowing a variety of methylation types to be generated. We simulate data based on an analysis of CpG methylation in *Arabidopsis* (see Additional file 1 for parameter details). Using these basic parameters, we evaluate the performance of each method, varying coverage, number of replicates, the magnitude of methylation difference between replicates.

We modified the standard WGBSSuite analysis by including the effects of sample specific non-conversion rates to the data. Non-conversion rates were estimated from the wild-type and various *dcl*-mutants in the Stroud et al. (2013) [12] dataset, as above. Parameters for a beta distribution approximating the distribution of observed rates were estimated by maximum likelihood. These parameters were then used to simulate non-conversion rates for each sample in a simulation.

The simulated data are evaluated using BSmooth/bsseq [24], MethylKit [25], MethylSig [26] and MethPipe [27]. BSmooth, MethylKit, and MethylSig are implemented as in the WGBSSuite benchmarking, as is a Fisher exact test. These methods primarily rely on the detection of differential methylation at the cytosine level, and construct DMRs from the identified differentially methylated cytosines. MethPipe offers two different implementations, the first similarly based on scores constructed at each cytosine supported by a two-state hidden Markov model used to identify regions of methylation (MethPipe-1), while the second uses a beta-binomial regression on the observed data and is recommended for larger sample sets (MethPipe-2).

Performance of the methods is evaluated primarily by constructing a ranked list of DMRs based on each method's test statistic. As in WGBSSuite's benchmarking, true postives are defined as the number of truly differentially methylated cytosines within identified DMRs, while false positives are the non-differentially methylated cytosines within the identified DMRs. Figure 5 shows a comparison between the methods for data simulated using parameters intended to produce data similar to those observed in CpG methylation in plant systems. Analyses are carried out using 1, 3, and 10 replicates, and with changes in the proportion of methylated cytosines between experimental groups of 0.05, 0.25 and 0.85.

In all simulations, the segmentSeq/baySeq methods described here perform as well or better than the other methods considered as assessed by the ROC (Receiver operating characteristic) curves. The segmentSeq approach failing to account for non-conversion on average performs well, but shows greater variation and some loss of performance compared to the segmentSeq-NC method which incorporates adjustments for non-conversion. This is particularly true for the experiments with few replicates; with higher numbers of replicates the effect of non-conversion will tend to average out across samples.

For large differences (a proportion shift of 0.85) in methylation all methods are able to detect differentially methylated cytosines in three and ten sample cases with almost perfect accuracy, with the exception of BSmooth and MethylKit. BSmooth shows reduced performance compared to other methods in the ten sample case and MethylKit is unable to make valid calls in any analysis. This is likely due to the design of MethylKit; it is primarily intended for the analysis of reduced representation bisulphite sequencing (RRBS) and does not appear suitable for the substantially lower coverage used in these simulations. For smaller shifts in methylation proportion the increase in performance through the segmentSeq/baySeq approach is more dramatic; this is to be expected as the increased data available to analyse methylation loci rather than individual cytosines gives greater power to detect small differences in methylation. In the analysis without replicates, the segmentSeq/baySeq approach shows substantially better performance over MethPipe and BSmooth for low and moderate differential methylation, with the MethPipe-1 analysis approaching this performance in the high differential case.

## Discussion

A number of methods have previously been developed to analyse high-throughput sequencing of the methylome [7]. These are predominantly focused on the identification of differential methylation and the discovery of differentially methylated regions from grouping differentially methylated cytosines. The methods presented here adopt an alternative strategy in which first methylated and un-methylated regions are identified, and differential methylation is subsequently evaluated. Comparisons on simulated data show that the approach developed here offers substantially more power to detect small changes in methylation across a region when compared to existing methods which operate on a cytosine-by-cytosine scale, without any loss of power in the detection of large shifts in methylation. Accounting for non-conversion rates, where possible, gives a small but consistent improvement in performance, particularly when replication or the level of change in methylation is low. This is perhaps of particular

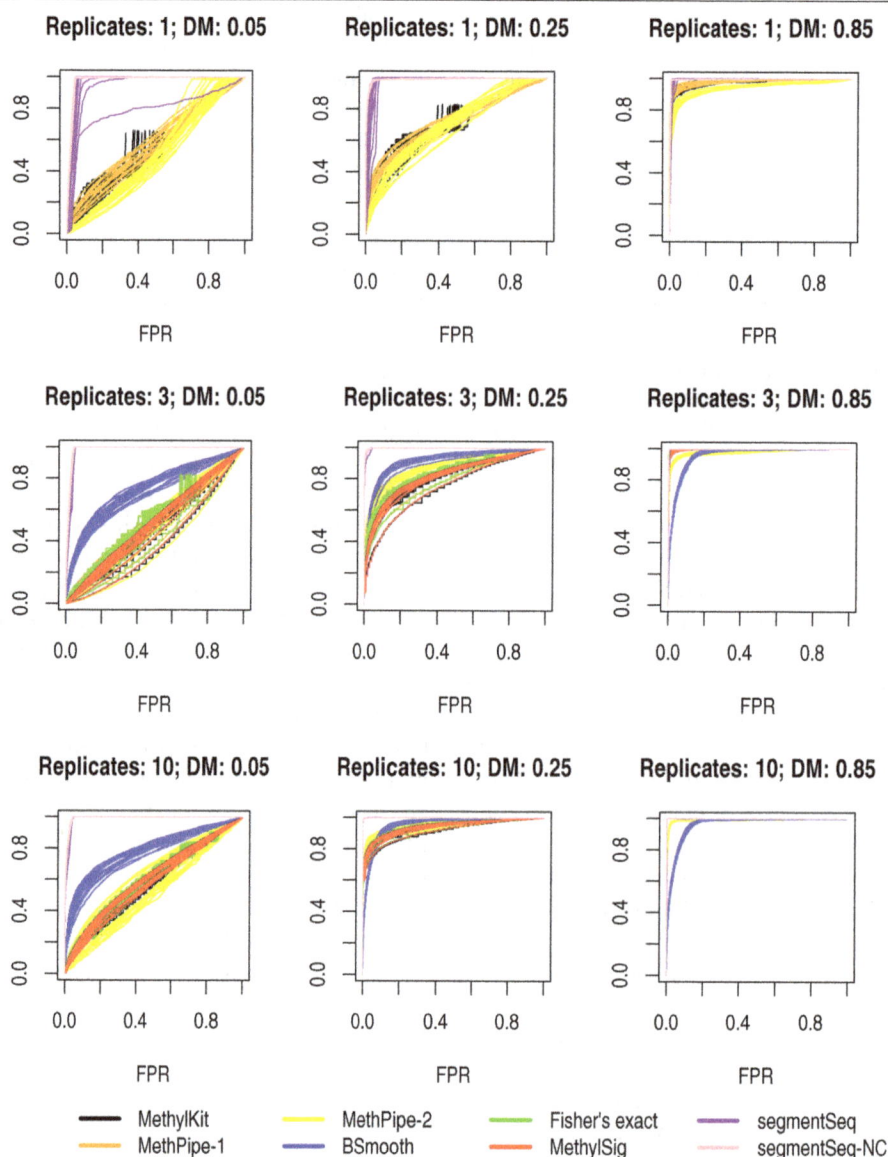

**Fig. 5** ROC curves from simulated data based on WGBSSuite analyses of CpG methylation in *Arabidopsis*. Analyses for 1, 3 and 10 replicates are shown, and for small (0.05), moderate (0.25) and large (0.85) changes in methylation in the differentially methylated regions. Twenty simulations were carried out for each choice of replicate number/difference in methylation and curves for each simulation are shown here

importance in analysing plant methylomes, in which wild-type levels of CHG and CHH context methylation are expected to be low, and consequently loss of methylation is marked by only a small shift in the observed data.

We demonstrate our methods on a subset of the Stroud et al. (2013) [12] dataset describing the *Arabidopsis* methylome. A primary strength of the approach presented here is its ability to analyse complex relationships between multiple replicate groups. We demonstrate this by the simultaneous analysis of all the *dcl* mutants, together with the wild-type samples contained in Stroud et al. (2013) [12]. Several novel patterns of methylation are identified through this analysis; most particularly a set of over a

thousand CpG loci which lose methylation in the *dcl2* mutant but not the *dcl2/4* double mutant; at somewhat fewer loci we identify similar patterns in CHG and CHH contexts. Similarly, we identify loci which show a reduction in methylation in the *dcl4* mutant but not the *dcl2/4* double mutant, and loci which show a reduction in the *dcl3* but not the *dcl2/3/4* triple mutant. The mechanisms associated with these loci are not directly explicable from the data but it seems likely that there is an antagonistic relationship between the DCL-proteins at at least some of the loci, as previously noted by Bouche et al. (2006) [28]. Support for these loci as biologically meaningful is demonstrated through comparisons with additional mutants of the methylome regulation pathways and through analysis of the genome localisation of the discovered loci.

## Conclusions

The methods described here allow for the identification of methylation loci from large sets of experimental conditions, the estimation of likelihoods for each condition that a region is truly methylated above background levels, and ultimately the detection of differential methylated regions. This approach allows downstream comparison between differentially methylated regions, non-differentially methylated regions, and non-methylated regions. Based on comparisons on simulated data, these methods also offer a number of significant performance advantages over existing methods for detection of differential methylation, particularly in the detection of small changes in methylation levels and in experiments with low numbers of replicates. These methods also allow for the analysis of complex experimental designs, as demonstrated on a reanalysis of methylation in a set of *dcl* mutants. This analysis demonstrates the potential utility of this method in identifying a variety of methylation loci demonstrating novel interactions between regulatory mechanisms of methylation. Though tested on methylation data derived from plant systems, there is no reason these methods should not be equally applicable to animal and human data given the conservation of CpG methylation between eukaryotes [29].

The methods are implemented and released within the segmentSeq [9] and baySeq [10], available on Bioconductor (www.bioconductor.org) [30]. In addition to usability and maintainence advantages, this ensures compatibility with the analyses of sRNA-seq, mRNA-seq and other high-throughput data already developed in these packages. Results acquired by high-throughput sequencing of methylation can thus be readily integrated with these other -omic data, allowing the differential methylome to be incorporated into in systems level analyses.

## Additional file

**Additional file 1:** Supplemental.pdf - supplementary materials. Description of simulation studies and parameter details; **Figure S1**: Boxplot of locus widths for each model/context; **Figure S2**: Profiles of cytosine methylation in additional mutants for identified differentially methylated loci; **Figure S3**: Profiles of model abundance across the genome; **Table S1**: Numbers of loci associated with models of differential methylation.

## Abbreviations
DCL: Dicer-like (protein); DMR: Differentially methylated regions; FDR: False discovery rate; sRNA: Small RNA; RdDM: RNA-directed DNA methylation; RRBS: Reduced representation bisulphite sequencing

## Acknowledgments
The author thanks Dr. Owen Rackham for his assistance in making WGBSSuite available for this study.

## Funding
This work was supported by European Research Council Advanced Investigator Grant ERC-2013-AdG 340642 - TRIBE.

## Authors' contributions
TJH is solely responsible for this study.

## Competing interests
The author declares that he has no competing interests.

## References
1. Zhang X, Yazaki J, Sundaresan A, Cokus S, Chan SW-L, Chen H, Henderson IR, Shinn P, Pellegrini M, Jacobsen SE, Ecker JR. Genome-wide high-resolution mapping and functional analysis of DNA methylation in arabidopsis. Cell. 2006;126(6):1189–201. doi:10.1016/j.cell.2006.08.003.
2. Berdasco M, Alcázar R, García-Ortiz MV, Ballestar E, Fernández AF, Roldán-Arjona T, Tiburcio AF, Altabella T, Buisine N, Quesneville H, Baudry A, Lepiniec L, Alaminos M, Rodríguez R, Lloyd A, Colot V, Bender J, Canal MJ, Esteller M, Fraga MF. Promoter DNA hypermethylation and gene repression in undifferentiated Arabidopsis cells. PloS ONE. 2008;3(10):3306. doi:10.1371/journal.pone.0003306.
3. Tsukahara S, Kobayashi A, Kawabe A, Mathieu O, Miura A, Kakutani T. Bursts of retrotransposition reproduced in Arabidopsis. Nature. 2009;461(September):3–7. doi:10.1038/nature08351.
4. Clark SJ, Statham A, Stirzaker C, Molloy PL, Frommer M. DNA methylation: bisulphite modification and analysis. Nat Protoc. 2006;1(5): 2353–64. doi:10.1038/nprot.2006.324.
5. Bock C. Analysing and interpreting DNA methylation data. Nat Rev Genet. 2012;13(10):705–19. doi:10.1038/nrg3273.
6. Hardcastle TJ. High-throughput sequencing of cytosine methylation in plant DNA. Plant Methods. 2013;9(1):16.
7. Robinson MD, Kahraman A, Law CW, Lindsay H, Nowicka M, Weber LM, Zhou X. Statistical methods for detecting differentially methylated loci and regions. Front Genet. 2014;5:324. doi:10.3389/fgene.2014.00324.
8. Zhao JH, Fang YY, Duan CG, Fang RX, Ding SW, Guo HS, Bird A, Chan SW, Henderson IR, Jacobsen SE, Zhang X, Lister R, Lindroth AM, Cao X, Jacobsen SE, Matzke M, Kanno T, Daxinger L, Huettel B, Matzke AJ, Zhang H, Zhu JK, Dalakouras A, Wassenegger M, Zhao M, Leon DS, Delgadillo MO, Garcia JA, Simon-Mateo C, Dalakouras A, Dadami E, Zwiebel M, Krczal G, Wassenegger M, Wierzbicki AT, Haag JR, Pikaard CS, Law JA, Jacobsen SE, Wu L, Mao L, Qi Y, Nuthikattu S, Stroud H, Greenberg MV, Feng S, Bernatavichute YV, Jacobsen SE, Lee TF, Wang H, Hao L, Shung CY, Sunter G, Bisaro DM, Wang H, Buckley KJ, Yang X, Buchmann RC, Bisaro DM, Buchmann RC, Asad S, Wolf JN, Mohannath G, Zhang Z, Yang X, Ivanov KI, Canizares MC, Li HW, Guo HS, Ding SW, Gonzalez I, Duan CG, Hamera S, Song X, Su L, Chen X, Fang R, Feng L, Duan CG, Guo HS, Cokus SJ, Li Y, Wang H, Mi S, Takeda A, Iwasaki S, Watanabe T, Utsumi M, Watanabe Y, Girard A, Hannon GJ, Sarazin A, Voinnet O, Creasey KM, Zhang X, Zhong X, Mosher RA, Schwach F, Studholme D, Baulcombe DC, Mirouze M, Mari-Ordonez A, Slotkin RK, Han BW, Wang W, Li C, Weng Z, Zamore PD, Mohn F, Handler D, Brennecke J, Siomi H, Siomi MC, Calvi BR, Gelbart WM, Dupressoir A, Heidmann T, Pasyukova E, Nuzhdin S, Li W, Flavell AJ, Ostertag EM, Lau NC, Brennecke J, Brower-Toland B, Carmell MA, Zahid K, Allen GC, Flores-Vergara MA, Krasynanski S, Kumar S, Thompson WF, Li R, Pathak RR, Lochab S, Mortazavi A, Williams BA, McCue K, Schaeffer L, Wold B, Kim KI, van de Wiel MA. Genome-wide identification of endogenous RNA-directed DNA methylation loci associated with abundant 21-nucleotide siRNAs in Arabidopsis. Sci Rep. 2016;6:36247. doi:10.1038/srep36247.

9. Hardcastle TJ, Kelly KA, Baulcombe DC. Identifying small interfering RNA loci from high-throughput sequencing data. Bioinformatics. 2012;28(4): 457–63. doi:10.1093/bioinformatics/btr687.

10. Hardcastle TJ. Generalized empirical Bayesian methods for discovery of differential data in high-throughput biology. Bioinformatics. 2016;32(2): 195–202. doi:10.1093/bioinformatics/btv569.

11. Rackham OJL, Dellaportas P, Petretto E, Bottolo L. WGBSSuite: simulating whole-genome bisulphite sequencing data and benchmarking differential DNA methylation analysis tools. Bioinformatics. 2015;31(14): 2371–3. doi:10.1093/bioinformatics/btv114.

12. Stroud H, Greenberg MC, Feng S, Bernatavichute Y, Jacobsen S. Comprehensive Analysis of Silencing Mutants Reveals Complex Regulation of the Arabidopsis Methylome. Cell. 2013;152(1):352–64. doi:10.1016/j.cell.2012.10.054.

13. Chapman EJ, Carrington JC. Specialization and evolution of endogenous small RNA pathways. Nat Rev Genet. 2007;8(11):884–96. doi:10.1038/nrg2179.

14. Gasciolli V, Mallory AC, Bartel DP, Vaucheret H. Partially Redundant Functions of Arabidopsis DICER-like Enzymes and a Role for DCL4 in Producing trans-Acting siRNAs. Curr Biol. 2005;15(16):1494–500. doi:10.1016/j.cub.2005.07.024.

15. Panda K, Slotkin RK. Proposed mechanism for the initiation of transposable element silencing by the RDR6-directed DNA methylation pathway. Plant Signal Behav. 2013;8(8). doi:10.4161/psb.25206.

16. Bond DM, Baulcombe DC. Epigenetic transitions leading to heritable, RNA-mediated de novo silencing in Arabidopsis thaliana. Proc Natl Acad Sci. 2015;112(3):917–22. doi:10.1073/pnas.1413053112.

17. Matzke MA, Kanno T, Matzke AJM. RNA-Directed DNA Methylation: The Evolution of a Complex Epigenetic Pathway in Flowering Plants. Annu Rev Plant Biol. 2015;66(1):243–67. doi:10.1146/annurev-arplant-043014-114633.

18. Hardcastle TJ, Kelly KA. Empirical Bayesian analysis of paired high-throughput sequencing data with a beta-binomial distribution. BMC Bioinforma. 2013;14(1):135. doi:10.1186/1471-2105-14-135.

19. Krueger F, Andrews SR. Bismark: a flexible aligner and methylation caller for Bisulfite-Seq applications. Bioinformatics. 2011;27(11):1571–2. doi:10.1093/bioinformatics/btr167.

20. Bond DM, Baulcombe DC. Small RNAs and heritable epigenetic variation in plants. Trends Cell Biol. 2014;24(2):100–7. doi:10.1016/j.tcb.2013.08.001.

21. Xu J, Pope SD, Jazirehi AR, Attema JL, Papathanasiou P, Watts JA, Zaret KS, Weissman IL, Smale ST. Pioneer factor interactions and unmethylated CpG dinucleotides mark silent tissue-specific enhancers in embryonic stem cells. Proc Natl Acad Sci U S A. 2007;104(30):12377–82. doi:10.1073/pnas.0704579104.

22. Zemach A, Kim MY, Hsieh PH, Coleman-Derr D, Eshed-Williams L, Thao K, Harmer S, Zilberman D. The Arabidopsis Nucleosome Remodeler DDM1 Allows DNA Methyltransferases to Access H1-Containing Heterochromatin. Cell. 2013;153(1):193–205. doi:10.1016/j.cell.2013.02.033.

23. Cao X, Aufsatz W, Zilberman D, Mette MF, Huang MS, Matzke M, Jacobsen SE. Role of the DRM and CMT3 methyltransferases in RNA-directed DNA methylation. Curr Biol. 2003;13(24):2212–7.

24. Hansen KD, Langmead B, Irizarry RA. BSmooth: from whole genome bisulfite sequencing reads to differentially methylated regions. Genome Biol. 2012;13(10):83. doi:10.1186/gb-2012-13-10-r83.

25. Akalin A, Kormaksson M, Li S, Garrett-Bakelman FE, Figueroa ME, Melnick A, Mason CE. methylKit: a comprehensive R package for the analysis of genome-wide DNA methylation profiles. Genome Biol. 2012;13(10):87. doi:10.1186/gb-2012-13-10-r87.

26. Park Y, Figueroa ME, Rozek LS, Sartor MA. MethylSig: a whole genome DNA methylation analysis pipeline. Bioinformatics. 2014;30(17):2414–2. doi:10.1093/bioinformatics/btu339.

27. Song Q, Decato B, Hong EE, Zhou M, Fang F, Qu J, Garvin T, Kessler M, Zhou J, Smith AD. A reference methylome database and analysis pipeline to facilitate integrative and comparative epigenomics. PloS ONE. 2013;8(12):81148. doi:10.1371/journal.pone.0081148.

28. Bouché N, Lauressergues D, Gasciolli V, Vaucheret H. An antagonistic function for Arabidopsis DCL2 in development and a new function for DCL4 in generating viral siRNAs. EMBO J. 2006;25(14):3347–56. doi:10.1038/sj.emboj.7601217.

29. Zemach A, McDaniel IE, Silva P, Zilberman D. Genome-wide evolutionary analysis of eukaryotic DNA methylation. Science. 2010;328(5980):916–9. doi:10.1126/science.1186366.

30. Gentleman RC, Carey VJ, Bates DM, Bolstad B, Dettling M, Dudoit S, Ellis B, Gautier L, Ge Y, Gentry J, Hornik K, Hothorn T, Huber W, Iacus S, Irizarry R, Leisch F, Li C, Maechler M, Rossini AJ, Sawitzki G, Smith C, Smyth G, Tierney L, Yang JYH, Zhang J. Bioconductor: open software development for computational biology and bioinformatics. Genome Biol. 2004;5(10):80. doi:10.1186/gb-2004-5-10-r80.

# Permissions

# List of Contributors

**Alencar Xavier and Katy Martin Rainey**
Department of Agronomy, Purdue University, 915 W. State St., Lilly Hall, West Lafayette, IN 47907, USA

**Shizhong Xu**
Department of Plant Science, University of California, 3134 Batchelor Hall, Riverside, CA 92521, USA

**William Muir**
Department of Animal Science, Purdue University, 915 W. State St., Lilly Hall, West Lafayette, IN 47907, USA

**Guoxian Yu, Chang Lu and Jun Wang**
College of Computer and Information Sciences, Southwest University, Chongqing, China

**Marcelo Rodrigo de Castro and Hermes Senger**
Computer Science Department, Federal University of São Carlos, Rod. Washington Luís, Km 235, 21040-900 São Carlos, Brazil

**Catherine dos Santos Tostes and Alberto M. R. Dávila**
LBCS-IOC, Oswaldo Cruz Foundation, Av Brasil 4365, 21040-900 Rio de Janeiro, Brazil

**Fabricio A. B. da Silva**
PROCC, Oswaldo Cruz Foundation, Av. Brasil 4365, 21040-900 Rio de Janeiro, Brazil

**Sheng Zhang, BoWang, LinWan and Lei M. Li**
National Center of Mathematics and Interdisciplinary Sciences, Academy of Mathematics and Systems Science, Chinese Academy of Sciences, 100190 Beijing, China
University of Chinese Academy of Sciences, 100049 Beijing, China

**Ksenia Khelik and Geir Kjetil Sandve**
Biomedical Informatics Research Group, Department of Informatics, University of Oslo, PO Box 1080, 0316 Oslo, Norway

**Karin Lagesen**
Biomedical Informatics Research Group, Department of Informatics, University of Oslo, PO Box 1080, 0316 Oslo, Norway
Norwegian Veterinary Institute, PO Box 750 Sentrum, 0106 Oslo, Norway

**Torbjørn Rognes**
Biomedical Informatics Research Group, Department of Informatics, University of Oslo, PO Box 1080, 0316 Oslo, Norway
Department of Microbiology, Oslo University Hospital, Rikshospitalet, PO Box 4950 Nydalen, 0424 Oslo, Norway

**Alexander Johan Nederbragt**
Biomedical Informatics Research Group, Department of Informatics, University of Oslo, PO Box 1080, 0316 Oslo, Norway
Centre for Ecological and Evolutionary Synthesis, Department of Biosciences, University of Oslo, PO Box 1066 Blindern, 0316 Oslo, Norway

**Pei Fen Kuan, Shuyao He and Kaiqiao Li**
Department of Applied Mathematics and Statistics, Stony Brook University, 100 Nicolls Road, 11794 Stony Brook, USA

**Scott Powers and Xiaoyu Zhao**
Department of Pathology, Stony Brook University, 100 Nicolls Road, 11794 Stony Brook, USA

**Bo Huang**
Oncology Business Unit, Pfizer Inc., 558 Eastern Point Rd, 06340 Groton, USA

**Daniel Amsel and André Billion**
Fraunhofer Institute for Molecular Biology and Applied Ecology, Department of Bioresources, Winchester Str. 2, 35394 Giessen, Germany

**Andreas Vilcinskas**
Fraunhofer Institute for Molecular Biology and Applied Ecology, Department of Bioresources, Winchester Str. 2, 35394 Giessen, Germany
Institute for Insect Biotechnology, Heinrich-Buff-Ring 26-32, 35392 Giessen, Germany

**Lili Zhao**
Department of Biostatistics, University of Michigan, 1415 Washington Heights, Ann Arbor, USA

**Mark T. Anderson and Harry L. T. Mobley**
Department of Microbiology and Immunology, School of medicine, University of Michigan, Ann Arbor, USA

**Weisheng Wu**
BRCF Bioinformatics Core, University of Michigan, Ann Arbor, USA

**Michael A. Bachman**
Department of Pathology, School of medicine, University of Michigan, Ann Arbor, USA

**Guoshuai Cai**
Department of Molecular and Systems Biology, Geisel School of Medicine at Dartmouth, Hanover, NH, USA
Department of Environmental Health Sciences, Arnold School of Public Health, University of South Carolina, Columbia, SC, USA

**Shoudan Liang and Xiaofeng Zheng**
Department of Bioinformatics and Computational Biology, The University of Texas MD Anderson Cancer Center, Houston, TX, USA

**Feifei Xiao4**
Department of Epidemiology and Biostatistics, Arnold School of Public Health, University of South Carolina, Columbia, SC, USA

**J. F. Mudge and J. E. Houlahan**
Department of Biology, Canadian Rivers Institute, University of New Brunswick, Saint John, NB E2L 4L5, Canada

**C. J. Martyniuk**
Center for Environmental and Human Toxicology & Department of Physiological Sciences, UF Genetics Institute, University of Florida, Gainesville, Florida 32611, USA

**Kit C. B. Roes and Marinus J. C. Eijkemans**
Biostatistics & Research Support, Julius Center for Health Sciences and Primary Care, University Medical Center Utrecht, 3508, GA, Utrecht, The Netherlands

**Putri W. Novianti**
Biostatistics & Research Support, Julius Center for Health Sciences and Primary Care, University Medical Center Utrecht, 3508, GA, Utrecht, The Netherlands

Department of Epidemiology and Biostatistics, VU University medical center, Amsterdam, The Netherlands
Department of Pathology, VU University medical center, Amsterdam, The Netherlands

**Victor L. Jong**
Biostatistics & Research Support, Julius Center for Health Sciences and Primary Care, University Medical Center Utrecht, 3508, GA, Utrecht, The Netherlands
Viroscience Laboratory, Erasmus Medical Center Rotterdam, 3015, CE, Rotterdam, The Netherlands

**Mahdi Heydari, Giles Miclotte, Piet Demeester and Jan Fostier**
Department of Information Technology, Ghent University-imec, IDLab, B-9052 Ghent, Belgium
Bioinformatics Institute Ghent, B-9052 Ghent, Belgium

**Yves Van de Peer**
Center for Plant Systems Biology, VIB, B-9052 Ghent, Belgium
Department of Plant Biotechnology and Bioinformatics, Ghent University, B-9052 Ghent, Belgium
Bioinformatics Institute Ghent, B-9052 Ghent, Belgium
Department of Genetics, Genome Research Institute, University of Pretoria, Pretoria, South Africa

**Shailesh Tripathi**
Predictive Medicine and Data Analytics Lab, Department of Signal Processing, Tampere University of Technology, Tampere, Finland

**Frank Emmert-Streib**
Predictive Medicine and Data Analytics Lab, Department of Signal Processing, Tampere University of Technology, Tampere, Finland
Institute of Biosciences and Medical Technology, Tampere, Finland

**Jason Lloyd-Price**
Department of Biostatistics, Harvard T.H. Chan School of Public Health, Harvard University, Boston, USA
Laboratory of Biosystem Dynamics, Department of Signal Processing, Tampere University of Technology, Tampere, Finland

**Andre Ribeiro**
Laboratory of Biosystem Dynamics, Department of Signal Processing, Tampere University of Technology, Tampere, Finland
Institute of Biosciences and Medical Technology, Tampere, Finland

**Matthias Dehmer**
Institute for Theoretical Informatics, Mathematics and Operations Research, Department of Computer Science, Universität der Bundeswehr München, Munich, Germany

**Olli Yli-Harja**
Computational Systems Biology, Department of Signal Processing, Tampere University of Technology, Tampere, Finland
Institute of Biosciences and Medical Technology, Tampere, Finland

**Marek Palkowski and Wlodzimierz Bielecki**
West Pomeranian University of Technology, Faculty of Computer Science, Zolnierska 49, 71-210 Szczecin, Poland

**Se-Ran Jun and Intawat Nookaew**
Department of Biomedical Informatics, College of Medicine, University of Arkansas for Medical Sciences, Little Rock, AR 72205, USA

**Loren Hauser**
Comparative Genomics Group, Biosciences Division, Oak Ridge National Laboratory, Oak Ridge, TN 37831, USA

**Andrey Gorin**
Computer Science and Mathematics Division, Oak Ridge National Laboratory, Oak Ridge, TN 37831, USA

**Aziz Khan**
Centre for Molecular Medicine Norway (NCMM), Nordic EMBL Partnership, University of Oslo, 0318 Oslo, Norway

**Anthony Mathelier**
Centre for Molecular Medicine Norway (NCMM), Nordic EMBL Partnership, University of Oslo, 0318 Oslo, Norway
Department of Cancer Genetics, Institute for Cancer Research, Oslo University Hospital Radiumhospitalet, 0310 Oslo, Norway

**Suyeon Kang and Jongwoo Song**
Department of Statistics, Ewha Womans University, Seoul, South Korea

**Thomas J. Hardcastle**
Department of Plant Sciences, University of Cambridge, Downing Street, CB2 3EA Cambridge, UK

# Index